SYSTEMATIC BOTANY MONOGRAPHS

VOLUME 9

The Genus Salix (Salicaceae) in the Southeastern United States

George W. Argus

THE AMERICAN SOCIETY OF PLANT TAXONOMISTS
20 January 1986

SYSTEMATIC BOTANY MONOGRAPHS
ISSN 0737-8211

Printed in the United States of America

Library of Congress Cataloging-in-Publication Data

Argus, George W. (George William)
The genus Salix (Salicaceae) in the southeastern United States.

(Systematic botany monographs ; v. 9)
Bibliography: p.
Includes index.
1. Willows—Southern States—Classification.
2. Willows—Southern States—Identification.
3. Willows—Southern States—Geographical distribution—
Maps. 4. Botany—Southern States—Classification.
5. Phytogeography—Southern States—Maps. I. Title.
II. Series.
QK495.S16A74 1986 583'.981 85-26871
ISBN 0-912861-09-6 (lib. bdg. : alk. paper)

THE GENUS SALIX (SALICACEAE) IN THE SOUTHEASTERN UNITED STATES

George W. Argus
Botany Division
National Museum of Natural Sciences
Ottawa, Ontario, Canada K1A 0M8

Abstract. The taxonomy of *Salix* in the southeastern United States was studied in the field and herbarium. Twelve species and one variety of native willows are recognized and seven species and one subspecies are shown to be introduced and naturalized. A quantitative study of variation in *S. caroliniana* and *S. nigra* supported species rank for these entities and showed that infraspecific variation was too diffuse to warrant taxonomic recognition. *Salix humilis* has been shown to consist of two varieties: var. *humilis* and var. *microphylla*. *Salix floridana* is an endemic species that belongs in sect. *Humboldtiana*. It may be a Tertiary relict that survived the Pleistocene glaciation in northern Florida. The phytogeographical patterns displayed by this and other species of *Salix* are related to Pleistocene events. During the Full-Glacial some northern willows presumably extended far south of their present ranges and some northern species had their ranges fragmented. Some northern species occur here now as peripheral species (*S. amygdaloides, S. bebbiana, S. discolor,* and *S. lucida*); others, such as *S. eriocephala,* left major disjunct populations behind following deglaciation. The disjunct Ozark population of *S. caroliniana* may also be related to these events, or it may have had its origin in Tertiary times. The recognition of these taxa of *Salix* is aided by keys to staminate, pistillate, and vegetative specimens, detailed descriptions, and comparison with related and similar species. Taxonomy, nomenclature, variation, and hybridization of each native species are discussed. Each species is illustrated; range maps are also provided.

INTRODUCTION

Salix is a genus of about 400 to 500 species distributed mainly in the northern parts of North America and Europe, and in the mountains of China. It is absent from or uncommon in tropical regions, and only three species are native to Central and South America. In the Western Hemisphere there are about 150 species. Of these, twelve species and one variety are native to the southeastern United States, and seven introduced species have become naturalized there. The southeastern United States, as defined in this study, includes Delaware, Maryland, Virginia, North Carolina, South Carolina, Georgia, Florida, Alabama, Mississippi, Louisiana, Arkansas, Tennessee, Kentucky, and West Virginia (Massey et al. 1981).

Willows occur throughout the Southeastern States, but relatively few species occur in any one place. Populations containing only one or two species are common. Some species in the region are wide-ranging (*S. nigra, S. caroliniana*), whereas others are localized (*S. floridana*). Between these extremes a number of species occupy relatively small areas; they may represent relicts of once widespread Pleistocene populations (*S. discolor, S. bebbiana*).

The taxonomy and phytogeography of the southeastern willows was studied by C. R. Ball and his collaborators during the first half of this century. During this time he produced a treatment of the species of *Salix* in the Southeastern

States (Ball 1938), as well as regional treatments for Maryland (Ball 1910), Virginia (Massey & Ball 1944), and West Virginia (Davis & Ball 1947), and a study of the endemic *S. floridana* (Ball 1943). *Salix* also has been treated in works on the trees and shrubs of the region, notably, *Trees of the Southeastern States* (Coker & Totten 1937) and *Trees of Northern Florida* (Kurz & Godfrey 1962), numerous regional guides to woody plants (see Little 1976b), and checklists. The ecology and biology of *S. caroliniana* in southern Florida have been treated by Craighead (1971) and by Tomlinson (1980). The most definitive taxonomic treatment of North American willows was made by Schneider between 1918 and 1921. His studies were handicapped, however, by a lack of opportunity for field study and a relative paucity of herbarium specimens, but his treatment of nomenclature and his discussion of historical material are invaluable.

The first North American willows to be described were species that occur in the Southeastern States: *S. nigra, S. humilis,* and *S. sericea* (Marshall 1785). The most recent species to be described from this region was the endemic *S. floridana* (Chapman 1860). The composition of the *Salix* flora of the Southeastern States has long been known, but many taxonomic questions have remained unsolved. What is the taxonomic relationship between *S. nigra* and *S. caroliniana*? Are they distinct species or are they subspecies that intergrade where their ranges overlap? To what degree do they hybridize, and, if they do hybridize freely, why do they not intergrade completely in the Ozark region where they are sympatric? Do the Ozark and interior populations of *S. caroliniana* constitute a separate race? Are *S. humilis* and *S. occidentalis* (*S. tristis*) distinct species or do they simply represent extremes of variation? What is the origin and taxonomic relationship of *S. floridana*? Is it a Tertiary relict related to the tropical Asian species *S. tetrasperma*? Or is it of hybrid origin, perhaps involving *S. caroliniana*? Exactly what effect did Pleistocene events have on the *Salix* flora of the Southeastern States? During the Pleistocene *Salix* appears to have flourished throughout the region, and many northern species extended their ranges far south of their present-day ranges. Did species such as *S. discolor* and *S. humilis* hybridize during the Pleistocene and leave behind introgressive populations? Could the unusual populations of *S. humilis* that occur today across the southern states from northern Florida to Louisiana have resulted from hybridization between that species and *S. occidentalis,* perhaps sometime during the Pleistocene? Questions of this kind are numerous. As pointed out by Kral (1981), there is no doubt that a lack of distributional information conceals real taxonomic and phytogeographical problems. The lack of information, however, lies not only in geographical ranges but extends to morphological variation, ecological requirements, chromosome numbers, hybridization, and general biology. The recent review of the biology of *S. caroliniana* by Tomlinson (1980) is a model for similar studies of other species.

The purposes of this study are to revise the taxonomy of the species of *Salix* of the southeastern United States, to elucidate the phytogeography of the species, and to make the identification of specimens comprehensible to the non-specialist. To this end I studied herbarium materials from numerous herbaria (see Acknowledgments) and conducted eight weeks of field study during three field seasons. The ranges of all taxa were mapped in detail, some for the first time, and data from specimens were computerized. Keys were written to aid in the identification of specimens in staminate, pistillate, and vegetative stages of development.

Since it is recognized that keys will not permit the identification of all specimens, detailed descriptions were prepared as well as illustrations and notes on recognition. Nomenclatural problems were studied, and the types of many names and much original material were examined. Finally, I attempted to solve the taxonomic problems presented by these species by both field and herbarium studies, some involving the statistical analysis of population samples.

Problems in the genus *Salix* are the result of a number of factors. First, the species are inherently variable. For example, in *S. caroliniana* and *S. nigra* variation in leaf and branchlet indumentum, which in other groups may be of taxonomic significance, occurs at random in many populations. Second, hybridization, although probably not as common as was suspected by some workers, does occur, and some of the blurring of lines between species may be due to present or past hybridization. The resolution of problems of hybridization may require the application of chemical studies, such as isozyme comparisons, as well as the more conventional cytogenetic studies. Third, the pattern of annual development, which varies from species that flower before the leaves emerge to those that flower at the same time as the leaves emerge (see below), makes it difficult to obtain all of the taxonomically important structures at any one time. It is therefore necessary to revisit the same populations or individuals at several times during the year. Much of the field work that I conducted yielded specimens in only one or another stage of development and emphasized the advantage of "living with the species" to come to understand them. It is hoped that botanists in the Southeastern States will find some of the problems described here sufficiently interesting to seek their own solutions.

MATERIALS AND METHODS

This study was based on more than 10,500 herbarium specimens borrowed from or studied at 59 herbaria (see Acknowledgments). All specimens were critically examined and label data were recorded for mapping; this documentation is available as a separate report (Argus 1985). A large number of these specimens was used to compile species descriptions and to conduct the morphological analyses.

Herbarium specimens were supplemented by field studies conducted in 1981, 1984, and 1985, during which time all of the Southeastern States, except Louisiana, were visited. Samples of field populations made in the area of overlap of *S. caroliniana–S. nigra* and of *S. humilis* var. *humilis–S. humilis* var. *microphylla* were supplemented by population samples collected in northern Florida by Robert K. Godfrey and by Kent D. Perkins. All of the native species with their main ranges in the Southeastern States were seen and studied in the field; the northern species previously had been studied in the Northern States and Canada.

An attempt was made to typify the names of all native taxa that occur in or near the study area. The application of names for which no types were seen was based on the protologue, especially the descriptions, and sometimes on the judgment of earlier students of the genus. Names, the application of which is still doubtful, are preceded by a question mark. Complete synonymy is given for most names except for introduced species and some peripheral species, including *S. bebbiana, S. discolor, S. exigua,* and *S. lucida.*

Naturalized taxa are not typified in this study. Linnaean species of *Salix* have never been adequately typified, and even European workers find it necessary to "employ these names in the traditional manner" (Skvortsov 1968: 33). In 1907 Enander published a lengthy study of *Salix* in the Linnaean Herbarium, but his inability to see "pure species" in but few of Linnaeus's specimens makes his work inadequate for typification of Linnaean names in *Salix* (Skvortsov 1968). The typification of these Linnaean names remains a task for future investigations.

Many of N. J. Andersson's infraspecific names were not assigned rank in their original publications. His usage of Greek letters, symbols, or the arrangement of names to indicate rank was inconsistent, sometimes even within the same publication, and it is inadvisable to assign rank automatically to them, as was suggested in Stafleu and Cowan (1976). Since all of Andersson's names that lack designated rank are synonyms, I have cited them here exactly as they appeared in the original publication and have omitted reference to the ranks attributed to them by other authors.

The descriptions of the taxa are based mainly on specimens from the Southeastern States. For species that are poorly represented in the area, such as *S. amygdaloides, S. bebbiana, S. discolor,* and *S. lucida,* southeastern specimens were supplemented by extraterritorial specimens or the literature. The description of naturalized species is often supplemented by specimens from Europe and Asia and from published descriptions, including Skvortsov (1968), Chmelar and Meusel (1976), Rechinger (1964), Roer and Elven (1975), and Paiero and Schoepf (1981).

The descriptive terminology used here is based mainly on Lawrence (1951), as required by the Flora Project (Massey et al. 1981), but some exceptions are made in accordance with recent literature (Stearn 1966; Argus 1973; Payne 1978); see section Morphology.

The descriptions of the taxa are based on mature, well-developed organs, except when immature organs are being described. Leaf descriptions are based on the largest mature leaf on a specimen; ament and flowering branchlet length are based on the largest flowering or fruiting inflorescence on a specimen; stipe and capsule length are based on one or more measurements taken from the lower third of an ament with fully open capsules. Stipe measurements in *S. caroliniana* and *S. nigra* were made on all specimens with mature fruit. The number of individuals measured or observed for descriptive purposes varied with the variability of the character. For statistical purposes a subset of the specimens was selected, usually on the basis of their well-developed foliage and/or aments.

The range of variation in quantitative characters is expressed in terms of mean, normal range of variation, and sometimes the extremes of variation. For example, in *S. caroliniana* leaf length is given as (5–) 7.5–11.5–15 (–22) cm. The value 11.5 is the mean, 7.5–15 is the normal range of variation, and 5 and 22 are extreme values.

The quantitative characteristics of leaves and stipes are analysed statistically by the use of Dice-grams (Hubbs & Hubbs 1953), which graphically represent some statistical parameters of the sampled population. For example, in Fig. 4 the long horizontal line represents the range; the short vertical line the sample mean; the long white bar one standard deviation on either side of the mean; and the short black bar two standard errors on either side of the mean. This graphical presentation permits the determination of the approximate statistical significance of the two means. In discussing this method of analysis Critchfield (1957) wrote, "the significance of a difference between the two sample means . . . may be determined graphically by aligning a straight-edge vertically at the end of either bar representing four standard errors. If the corresponding bar of the other sample does not touch, or just touches, the straight edge, the difference is significant (P value: .01 to .005). If the two bars overlap, the difference is not significant (same P value)." He also points out that statistical problems may be caused by differences in sample size. In the data presented here the differences in sample size and standard error are not large enough to be a major source of error.

Phenological data were recorded for all specimens with flowering aments. Staminate aments deemed to be in anthesis had stamens with unopened anthers or loose pollen and pistillate aments had stigmas that appeared to be receptive.

Distribution maps were prepared from computerized data of specimen labels; the data were machine-sorted geographically and hand-plotted by county. Literature records were plotted only for extraterritorial states and were usually based on the following sources: Great Plains (McGregor & Barkley 1977), Illinois (Mohlenbrock 1980), Indiana (Deam 1940), Missouri (Steyermark 1963), Ohio (Braun 1961; Cusick & Silberhorn 1977), Pennsylvania (Wherry et al. 1979), and Texas (Ball 1961). The North American range maps were based on published literature and on herbarium specimens as indicated in the captions.

Herbarium labels were recorded and computerized for all specimens examined in the study. The data recorded include: taxon, state, county, principal collector, collection number, date of collection, and herbarium acronym. A printout of this computerized catalogue (Argus 1985) is available free-of-charge from: Information Centre, Museum of Natural Sciences, Ottawa, Ontario, Canada, K1A 0M8. Copies of the catalogue have been distributed to major North American botanical libraries. Representative specimens from the catalogue are cited in this paper and include, at least, one specimen from each county of occurrence and all specimens cited in the text. The specimens selected were those most widely distributed in herbaria or the most recent collection.

The physiographic zones in which the species occur was determined by a comparison of the distribution map with a map of physiographic zones (Fig. 1).

MORPHOLOGY

The general morphological features of *Salix* have been discussed previously (Argus 1965; Skvortsov 1968); the comments here will apply mainly to the species as they occur in the Southeastern States. Many of these features are illustrated in the figures.

Habit. The habit of *Salix* varies from trees 20 m or more tall to decumbent shrubs that may be only 3 dm tall. Only *S. nigra, S. caroliniana, S. amygdaloides, S. floridana, S. lucida, S. pentandra, S. alba, S. fragilis,* and *S. babylonica* are trees; some of the other species in the region may grow as tall as 6 m, but they branch from the base and are considered to be shrubs. The only "dwarf" species is *S. humilis* var. *microphylla.* Two species are clonal, *S. exigua* by root shoots, and *S. humilis* by layering. The production of root shoots, termed soboliferous by Eckenwalder (pers. comm.), is very unusual in *Salix,* and only the members of

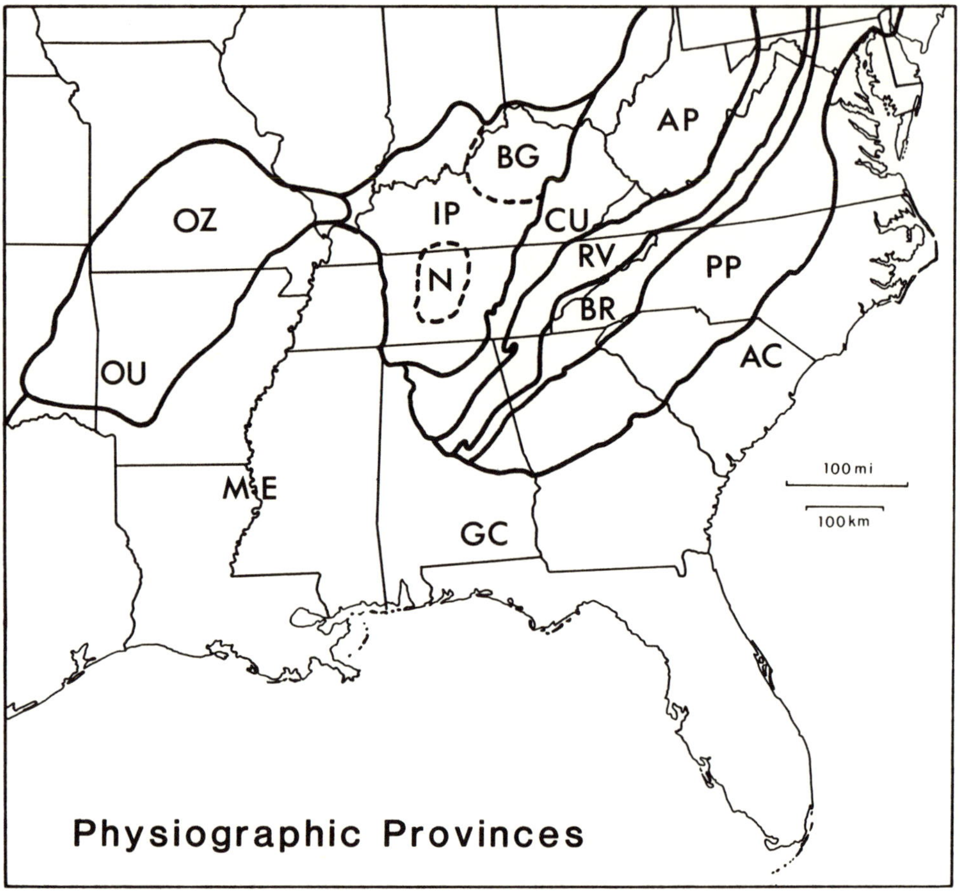

FIG. 1. Physiographic provinces and subdivisions of the Southeastern United States. *Coastal Plain:* AC–Atlantic Coastal Plain, GC–Gulf Coastal Plain, ME–Mississippi Embayment. *Piedmont:* BG–Blue Grass Basin, IP–Interior Low Plateaus, N–Nashville Basin, PP–Piedmont Plateau. *Mountains:* AP–Allegheny Plateau, BR–Blue Ridge, CU–Cumberland Plateau, OU–Ouachita Mountains, OZ–Ozark Mountains, RV–Ridge and Valley. Based on Atwood (1940), Lobeck (1950), and Massey et al. (1981).

sect. *Longifoliae* exhibit this form of reproduction. Layering occurs in *S. humilis* as the decumbent branches are buried in moss or litter and take root. Sometimes patches several meters across may be derived in this way from a single individual. A third form of vegetative propagation takes place when branches break from an individual and take root separately. This mode of propagation is common in species that occur on alluvial substrates, many of which are adapted to it by possessing very brittle branches that readily break free and can be dispersed by water (see below).

Branches. The branches of *Salix* vary from pubescent to glabrous. The young branches or branchlets may be clothed with white, gray, brown, or ferruginous hairs that are usually lost progressively throughout the season. For example, by the end of the growing season hairs may be retained by *S. sericea* only behind the buds. Infraspecific variation in branchlet indumentum is common in some species; in *S. caroliniana* the branchlets of some plants may be glabrous whereas in others they may be densely villous. The terminology used here to describe indumentum types is discussed below. The branches and branchlets of members of sect. *Humboldtiana,* sect. *Salix,* and in *S. sericea* are usually brittle at the base and break cleanly from the stem when flexed at the base or when bent by strong winds or flood waters. *Salix amygdaloides* is exceptional in that its branches are sometimes tenacious; the branches in some other members of sect. *Humboldtiana* may be variously brittle. The branches of *S. babylonica* and one cultivar of *S. alba* are pendulous; other species have spreading or erect branches. The wood under the bark of most species is smooth, but in some members of sect. *Vetrix* the wood has prominent ridges that vary from less than 5 mm long to several centimeters (see *S. cinerea,* no. 15, for discussion).

Buds. The buds of *Salix* vary in shape, size, color, indumentum, and in the degree of coalescence of the bud scale itself. The bud scale is formed by the coalescence of two lateral prophylls. Most species have a single calyptrate or caplike bud scale. In sect. *Humboldtiana,* other primitive extraterritorial species of *Salix,* and in *Chosenia* Nakai the margins of the bud scale are free and overlapping on the adaxial side of the bud. This characteristic is useful to distinguish *S. caroliniana, S. nigra, S. amygdaloides,* and *S. floridana* from other southeastern *Salix* species and indicates a relationship between these relatively primitive tropical and subtropical species and the eastern Asian sect. *Urbanianae* C. Schneider and the genus *Chosenia.* This feature is constant (Chmelar 1978), and it has been used by Nakai (1928) and Kimura (1928) to erect subgenera (see below).

Leaves. There are three kinds of leaves in *Salix:* (1) ordinary leaves that appear toward the end of the growing season from about the lower third of the branchlet to the tip, (2) the reduced leaves that occur on the lowermost part of the vegetative branchlets and on the flowering branchlets, and (3) sprout-shoot leaves that are usually more vigorous and more prominently stipulate.

The leaves at the base of vegetative branchlets and those on the flowering branchlets are similar to each other in morphology and indumentum, and both differ in size and shape from the ordinary leaves. On a branchlet they range from bractlike cataphylls to larger structures that intergrade into the typical leaves. In some cases these leaves provide useful taxonomic characters (Argus 1965).

Leaf blades. The leaf blades range in shape from narrowly elliptic, or linear, as in *S. exigua,* to broadly elliptic or obovate. The shape as expressed by the ratio of leaf length to width is as low as 2 in *S. pentandra, S. bebbiana, S. caprea,* and *S. cinerea* and as high as 31 in *S. exigua.* The narrow-leaved species have a mean length/width range from 4.5 to 9 and the broad-leaved species a range from 2.5 to 5. The leaves are often covered beneath, and rarely above, with a whitish, waxy bloom, termed here glaucescence. Leaf glaucescence may be lost if excessive heat is used in drying herbarium specimens. The presence or absence of this bloom is usually constant in southeastern species and is taxonomically useful (see *S. nigra,* no. 2). In the Southeastern States the only species that lack leaf glaucescence are *S. nigra, S. lucida,* and *S. pentandra.* In *S. humilis* var. *microphylla* the leaves are densely tomentose beneath, and the wax is not evident. The leaf margins are flat in most species, but in *S. humilis, S. caprea,* and *S. cinerea* they are revolute. Leaf margins vary from entire to serrate. In *S. exigua,* and to a certain extent in

S. babylonica, the teeth are elongate and widely spaced. In most other species they are relatively uniformly spaced, and the margins are serrate to serrulate (sections *Humboldtiana, Salicaster, Salix, Vimen,* and *Cordatae*). In sect. *Vimen* the denticulations are less well-developed, and the leaves are entire to crenate or, as in *S. bebbiana,* coarsely serrate. In this section the leaf margins, even when entire, are usually distantly glandular-dotted. The glands are not borne on the edge of the leaf, as in other sections, but are slightly inset from the edge. In *S. purpurea* the leaves are entire at the base, distantly serrate in the midsection, and closely serrulate toward the apex.

The lower surface of the leaves is always covered with numerous stomata. Stomata also appear on the upper surface of the cataphylls, but they are uncommon on the upper surface of ordinary leaves. When stomata do occur on both leaf surfaces (leaves amphistomatous), this condition is usually of taxonomic importance. Amphistomatous leaves occur in *S. exigua, S. alba, S. babylonica,* and *S. fragilis* and in about 50% of specimens of *S. nigra* (which see for discussion). All other species in our flora have stomata only in the lower epidermis (leaves hypostomatous).

The pubescence of leaves, and also of branchlets, pistils, and floral parts, varies from species to species. The hairs or trichomes are simple, unicellular, and usually tubular, tapering to a point. In some species, e.g., *S. caroliniana,* the trichomes are flattened tubes with thickened edges, and in members of sect. *Myrtosalix* A. Kerner they are flat (ribbonlike), without thickened edges, and crinkled. The hairs may be short or long, straight or wavy, erect or appressed, dense or sparse, and colored white, gray, yellowish, or ferruginous. These differences are useful to the taxonomist who, through experience, may come to recognize specific patterns. The description of these differences, however, is difficult because of the imprecision and inconsistent usage of the available nomenclature. An attempt has been made to relate indumentum terminology to specific examples (Argus 1973), but the coverage is not comprehensive. In this paper the terminology for indumentum types generally follows Lawrence (1951) but as modified by Payne (1978). The terms are defined in the following way:

A. Indumentum of short hairs.
Lanulose (woolly). Densely curly.
Pubescent. Soft, dense, or generally hairy.
Puberulent. Soft, very short, sparse.
Sericeous (silky). Straight, appressed, glossy.
Tomentose. Rigid, densely lanate.
Velutinous. Dense, erect, moderately rigid.

B. Indumentum of long hairs.
Hirsute. Stiff.
Lanate (woolly). Dense curly, matted.
Pilose. Soft, shaggy, sparse.
Sericeous (silky). Straight, appressed, glossy.
Villous. Soft, curly but not matted.

Indumentum referred to as sericeous because of its shiny, silky appearance is usually composed of long trichomes (ca 0.3–0.6 mm long), but in some species, cf. *S. sericea,* the hairs are very short (ca 0.2–0.3 mm long). I refer to this indumentum as short-sericeous.

The type of pubescence on the lower surface of leaves is characteristic of some species. The leaves of *S. sericea* are short-sericeous beneath with short, appressed hairs, in *S. humilis* and *S. cinerea* the leaves are densely tomentose beneath with rigid curly hairs, and in *S. bebbiana* the pubescence is sericeous-tomentose. In most other species the mature leaves are glabrous or soon glabrate. Ferruginous, or rust-colored, hairs occur in *S. lucida, S. sericea,* and in species of sect. *Vetrix,* except for *S. bebbiana* and *S. cinerea* subsp. *cinerea.*

Petioles. Petioles vary in length from species to species. In *S. exigua* and *S. humilis* var. *microphylla* the petioles are characteristically very short (1–5 mm long), whereas in *S. amygdaloides* and *S. floridana* they are long (7–21 mm long). In members of sections *Humboldtiana, Salicaster,* and *Salix* glandular dots or processes occur at the distal end of the petiole near the base of the leaf blade (Fig. 22). In *S. babylonica* the glands may be absent or may appear only as two minute dots; all other species are eglandular.

Stipules. Stipules are characteristic of most species of *Salix* but in some they are lost early (caducous), whereas in others, such as *S. amygdaloides, S. humilis* var. *microphylla,* and *S. sericea,* they are undeveloped or minute. The largest stipules occur on sprout shoots or on other vigorous shoots. The shape of stipules is often more useful taxonomically than size. Features that are useful are the curvature of the axis, the symmetry of the stipule, the toothing on the margin, and the presence of glandular processes on the upper surface; but even these features may display considerable variation (cf. *S. caroliniana* and *S. nigra*).

Flowering. The relative time of flower and leaf emergence is described by the following terms: precocious—flowers appear before the leaves emerge; subprecocious—flowers appear just before the leaves emerge; coetaneous—flowers appear at the same time as the leaves emerge; serotinous—flowers appear after the leaves emerge. The aments of precocious flowering plants are usually sessile on the branch. They are subtended by two or three greenish or leafy bracts that grade into the floral bracts that subtend the flowers (Figs. 34, 41). In coetaneous and serotinous plants the aments are borne on distinct flowering branchlets, which bear leaves corresponding to those on the lower portion of vegetative branchlets (Figs. 18, 30). These short branchlets have been termed spur shoots by Tomlinson (1980).

Aments. The staminate aments or catkins are shed soon after anthesis, the pistillate ones after the seeds are shed. In species in which the aments are borne on flowering branchlets the aments and branchlets may either fall together or they may separate and fall separately (e.g., *S. caroliniana*). In some species the flowering branchlets may persist and, in *S. caroliniana* (according to Tomlinson 1980), may form a considerable part of the renewal growth of the season by sylleptic branching (see Figs. 3, 11). In this condition a branch emerges from the axil of the distal leaf on the flowering branchlet contemporaneously with the ament without bud formation or an initial period of dormancy. This new shoot grows alongside the ament and makes the ament appear to be lateral. Tomlinson noted this development in *S. caroliniana* and in about 30% of South Florida trees. It has been observed also to occur in *S. nigra, S. gooddingii* C. Ball, and *S. humboldtiana* Willdenow of sect. *Humboldtiana,* and in *S. exigua* of sect. *Longifoliae.* It may occur in other members of sect. *Humboldtiana,* but it has not yet been observed in *S. amygdaloides.*

Floral bracts. A single floral bract subtends each flower. In the staminate flowers of all species these bracts are persistent, but they are deciduous in the pistillate flowers of sections *Humboldtiana* (except in *S. floridana,* no. 4, see for discussion), *Salicaster, Salix,* and *Longifoliae.* They are persistent in the pistillate flowers of the other sections. The color of the bracts varies from pale tawny to dark brown, or sometimes brown at the tip and pale below. Unpigmented bracts, described as tawny, occur in all of the sections that have deciduous pistillate bracts as well as in sect. *Subalbae* and in *S. bebbiana* of sect. *Vetrix.* In some plants of *S. bebbiana* they are pinkish or brownish at the tip; the bracts of this species are not strictly unpigmented. The floral bracts are sometimes glabrate, but commonly they are clothed with hairs that may be long or short, straight, wavy, or crispate, and white or ferruginous.

Staminate flowers. In members of sections *Humboldtiana* and *Salicaster* the number of stamens per flower averages about six with a range of three to twelve (Figs. 2, 18). The flowers in all other sections have two stamens; in sect. *Helix* they are fused into one (Fig. 50). The nectaries in *S. caroliniana, S. nigra,* and *S. amygdaloides* vary from one or two to several slender lobes surrounding or intermixed with the stamens. In *S. floridana,* and in sections *Salicaster, Salix,* and *Longifoliae* there are two nectaries, one located adaxially and one abaxially. All other species have only one abaxial nectary.

Pistillate flowers. Each pistillate flower consists of a single, bicarpellate, unilocular pistil that may be sessile (*S. purpurea* and in some plants of *S. babylonica*) or borne on a stipe or gynophore ("pedicel" of authors). The long stipes of *S. bebbiana* are often diagnostic of the species, and the long stipes of *S. caroliniana, S. amygdaloides,* and *S. floridana* separate them from the related *S. nigra.* The pistils are glabrous in sections *Humboldtiana, Salicaster, Salix,* and *Subalbae.* In sect. *Longifoliae* and in subg. *Vetrix* the pistils are usually pubescent. The styles are usually adnate or only slightly bifurcate. At the base of the stipe is a nectary, usually square or oblong and located abaxially. In *S. caroliniana* and *S. amygdaloides* the nectary may partly surround the stipe, or there may be lateral nectaries. *Salix pentandra* and *S. fragilis* occasionally have flowers with two nectaries, abaxially and adaxially.

Capsules. The ripened fruit is a capsule. The indumentum of the capsule is usually less dense than in the flowering pistil, but it is rarely lost entirely. The two capsule valves separate along the suture, and, remaining attached at the base, spread away from each other and expose the seeds borne near their base. In some species the valves curl back on themselves (e.g., *S. bebbiana*), but in most they are simply arcuate.

Ovules. In our species the number of ovules per carpel ranges from 2 to 15. There is some infraspecific variation, but in general each species has a basic number of ovules per carpel around which the variation may occur; for example, *S. humilis* usually has 3 ovules per carpel but occasionally may have 2, 4, or 5 ovules per carpel. Ovule number is potentially of taxonomic importance (Chmelar 1978) (see also *S. floridana,* no. 4), but more observations need to be made of this characteristic. In the Southeastern States *S. floridana* and *S. babylonica* have 2 ovules per carpel; *S. humilis, S. sericea,* and *S. purpurea* have 3; *S. alba, S.*

fragilis, S. cinerea have 3 to 6; *S. caroliniana, S. nigra, S. amygdaloides, S. bebbiana, S. discolor, S. caprea,* and *S. eriocephala* have 6 to 9; *S. lucida* and *S. pentandra* have 8 to 12. The highest number of ovules, in a southeastern willow, occurs in *S. exigua* which has 9 to 15.

Seeds. The seeds of *Salix* are green or reddish, narrowly oblong or obovate, and vary in length from about 0.8 to 3 mm. They are surrounded by a spreading coma of fine, silky hairs, which exceed the length of the seed. The hairs are attached at one end to form a ring in which the radical end of the seed is seated. The fine hairs surrounding each seed facilitate dispersal by wind. The seeds of *Salix* are generally very similar, except for size, and they do not seem to provide differences of taxonomic significance.

CYTOLOGY

Salix is one of the few woody genera with a large number of polyploid taxa (Suda 1963). The base chromosome number in the genus is 19, and, based on this number, ploidal levels range from diploid (2x) to dodecaploid (12x) (Suda & Argus 1968). Species with a base number of 22 have been reported by a number of workers (cf. Suda 1963), but these may be due to fragmentation or to inaccurate counts. Chromosome numbers have not yet played a significant role in the taxonomy of the genus (Skvortsov 1968), although theoretically they should be of fundamental importance in understanding the evolution of the group. They have been useful in recognizing the parentage of some hybrids (Chmelar 1979), but the variability in chromosome numbers that occurs within some sections and even within some species is still perplexing. Chromosomes of *Salix* are very small (about 1 micron long) and are difficult to count accurately and rapidly enough for a sufficiently large sample of both species and populations to be available for analysis. A secondary problem is the failure of some early workers to preserve voucher specimens, so that the identity of their material cannot be verified.

No chromosome counts have been made on southeastern material, and no counts are available for *S. caroliniana, S. floridana, S. humilis* var. *microphylla,* or *S. sericea.* A summary of the available counts is given in Appendix I.

CLASSIFICATION

The family Salicaceae consists of three genera: *Populus* Linnaeus, *Chosenia* Nakai, and *Salix* Linnaeus. *Populus* is anemophilous and *Salix* is basically entomophilous, although wind pollination in *Salix* may be more common than generally thought (Argus 1974). The reproductive morphology of these genera generally reflects this basic difference. In *Populus* the aments are loose and pendulous, there are numerous stamens (5–80), the flowers are eglandular, and the pollen grains are thin-walled and non-aperturate. In *Salix* the aments, with few exceptions, are compact and erect, there are usually two stamens (some groups have up to seven stamens and in others they are reduced by fusion to one), the flowers have nectaries, and the pollen grains are thick-walled and tricolpate. In addition, poplars are trees with monopodial branching, their wood has homogeneous rays, their shoots are often heterophyllous, and the buds have three to ten bud scales.

The willows, on the other hand, are usually shrubs or less commonly trees with sympodial branching, their wood has heterogeneous rays, their shoots are homophyllous, and the buds are enclosed by a single bud scale. The differences between these genera are considered to be so great that some authors (Kuprianova 1965; Rouleau 1970) have proposed placing them in separate families; but this unduly overemphasizes their differences and does not recognize their similarities and the intermediate position of *Chosenia.*

The monotypic genus *Chosenia* is very similar to *Salix* and may have been derived from a very early *Salix* ancestor (Skvortsov 1968). *Chosenia arbutifolia* (Pallas) A. Skvortsov resembles *Populus* in being an anemophilous tree with pendulous aments, and in having wood with homogeneous rays. Nectaries are absent or rudimentary, and the stigmas are deciduous. Its pollen grains have an entomophilous character, but the colpi are reduced as is typical of anemophilous plants (Kuprianova 1965). With respect to flowers and floral development *Chosenia* resembles *Salix* (Helmqvist 1948), especially members of sect. *Humboldtiana,* which also have buds with an imbricate bud scale, caducous floral bracts, glabrous pistils, and several stamens. There seem to be sufficient grounds for recognizing *Chosenia* as a distinct genus, but it is very similar to some species of *Salix,* and the features in which it resembles *Populus* may be secondarily derived (Skvortsov 1968).

The infrageneric classification of *Salix* in the southeastern United States is summarized in Appendix II. The genus *Salix* consists of three subgenera, *Salix, Vetrix* (Dumortier) Dumortier, and *Helix* Dumortier. A fourth subgenus, *Chamaetia* (Dumortier) Nasarov, is recognized by Skvortsov (1968), but it is best included in subg. *Vetrix* (Dorn 1976). The recognition of only two subgenera based on the nature of the bud scale, as proposed in 1928 by Kimura (*Euitea* with a free overlapping margin and *Protitea* with a coalescent bud scale) and in the same year by Nakai (*Pleiolepis* and *Calyptrolepis,* respectively), may prove to be more natural than the classification adopted here, but it is not yet generally accepted.

Subgenus *Salix* is a heterogeneous group in which some sections have flowers with several stamens and others have flowers with only two; the buds in some sections have a bud scale with a free, overlapping margin and in others it is calyptrate; and in some sections the bracts of pistillate flowers are deciduous after flowering, whereas in others the bracts are persistent. All members of subg. *Salix* have glands on the petiole near the base of the blade (although these are poorly developed in sect. *Subalbae*) and have pale non-pigmented bracts (which also occur in some members of subg. *Vetrix*). In the Southeastern States sections *Humboldtiana, Longifoliae, Salicaster,* and the naturalized species of sections *Salix* and *Subalbae* are all placed in subg. *Salix.*

Subgenus *Vetrix* is made up of a large number of species that have two stamens (sometimes reduced to one), petioles that lack glands at the base of the blade, floral bracts that are usually pigmented and persistent, and buds that have a single calyptrate bud scale. In the Southeastern States this subgenus includes sections *Vetrix, Vimen,* and *Cordatae.*

Subgenus *Helix* is represented in our flora only by sect. *Helix,* which includes the naturalized *S. purpurea.* The subgenus is characterized by staminate flowers with one short, almost square, nectary, one stamen derived from two connate filaments and anthers, and leaves that are mostly narrow, entire or serrulate, and mostly amphistomatous.

ECOLOGY

The indigenous species of *Salix* in the Southeastern States fall into two general ecological classes, riparian and non-riparian. The species that occur predominantly in riparian habitats are *S. caroliniana, S. nigra, S. amygdaloides, S. floridana, S. lucida, S. exigua, S. sericea,* and *S. eriocephala.* Those that occur predominantly in non-riparian habitats are *S. bebbiana, S. discolor,* and *S. humilis.* All of the latter species are in sect. *Vetrix.*

It was noted by Skvortsov (1968) that riparian species require a well-aerated substrate and flowing moisture, and that these species are usually trees with high energy requirements. The non-riparian species are less exacting with respect to soil aeration and frequently can tolerate even stagnant moisture conditions. Riparian species are most common on the margins of rivers and streams, but they are not restricted to these habitats and may occur also in other wetland habitats, such as swamps, marshes, meadows, and roadside ditches. They do not, however, occur in upland woods or in the drier habitats sometimes occupied by some non-riparian species. The range of habitats occupied by species of *Salix* is controlled more by requirements for seed germination and ecesis than by the moisture requirements of adult plants. This has been shown for *S. nigra* (McLeod & McPherson 1973), and it holds for other species as well.

Willows generally require two basic conditions for colonization of a habitat: moist substrate at the time that the seeds are shed (usually late spring and early summer) and ample sunlight (plants are heliophilic). Seeds do not generally survive long in the absence of moisture, and seedlings will not survive or thrive under shade. Suitable conditions for growth are often provided by man-made habitats, and willows may occupy a wide variety of disturbed habitats including roadside ditches, railroad rights-of-way, old fields, old mine tailings, gravel pits, etc.

PHYTOGEOGRAPHY

The geographical distributions of the 12 species of *Salix* indigenous to the Southeastern States form a cline ranging from species with their main ranges in the south, or with minor extension into the northern peripheral states, to species with their main ranges in the northern glaciated region and only peripheral occurrences in the Southeastern States. It is possible to recognize in these ranges six general patterns.

1. *Southern Coastal Plain Species. Salix caroliniana* (Fig. 2) and *S. floridana* (Fig. 17) have their main ranges in the Southeastern States; the former has outliers in the northern states and on the Great Plains.

2. *Eastern Temperate Species. Salix nigra* (Fig. 11) occurs from the Great Lakes south to the Gulf of Mexico and from the Atlantic coast to the eastern Great Plains. It is generally absent from the Atlantic Coastal Plain and from peninsular Florida.

3. *Appalachian-Northeastern Temperate Species. Salix humilis* var. *microphylla* (Fig. 39) and *S. sericea* (Fig. 48) range from the Great Lakes to the Appalachian Mountains.

4. *Appalachian-Southern Boreal Species. Salix humilis* var. *humilis* (Fig. 37) and *S. eriocephala* (Fig. 47) are northeastern taxa that extend westward to the Great Plains and southward through the Appalachian Mountains, sometimes with disjunct southern populations.

5. *Great Plains Species. Salix amygdaloides* (Fig. 14) and *S. exigua* (Fig. 30) are widespread in central North America. They are peripheral to the Southeastern States, where they occur mainly in the Mississippi and Ohio river valleys.

6. *Southern Boreal Species. Salix lucida* (Fig. 18), *S. bebbiana* (Fig. 32), and *S. discolor* (Fig. 34) occur mainly in formerly glaciated regions, but they still retain isolated populations on the northern edge of the Southeastern States.

The possible origin of these distributional patterns may be inferred by comparing them with the vegetational and climatic history of the Wisconsinin and Holocene Ages in eastern North America and the ecological behavior of the species.

During the Pleistocene Period climatic changes resulted in major disruptions of the vegetation of eastern North America (Maxwell & Davis 1972), which permitted northern species to penetrate the southeastern deciduous forest region and to become established far south of their present areas of occurrence. Evidence of the migration of northern floras well south during the Pleistocene was anticipated by Core (1932, 1938), who supposed that tundra conditions would have prevailed in West Virginia, southern Ohio, and southern Pennsylvania, and that northern species would have been forced hundreds of miles farther south. In 1949 Deevey, on the basis of biogeographical data, concluded that "glacial chilling in the Southeastern States must have been fairly extensive" and that "the warmth-loving species survived in peninsular Florida and in Mexico" (p. 1375). These views have been supported by recent palynological investigations that show, furthermore, that the climatic changes that led to this migration did not result simply in a southward shift of intact vegetation zones, but in their disassociation and migration as individual species (Maxwell & Davis 1972). This concept is very different from that proposed by Braun (1955) and by Graham (1965, p. 581), who concluded that "although the ranges of certain species were modified during the Pleistocene the general effect of climatic changes on the biota of Southeastern United States was not extensive." Studies of Late-Pleistocene palynology reveal very extensive biotic effects and species assemblages for which there are no modern analogs (Whitehead 1967; Watts 1971; and others). The use in the following discussion of names for vegetational assemblages, such as spruce-pine and oak-hickory, does not imply that they are the same as modern assemblages, for they differed from them both in species composition and species frequency (Maxwell & Davis 1972).

Salix species are adapted to the colonization of disturbed areas and open vegetation (see section Ecology). The dispersal of these species would have been facilitated by the disruption of the vegetation by climatic changes and the increased flow of water resulting from glacial melting. It is very likely that willows greatly expanded their ranges in the Southeastern States during this period of climatic and vegetational change. The present distribution of *Salix* in the Southeastern States is a reflection of past climatic events, the individual species' response to climatic changes, competition with other species, and the availability of suitable habitats.

Outline of the Vegetational History of the Southeastern States During the Wisconsinin and Holocene Ages.

The following history of the vegetation of the Southeastern States during the late Pleistocene is based mainly on the summary by Delcourt and Delcourt (1981). In their paper they present vegetation maps for seven time periods: 40,000 yr BP; 25,000 yr BP; 18,000 yr BP (midpoint of the Full-Glacial Interval); 14,000 yr BP (Late-Glacial Interval); 10,000 yr BP (early Holocene); and 200 yr BP (late Holocene). The vegetational changes that took place during this time period are particularly pertinent to the present distribution of *Salix.*

The most significant events in this regard began about 40,000 years ago with the extension of the Laurentide Ice Sheet into the Great Lakes region and the formation of discontinuous patches of tundra in the north-central Appalachian Mountains and of boreal forest, dominated by spruce (*Picea*) and jack pine (*Pinus banksiana*), from southern Pennsylvania to eastern Tennessee and central North Carolina. Peninsular Florida was covered by xeric sand dune scrub, and cypress-gum (*Taxodium-Nyssa*) swamp forest probably occured as far north as southeastern Mississippi (Delcourt & Delcourt 1981). Presumably this forest also occurred along the Gulf of Mexico coast, especially on peninsular Florida. In northern Florida pollen of *Salix,* "along with that of *Myrica, Gordonia, Ilex, Nyssa,* and Ericaceae" was conspicuous, which suggests the occurrence of grass-swamp vegetation (Watts 1969, p. 638). Deciduous taxa of the modern mesic cool-temperate forest may have found edaphically suitable conditions on the Blufflands (loess-mantled uplands) adjacent to the Mississippi Valley and on hill slopes and ravines adjacent to the major river systems (Delcourt & Delcourt 1977).

About 25,000 yr BP there was a period of mild warming followed by major climatic cooling. During the Full-Glacial Interval (18,000 yr BP) a belt of tundra adjacent to the ice sheet covered the northern tip of Kentucky and extended along the crest of the Appalachians south to the Great Smoky Mountains. Boreal forest, composed of spruce and jack pine, occupied the Great Plains, the Ozark Plateau, and southern Illinois and Indiana. Open jack pine forest, with subdominants of white spruce (*Picea glauca*) and fir (*Abies*), extended from southern Pennsylvania westward to the Mississippi valley and south to northern South Carolina and Mississippi. At this time the watertable in southeastern Virginia was high, but the regional vegetation was xeric (Whitehead 1967). White spruce forest occupied the entire Mississippi Valley and separated Appalachian populations from those in the Ozarks (Delcourt 1979). The flow of glacial meltwater down the Mississippi Valley ameliorated the climate by reducing extremes in temperature, which permitted mixed hardwood forest to exist on the uplands flanking the major drainage systems. A narrow band of cool-temperate conifers and northern hardwoods may have occurred from northern Mississippi to southern South Carolina (Delcourt & Delcourt 1981). A warm-temperate forest of oak (*Quercus*), hickory (*Carya*), and southern pine (*Pinus* spp.) covered a wide band from Texas to southern South Carolina south to northern peninsular Florida. No pollen evidence of Full-Glacial cypress-gum forest has been found, but Delcourt and Delcourt (1981) speculate that this forest may have occurred along the Gulf Coastal Plain where, during the Full-Glacial, sea level was about 100 m lower than it is now, and an extensive area of coastal plain would have been available for colonization.

The disintegration and northerly retreat of the Laurentide Ice Sheet started during the Late-Glacial, about 14,000 yr BP. Open forests of spruce and jack pine extended eastward across Kentucky and northern Tennessee into western West Virginia. Mixed conifer-hardwood forest expanded northward replacing jack pine-spruce forest. The vegetational assemblages to the south remained relatively stable. With the onset of major climatic amelioration at about 12,500 yr BP the Appalachian tundra declined slightly, and a band of spruce forest formed between the tundra and the boreal forest dominated by spruce and jack pine; spruce forest within the lower Mississippi Valley was replaced by cypress-gum forest. From about 8500 to 5000 yr BP lowered watertables probably were critical in controlling the distribution of Coastal Plain tree species (Watts 1971); but even as watertables fell in the region *Salix* was present (King & Allen 1977). Sea level rose rapidly during the period from about 14,000 to 7000 yr BP and then slowly and steadily to modern levels (Watts 1971).

The climatic warming that took place during the Hypsithermal Interval, about 5000 yr BP, resulted in major changes in the vegetation of the Southeastern States. The mixed hardwood forest was reduced in extent and was restricted to favorable gorge and slope habitats in the Cumberland and Allegheny plateaus and to the eastern margins of the Mississippi Valley. With the rise of watertables on peninsular Florida, southern pine forest started to replace the oak savannah and led to the development of "hammocks" of mesic deciduous trees (Watts 1971); extensive swamps and marshes developed along the present coasts.

During the late Holocene (200 yr BP) major deltaic systems were formed by the Mississippi River. With the advent of human settlement vegetational disruption was renewed and man-made habitats became available for *Salix*.

The climatic conditions that prevailed in the Southeastern States during the climatic deterioration that started about 40,000 yr BP and continued through the Full-Glacial Interval (18,000 yr BP) up to the Hypsithermal Interval (5000 yr BP) would have created environmental conditions favorable for the expansion of some northern species of *Salix* into the Southeastern States. Conditions necessary for the establishment of *Salix,* namely cool, moist climate, mineral soil, and open habitats, probably were common during this time. In a study of the unusual winter activity of some flightless genera of scarabaeid beetles Howden (1963) inferred that the summer climate of central Florida during the Full-Glacial was possibly equivalent to the summer climate of Marquette, Michigan, today. Riparian species would have found ideal habitats along the streams that drained glacial melt waters, particularly the braided streams that occupied the Mississippi Valley. Upland species would have found the open savannahlike vegetation and high watertables suitable for seedling germination and ecesis.

The open forests of jack pine, or red pine (*Pinus resinosa*), and spruce that extended as far south as central South Carolina and west to the Mississippi Valley during the Full-Glacial probably provided habitats very similar to those that prevail today in central Quebec, where marsh habitats are available for *S. discolor,* open, xeric forest habitats for *S. humilis* var. *humilis* and *S. bebbiana,* and riverine habitats for the riparian species *S. eriocephala* and *S. lucida.* It is probable that during the Full-Glacial a large number of *Salix* species, some now no longer represented in the region, occurred throughout the Southeastern States.

Species of southern affinities that occupy riparian habitats, such as *S. caroliniana* and *S. nigra,* may have been part of the Arcto-Tertiary flora, and may have occurred widely in the Southeastern States prior to the Pleistocene Age. These species may have occurred in Tertiary lowland forest similar to that which occurs in southeastern Missouri today (Steyermark 1959). During the Sangamonian Stage (ca 70,000 yr BP), when warm, dry climatic conditions prevailed in the Great Plains and cypress vegetation occurred in south-central Illinois (Grüger 1972), *S. caroliniana* and *S. nigra* may have ranged from the Coastal Plain through the Mississippi Valley to the Ozark Plateau. This connection between the Ozarks and the Coastal Plain probably remained open during the Altonian and Farmdalian Subages (60,000–22,000 yr BP) (Delcourt et al. 1980). By the Full-Glacial (18,000 yr BP), as the northern forest species moved southward and white spruce and larch (*Larix laricina*) occupied the Mississippi Valley (Delcourt et al. 1980), the cold-intolerant *S. caroliniana* may have become isolated in the Ozark region or on the Blufflands along the Mississippi Valley. From these areas it may have entered the Ozark Plateau, where it presently occurs as a major disjunct population. The morphological changes evident in the Ozark population of *S. caroliniana* (see below) probably originated during this time. Ross (1965) reports similar changes in the insect genus *Allocopina* and in other genera where sister species evolved during the separation of populations on either side of the Mississippi Valley during the Full-Glacial. At about the same time *S. nigra* may have spread throughout the mixed oak-hickory-southern pine vegetation from the Mississippi Valley to the Georgia coastal plain and northern Florida. *Salix nigra* was probably unable to enter the dry, warm, sand dune scrub vegetation that covered peninsular Florida or the cypress-gum forest on the Gulf Coastal Plain, and it is absent from those regions today.

During the Full-Glacial *S. caroliniana* probably occurred also on the Gulf Coastal Plain of peninsular Florida and from there spread northward along the Atlantic Coastal Plain. As the Gulf Coastal Plain became restricted (ca 10,000 yr BP), *S. caroliniana* may have become less common there. The climate of peninsular Florida during the Full-Glacial was dry, and many lakes dried up because watertables were lower than at present (Watts & Stuvier 1980). At this time habitats for *Salix* may have been limited. *Salix floridana,* however, is presumed to have survived in refugia in north-central Florida, and it is possible that *S. caroliniana* did the same.

Disjunct populations of *S. eriocephala* and *S. humilis* var. *humilis* in Georgia and Alabama suggest that these species, and probably others, entered the watershed of the Gulf of Mexico during the climatic deterioration or during the Full-Glacial. As the climate became warmer and drier they may have been eliminated from the southern Appalachian region by the stabilization of alluvial habitats and by the establishment of closed vegetation to the water's edge. The warmer climate itself was probably not inimical to these species, for naturalized populations of *S. lucida* and *S. discolor* thrive today more than 250 km south of their present natural ranges; but as these and other northern species moved northward into the newly deglaciated region, they were all but eliminated from the Southeastern States. *Salix discolor* survives in northeastern West Virginia only in pockets of relict boreal forest (Allard & Leonard 1952).

The Appalachian-Northeastern Temperate species *S. sericea* and *S. humilis* var. *microphylla* probably did not penetrate southward to the Gulf watershed during the Full-Glacial and survive today mainly in the Appalachian region. Since the Hypsithermal Interval they have not been as successful in their northward migration as other willows have been. *Salix sericea* may not have been able to compete successfully with the closely related *S. petiolaris* for wetland habitats.

During the Full-Glacial the Great Plains species *S. exigua* may have occurred on the sandy and silty alluvial floodplains of major rivers and extended its range eastward to western North Carolina. It is probable that *S. amygdaloides* did not occur in the Southeastern States during the Wisconsinin Age, but has migrated into the Mississippi and Ohio river valleys only in recent times.

Since the advent of European settlement in the Southeastern States new habitats for *Salix* have become available. The ranges of many species are expanding along roadsides, railways, on abandoned farm land, and in other man-made habitats (see *S. caroliniana,* no. 1). Some minor range disjunctions may be related to this modern expansion of *Salix* into habitats created by human activity.

TAXONOMY

SALIX Linnaeus, *Sp. pl.* 2: 1015. 1753.—Type: *Salix alba* Linnaeus.

Dioecious shrubs or trees with sympodial branching. Bud scales single, nonresinous, fused into a cap or with free overlapping margins. Leaves deciduous, simple, alternate to subopposite, glandular-toothed to subentire, petiolate, stipulate. Aments ascending to spreading, developing before (precocious), with (coetaneous), or, rarely, after (serotinous) the leaves emerge, solitary, sessile, or terminating a short, leafy branchlet. Flowers simple, lacking a perianth, each subtended by a bract with an entire or erose apex. Stamens 1, 2, 3–6, or 7, filaments free, sometimes connate below. Pistils bicarpellate, unilocular, stipitate or sessile, style 1, stigmas 2, entire or bifid, placentae two, parietal, ovules (1–) 2–12 (–21) per carpel, nectary usually a single short, adaxial protuberance, sometimes with a second abaxial gland or several lateral glands (in some a cuplike nectary may surround the stipe). Capsules bivalvate. Seeds surrounded by an arillate coma.

Keys to the Species of Salix in the Southeastern United States

Key to Staminate Specimens

1. Stamens three or more.
    2. Aments slender, loosely flowered.
        3. Leaves glaucous beneath.
            4. Leaves narrowly to broadly lanceolate, margins serrulate.
                5. Leaves lanceolate to very narrowly so, length/width 5–7.5–10; leaf apex acuminate; older branchlets dark to light brown, usually pubescent; stipules usually prominent and persistent. 1. *S. caroliniana.*
                5. Leaves broadly to narrowly lanceolate, length/width 2.8–4–6; leaf apex acuminate to caudate; older branchlets yellowish, tan, or grayish, glabrous; stipules small or absent. 3. *S. amygdaloides.*

4. Leaves elliptical to oblong, margin serrate to denticulate. 4. *S. floridana.*
3. Leaves non-glaucous beneath, rarely thinly glaucous. 2. *S. nigra.*
2. Aments thickish, densely flowered.
6. Young leaves pubescent with caducous, ferruginous hairs; stipules prominent. 5. *S. lucida.*
6. Young leaves glabrous; stipules minute or absent. 6. *S. pentandra.*
1. Stamens two.
7. Aments coetaneous, borne on short, leafy branchlets.
8. Petioles glandular near base of blade.
9. Flowering branchlets 1–1.5 cm long; branches usually spreading, sometimes pendulous.
10. Aments 3–3.5 cm long; leaves narrowly lanceolate; young leaves densely long-sericeous. 7. *S. alba.*
10. Aments 5–6 cm long; leaves lanceolate; young leaves very sparsely sericeous or glabrous. 8. *S. fragilis.*
9. Flowering branchlets very short, ca 0.3 cm long; branches long and pendulous, glabrate. 9. *S. babylonica.*
8. Petioles eglandular.
11. Floral bracts tawny; aments on leafy branchlets.
12. Aments often branched; filaments pilose at base; leaves linear. 10. *S. exigua.*
12. Aments unbranched; filaments glabrous; leaves elliptical. 11. *S. bebbiana.*
11. Floral bracts dark brown; aments sessile with green bracts at base. 17. *S. eriocephala.*
7. Aments precocious, sessile, sometimes with a few leafy bracts at base.
13. Filaments and anthers distinct; aments alternate.
14. Leaves entire, crenate or coarsely serrate, usually pubescent or glabrate, but not sericeous.
15. Floral bracts tawny, sometimes pinkish at tip. 11. *S. bebbiana.*
15. Floral bracts dark brown.
16. Aments 0.5–2 cm long; branchlets tomentose or velutinous-tomentose. 13. *S. humilis.*
16. Aments 2–5 cm long; branchlets villous to velutinous, sometimes glabrate.
17. Young leaves sparsely pubescent; native shrubs. 12. *S. discolor.*
17. Young leaves densely villous; introduced trees or shrubs.
18. Decorticated wood smooth or with a few short ridges; year-old branchlets almost completely glabrous. 14. *S. caprea.*
18. Decorticated wood with prominent, long ridges; year-old branchlets densely pubescent. 15. *S. cinerea.*
14. Leaves serrulate, short-sericeous beneath. 16. *S. sericea.*
13. Filaments and anthers united; at least some aments opposite or subopposite. 18. *S. purpurea.*

Key to Pistillate Specimens

1. Pistils glabrous.
2. Floral bracts deciduous after flowering; petioles with glandular lobes or dots near base of blade.
3. Leaves glaucous beneath.
4. Aments loosely flowered; stipes 1.25–5.25 mm long; leaves usually lacking stomata above (hypostomatous).
5. Leaves lanceolate to very narrowly so, apex acuminate; older branchlets dark to light brown, usually pubescent; stipules usually prominent and persistent. 1. *S. caroliniana.*
5. Leaves broadly to narrowly lanceolate, apex acuminate to caudate; older branchlets yellowish, tan, or grayish, glabrous; stipules small or absent. 3. *S. amygdaloides.*
4. Aments densely flowered; stipes less than 1 mm long; leaves usually with stomata above (amphistomatous).

6. Styles about 0.2 mm long; leaves long-sericeous beneath, margin serrulate. 7. *S. alba.*
6. Styles 0.6–0.8 mm long; leaves glabrous beneath, margin coarsely serrate-crenate. 8. *S. fragilis.*
3. Leaves non-glaucous beneath (rarely thinly glaucous in *S. nigra*).
7. Leaf margins closely serrate or serrulate; aments unbranched.
8. Capsules 3–5 mm long; styles up to 0.4 mm long. 2. *S. nigra.*
8. Capsules 5–9 mm long; styles more than 0.4 mm long.
9. Capsules 5–7 mm long; nectaries 0.2–0.5 mm long, ⅕ to ⅓ as long as stipe. 5. *S. lucida.*
9. Capsules 8–9 mm long; nectaries 0.4–0.8 mm long, ½ to ¾ as long as stipe. 6. *S. pentandra.*
7. Leaf margins distantly glandular-denticulate; aments often branched. 10. *S. exigua.*
2. Floral bracts persistent after flowering; petioles mostly eglandular (glandular in *S. floridana,* sometimes so in *S. babylonica*).
10. Aments slender, up to 2.2 cm long; pistils sessile, or stipes up to 0.3 mm long. 9. *S. babylonica.*
10. Aments stout, 2.5–7.5 cm long; pistils stipitate, stipes more than 1 mm long.
11. Stipes 3–6 mm long; petioles 13–20 mm long; branches very brittle. 4. *S. floridana.*
11. Stipes 1.2–1.6 (–2.8) mm long; petioles 3–14 mm long; branches usually flexible. 17. *S. eriocephala.*

1. Pistils pubescent.
12. Pistils stipitate; aments alternate.
13. Floral bracts deciduous after flowering; aments coetaneous or serotinous, borne on leafy branchlets. 10. *S. exigua.*
13. Floral bracts persistent; aments precocious, sessile or subsessile.
14. Pistils and capsules long-beaked; leaves glabrate, villous, or tomentose beneath.
15. Floral bracts tawny, sometimes pinkish at tip. 11. *S. bebbiana.*
15. Floral bracts dark brown to blackish.
16. Aments 1–3.5 cm long; stipes mostly less than 2 mm long. 13. *S. humilis.*
16. Aments 3–8 cm long; stipes 2 mm long or longer.
17. Styles 0.3–1 mm long; unfolding leaves sparsely pubescent, usually with some ferruginous hairs. 12. *S. discolor.*
17. Styles 0.2–0.5 mm long; unfolding leaves densely white-villous, rarely with ferruginous hairs.
18. Trees or tall shrubs; decorticated wood smooth or with a few short ridges; buds coarsely pubescent to glabrate. 14. *S. caprea.*
18. Shrubs; decorticated wood with prominent, long ridges; buds coarsely pubescent. 15. *S. cinerea.*
14. Pistils and capsules short-beaked; leaves short-sericeous beneath. 16. *S. sericea.*
12. Pistils sessile; at least some aments opposite or subopposite. 18. *S. purpurea.*

Key to Vegetative Specimens

1. Leaves always alternate.
2. Bud apex sharp-pointed; bud scale margin free and overlapping.
3. Leaves glaucous beneath.
4. Leaves narrowly to broadly lanceolate, margin serrulate.
5. Leaves lanceolate to very narrowly so, length/width 5–7.5–10; leaf apex acuminate; older branchlets dark to light brown, usually pubescent; stipules usually prominent and persistent. 1. *S. caroliniana.*
5. Leaves broadly to narrowly lanceolate, length/width 2.8–4–6; leaf apex acuminate to caudate; older branchlets yellowish, tan, or grayish, glabrous; stipules small or absent. 3. *S. amygdaloides.*
4. Leaves elliptical to oblong, margin serrate to denticulate. 4. *S. floridana.*
3. Leaves non-glaucous beneath (rarely thinly glaucous). 2. *S. nigra.*
2. Bud apex blunt; bud scale margin fused.

6. Leaves green or pale beneath.
   7. Leaves lanceolate or elliptic-lanceolate, margin serrate, stomata usually absent above (hypostomatous).
      8. Stipules glandular at base of blade.
         9. Leaf apex caudate; stipules persistent and prominently glandular; young leaves with caducous, ferruginous hairs. 5. *S. lucida.*
         9. Leaf apex short-acuminate; stipules minute or absent; young leaves glabrous. 6. *S. pentandra.*
      8. Stipules not glandular at base of blade. 17. *S. eriocephala.*
   7. Leaves linear, margin distinctly glandular-denticulate, stomata present above (amphistomatous). 10. *S. exigua.*
6. Leaves glaucous beneath.
   10. Leaf margin serrulate or serrate.
      11. Native shrubs; leaves lacking stomata above (hypostomatous).
         12. Leaves densely short-sericeous beneath; stipules of small glands or absent. 16. *S. sericea.*
         12. Leaves glabrous or glabrate beneath; stipules prominent. 17. *S. eriocephala.*
      11. Introduced trees; leaves with stomata above (amphistomatous).
         13. Leaf margin coarsely and irregularly serrate-crenate, blades glabrous beneath. 8. *S. fragilis.*
         13. Leaf margin minutely and uniformly serrulate or spinulose-serrulate, blades long-sericeous or glabrate beneath.
            14. Leaves narrowly lanceolate, margin serrulate, blades long-sericeous beneath. 7. *S. alba.*
            14. Leaves very narrowly lanceolate, margin spinulose-serrulate, blades glabrate beneath. 9. *S. babylonica.*
   10. Leaf margin entire or crenate.
      15. Leaves glabrate, margin flat; young leaves usually sparsely pubescent.
         16. Branches persistently pubescent; leaves rugose-veiny beneath, lacking ferruginous hairs. 11. *S. bebbiana.*
         16. Branches glabrate; leaves plane beneath; young leaves with some ferruginous hairs. 12. *S. discolor.*
      15. Leaves permanently pubescent, at least beneath, margin usually revolute; young leaves densely villous or tomentose.
         17. Leaf margin entire and undulate; leaves oblong to oblanceolate or narrowly so. 13. *S. humilis.*
         17. Leaf margin crenate or irregularly serrate, rarely subentire; leaves elliptic, obovate to oblanceolate.
            18. Leaf margin flat; stipules small, less than 2 mm long, caducous. 11. *S. bebbiana.*
            18. Leaf margin revolute; stipules conspicuous, up to 15 mm long, persistent.
               19. Decorticated wood smooth or with a few short ridges; trees or tall shrubs; year-old branchlets almost glabrous. 14. *S. caprea.*
               19. Decorticated wood with many long, prominent ridges; shrubs; year-old branchlets densely pubescent. 15. *S. cinerea.*

1. Leaves mostly alternate, but at least some opposite or subopposite. 18. *S. purpurea.*

I. Salix subgenus **Salix** section **Humboldtiana** Pax in Engler & Prantl, *Nat. Pflanzenfam.* 3 (1): 36. 1888.—TYPE: *S. humboldtiana* Willdenow.

1. Salix caroliniana Michaux, *Fl. bor.-amer.* 2: 226. 1803.—TYPE: "Carolinea," *Michaux s.n.* (lectotype, here designated: P! (Michaux herb.), pistillate specimen; syntype: sheet with no locality given and bearing staminate and vegetative branches: P!).—Carolina willow, Coastal Plain willow.

Salix occidentalis Bosc ex Koch, *Salic. eur.* 16. 1828, non Walter, 1788.—TYPE: CUBA. *Sieber s.n.* (possible fragment at S!).

Salix pitcheriana J. Barratt, *Salic. amer.,* Sp. 22. 1840, nomen nudum.

Salix longipes Shuttleworth ex Andersson, Öfvers. Förh. Kongl. Svenska Vetensk.-Akad. 15: 114. 1858. *Salix nigra* [var.] ****Salix longipes* (Shuttleworth ex Andersson) Andersson, *Monogr. Salicum* 22. 1867. *Salix occidentalis* var. *longipes* (Shuttleworth ex Andersson) Bebb, Gard. & Forest 8: 363. 1895.—TYPE: FLORIDA. Wakulla Co., St. Marks, Apr 1843, *Rugel s.n.* (holotype: K!; isotypes: FLAS! MO! NA!).

Salix longipes var. *pubescens* Andersson, *Salices bor.-amer.* 7. 1858. *Salix nigra* [f.] – *gongylocarpa* Shuttleworth ex Andersson, *Monogr. Salicum* 22. 1867.—TYPE: FLORIDA. Wakulla Co., St. Marks, Apr 1843, *Rugel s.n.* (holotype: K, fragments: A!; isotypes: MO! S!).

Salix nigra [f.] – *venulosa* Andersson, *Monogr. Salicum* 22. 1867. *Salix nigra* var. *venulosa* (Andersson) Bebb in Coville, Contr. U.S. Natl. Herb. 4: 199. 1893. *Salix longipes* var. *venulosa* (Andersson) C. Schneider, Bot. Gaz. 65: 14. 1918.—TYPE: TEXAS. Uvalde Co., Nueces R. (not New Mexico as on label), 15 May 1851, *Wright 1879* (lectotype, here designated: GH!; isolectotypes: NY! US!).

Salix cordata [f.] – *discolor* Andersson, *Monogr. Salicum* 159. 1867.—TYPE: no specimen cited, but a specimen so identified by N. J. Andersson at GH is annotated as a synonym of *Salix longipes* var. *wardii* by Schneider (1919: 27); Bebb (1895a) comments that the description is of "*Salix wardii.*"

Salix nigra var. *wardii* Bebb in Ward, Bull. U.S. Natl. Herb. 22: 114. 1881. *Salix wardii* (Bebb) Bebb, Gard. & Forest 8: 363. 1895. *Salix longipes* var. *wardii* (Bebb) C. Schneider, Bot. Gaz. 65: 22. 1918.—TYPE: DISTRICT OF COLUMBIA. Washington, 4 June 1881, *Ward s.n.* (lectotype, here designated: F!; isolectotypes: A! CM! GH! NA!).

Salix amphibia Small, *Fl. Miami.* 61. 1913.—TYPE: FLORIDA. Dade Co., Long Key, 18–26 Jan 1909, *Small & Carter 2898* (lectotype, here designated: NY!; isolectotypes: DUKE! FLAS! GH! MTJB! MO! NCU! PH! US! WVA!).

Salix harbisonii C. Schneider, J. Arnold Arbor. 1: 29. 1919.—TYPE: GEORGIA. Chatam Co., Savannah, 5 Apr 1918, *Harbison 5* (holotype: A!; isotypes: NCU! PAC-(*3984*)! US!).

Trees less than 10 m tall. Branches dark to light brown, glabrous or sparsely pubescent; branchlets reddish brown to yellowish brown, glabrous, villous or tomentose with white and ferruginous hairs, becoming glabrate, brittle at base or sometimes somewhat tenacious; buds reddish brown, glabrous to sericeous, conical, apex sharply pointed, bud scale margins free and overlapping. Leaf blades lanceolate to very narrowly so, the larger blades (5–) 7.5–11.5–15 (–22) cm long, (0.6–) 1–1.6–2.2 (–3.5) cm wide, length/width 5–7.5–10, apex acuminate to short acuminate, base usually acute sometimes rounded or auriculate, margin glandular-serrulate; immature leaves reddish green or yellowish green, translucent, blades glabrous or tomentose to sericeous-tomentose with white hairs sometimes intermixed with ferruginous hairs, midrib pilose to tomentose; mature leaf blades glabrate and glossy above, midrib usually pilose, lower surface thickly glaucous, glabrous, midrib sparsely tomentose to glabrate, leaf blades rarely persistently pubescent on both sides, hypostomatous; petioles yellowish to reddish brown, (3–) 4.5–8–14 (–22) mm long, tomentose to sparsely pilose adaxially, glandular dots or processes at distal end; stipules usually prominent, curved obliquely, apex obtuse to acute, usually glandular adaxially, maximum length about 15 mm. Aments coetaneous, borne on leafy flowering branchlets, appearing lateral if bud in axil of distal leaf emerges (syllepsis), leaves deciduous after flowering, aments finally deciduous, abscissing above distal leaf. Staminate aments 3–5–9.5 cm long, borne on flowering branchlets 0.4–1.3–2.5 cm long; stamens 6, rarely 4, 5, or 7–12, filaments pilose on proximal half, dry anthers 0.4–0.6 mm long; nectaries 1, 2, or several slender lobes, adaxial lobe often broadest, others slender and located abaxially or intermixed with stamens; bracts tawny, 1–3 mm long, obovate to broadly ovate, apex acute to rounded, entire or erose, densely crisp-villous on both sides or glabrate abaxially. Pistillate aments loosely flowered, 2–5–9.5 cm long, borne on flowering branchlets 0.6–1.9–3.5 cm long; pistils greenish brown or reddish brown, white-punctate with stomata, pyriform to obclavate, glabrous, capsules brownish, 4–6 mm long, often deciduous after dehiscence, styles (including the 2-branched stigmas) 0.3–0.8 mm long, stipes 1.25–2–5.25 mm long; nectary 1, adaxial, short and broad, often half surrounding stipe, 0.3–0.4 mm long, ⅙ as long as stipe; ovules 6–8 per carpel; bracts as in staminate flowers but often oblong, with an acute or blunt apex and deciduous after flowering. Chromosome number unknown. Fig. 2.

Phenology. Flowering in the South (Georgia to Florida and Louisiana) from December to early May with a mean in early March, and in the North and West from mid-March to early June with a mean in late April; based on 424 records.

Distribution. (Figs. 2-inset, 3). *Salix caroliniana* is a southern species with two main centers of occurrence, one on the Coastal Plain and the other on the Ozark Uplands. The Coastal Plain populations extend from Delaware southward to Florida, and westward along the Gulf Coast to Alabama, with occurrences also in western Cuba and Guatemala. On the Atlantic Coastal Plain north of Florida *S. caroliniana* occurs most commonly on the outer or lower Coastal Plain. The Ozark center covers the northwestern half of Arkansas, most of Missouri, and extends into the adjacent states. Secondary centers occur in the Nashville Basin, and the surrounding Highland Rim in Tennessee and northern Alabama, the Blue Grass Basin of Kentucky, and the Appalachian Ridge and Valley Province of eastern West Virginia and adjacent Virginia. Along its northern boundary the species occurs sporadically in Indiana, Ohio, and Pennsylvania. Many of the isolated occurrences on the Appalachian Upland are associated with limestone or other calcareous substrates.

Salix caroliniana occurs primarily in alluvial woods on the floodplains of rivers and streams. On the Coastal Plain it also grows in swamps, hammocks, marshes, wet depressions between sand dunes, and along ditches and canals. In the Ozark region and at other interior localities it occurs primarily on rocky or gravelly stream beds with secondary occurrences in moist, open fields and woods. These populations are usually associated with calcareous substrates. The occurrence of the species in northeastern West Virginia is correlated with limestone outcrops as mapped by Gillespie (1963, 1964).

On disturbed freshwater wetlands in Dade Co., Florida, *S. caroliniana* forms

FIG. 2. *Salix caroliniana*. a. Branch bearing a vegetative branchlet and pistillate aments in late flowering (×0.75). b. Mature leaf of an Ozark specimen (×0.75). c. Staminate ament with a sylleptic side branch, arrow (×1). d. Staminate flower (×10). e. Young pistillate flower (×10). Inset: map of total distribution, based on herbarium specimens.

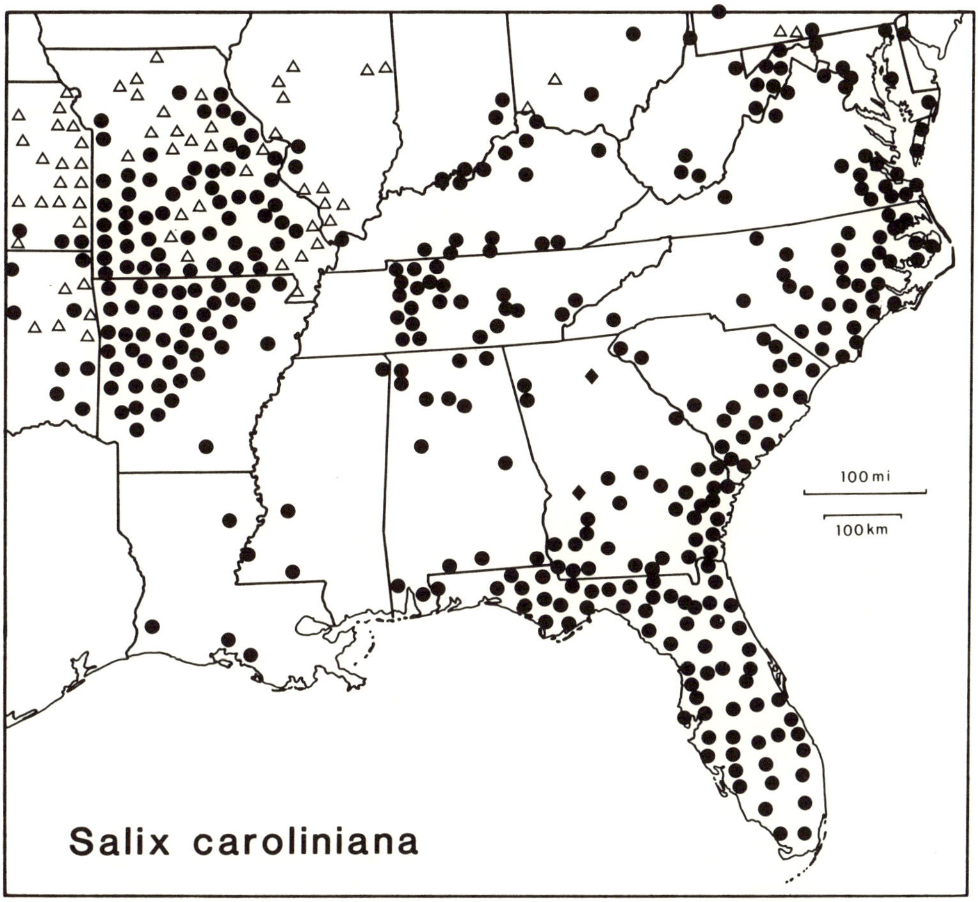

FIG. 3. Distribution of *Salix caroliniana* in the Southeastern and peripheral states. Solid dots, herbarium specimens; diamonds, W. Duncan (pers. comm.); open triangles, literature reports.

islands of trees called willow heads (Craighead 1971). Nearly pure stands of *S. caroliniana* and sea-myrtle (*Baccharis halimifolia*) occur on abandoned farm fields in glades and sloughs traversing the pinelands. During the past 30 years the occurrence of *S. caroliniana* has increased greatly in southern Florida (Craighead 1971).

Salix caroliniana is rare in three Southeastern States: Delaware, Louisiana, and Mississippi. In Delaware it is known only from New Castle Co., where it was last collected in 1923. In Louisiana it is represented by four collections from four parishes; three of these parishes were last collected between the 1890's and 1925 and one in 1969. In Mississippi it is known from eight collections from four counties; these collections were made between 1957 and 1978. Outside of the Southeastern States *S. caroliniana* appears to be rare also in Pennsylvania, Ohio, Indiana, and Kansas. It is listed as endangered, however, only in Indiana and in Ohio (Bacone & Hedge 1980; Cooperrider 1982). It seems certain that it is endangered, if not extirpated, in Delaware, and it is probably threatened or endangered in Louisiana.

Representative Specimens. U.S.A. Alabama: Baldwin Co., 1935, *Smith s.n.* (NA); Blount Co., *Clark 9407* (NCU); Colbert Co., *Harper 3891* (GH, MO, PH, US); Covington Co., *Kral 38736* (VDB); Cullman Co., *Clark 4031* (NCU); Escambia Co., *Clark 9928* (NCU); Franklin Co., 19 Mar 1898, *Mohr s.n.* (US); Houston Co., *Clark 7419* (NCU); Jackson Co., *Whetstone 4265* (NCU); Madison Co., *Terrell 3496* (FSU, US); Mobile Co., *Meyer 10388* (NA); Tallapoosa Co., *Clark 15170* (NCU); Tuscaloosa Co., *Deramus 304* (UNA); Winston Co., *Clark 13074* (NCU).—Arkansas: Ashley Co., 13 Apr 1924, *Ashe s.n.* (NCU); Baxter Co., *Demaree 23591* (MO, NY); Benton Co., 28 May 1975, *Bailey s.n.* (UARK); Boone Co., *Demaree 30209* (UARK); Bradley Co., *Demaree 21916* (NA, NY); Carroll Co., *Palmer 5588* (A, MO, US); Clark Co., *Demaree 9983* (A, MO); Clay Co., *Demaree 27000* (NA, UNA); Cleburne Co., *Demaree 30498* (DUKE, NCU, TENN, VDB, WILLI); Conway Co., *Demaree 23069* (MO, NA, NY); Crawford Co., *Demaree 64012* (WILLI); Cross Co., *Demaree 3929* (UARK); Faulkner Co., 23 Sep 1925, *Ashe s.n.* (NCU); Franklin Co., *Barber 204* (UARK); Fulton Co., *Demaree 27367* (NCU, TENN, WILLI); Garland Co., *Demaree 57703* (APCR, NCU, NLU, TENN, VDB, WILLI); Hempstead Co., *Demaree 9924* (A); Hot Springs Co., *Demaree 14848A* (A, MO, NA, NY); Howard Co., *Demaree 9878* (A, MO, NY); Independence Co., *Thomas 8013* (NLU); Izard Co., *Demaree 23393* (MO, NY); Johnson Co., *Demaree 20255* (MO, NA); Lawrence Co., *Demaree 30456* (NA); Logan Co., *Demaree 55480* (AUA, NCU, TENN, VDB); Madison Co., 19 Apr 1930, *Henbest s.n.* (UARK); Marion Co., *Palmer 14308* (A, MO, US); Montgomery Co., *Demaree 9811* (NY, MO); Newton Co., *Demaree 25001* (GH, NA); Perry Co., *Demaree 20164A* (MO, NA); Pike Co., *Demaree 15959* (NA, NY); Polk Co., *Demaree 15708* (MO, NA, NY); Pope Co., *Merrill 95* (A, MO, NY); Pulaski Co., *Demaree 8673* (A, NY, MO, US); Randolph Co., *Demaree 69649* (MO); Saline Co., *Demaree 35327* (FSU, GH, MO, NCU, NO, USF); Scott Co., 8 Aug 1974, *Lynn s.n.* (UARK); Searcy Co., *Demaree 22259* (NA, NY, MO); Sebastian Co., *Palmer 39306* (A, MO, NY); Sharp Co., *Demaree 27729* (DUKE, NA, NY); Stone Co., *Thomas 7881* (NLU, TENN); Van Buren Co., *Demaree 25461* (FSU, GH); Washington Co., *Moore 850* (LTU, UARK); White Co., *Lawson 602* (NLU); Yell Co., *Demaree 18230* (MO, NA, NY).—Delaware: New Castle Co., *Tatnall 1768* (DOV).—District of Columbia: *Martin 938* (DAO, DUKE, GH, MO, MUHW, NA, NCU, NY, PH, UNA, USF, WVA).—Florida: Alachua Co., *D'Arcy 1483* (FLAS, GH, VDB); Bay Co., *Godfrey 54481* (DUKE, FLAS, FSU, GH, NA, NCU, NY, USF, VDB); Bradford Co., *Wyman 169* (NA); Brevard Co., *Kral 5499* (FSU, VPI); Broward Co., *Seibert 1192* (A, MO); Calhoun Co., 7 Apr 1969, *Whatley s.n.* (NLU); Charlotte Co., *Deam 58738* (NA); Citrus Co., *Cooley 5873* (GH, USF); Clay Co., *Ward 5496* (FLAS, NCU, NY, US); Collier Co., *Lakela 29432* (NY, USF); Columbia Co., *Godfrey 62655* (FLAS, NCU); Dade Co., *Demaree 10197* (FSU, MO, US); Desoto Co., *Baltzell 6092* (FLAS); Dixie Co., *Godfrey 56490* (DUKE, FSU, GH, NA); Duval Co., *Curtiss 5826* (DOV, GH, FLAS, MO, NCU, NY, US); Escambia Co., 20 Aug 1955, *Kurz s.n.* (FSU); Flagler Co., 18 Apr 1940, *West s.n.* (FLAS); Franklin Co. *Godfrey 54848* (DUKE, FLAS, FSU, GH, NA, NCU, NY, USF, VDB); Gadsden Co., *Godfrey 54692* (FSU, NA, USF); Glades Co., *Brass 14493* (FLAS, FSU, GH, US); Gulf Co., *Harbison 10* (A, S); Hamilton Co., *Godfrey 53975* (DUKE, FSU, GH, NY, USF); Hardee Co., *Baltzell 9665* (FLAS); Hendry Co., *Lakela 25637* (GH, USF); Hernando Co., *Godfrey 57260* (FSU, GH, NA, USF); Highlands Co., *Brass 14565* (GH, US); Hillsborough Co., *Blanton 6945* (A, CM, NCU, US); Holmes Co., *Godfrey 54681* (FLAS, FSU, NA, NY, USF); Indian River Co., *Lemaire 117* (FSU, GH, NA, NCSC, NCU); Jackson Co., *Mitchell 1018* (FSU); Jefferson Co., *Godfrey 54511* (FSU, NA, NCU, UNA, USF, VDB); Lafayette Co., *Godfrey 62590* (DUKE, FLAS, FSU, GH, LTU, NA, NCU, VDB); Lake Co., *Harper 18* (GH, MO, NY, US); Lee Co., *Cooley 2494* (FLAS, GH, NY, USF); Leon Co., *Godfrey 54464* (FLAS, NA, UNA, USF, VDB); Levy Co., *Ward 1295* (FLAS, GH, NCU, NLU, USF); Liberty Co., *Godfrey 284* (FSU); Madison Co., *Godfrey 54521* (DUKE, FLAS, FSU, GH, NY); Manatee Co., *Dodson 4965* (USF); Marion Co., *Argus 11460* (CAN); Martin Co., *Smith 916* (FLAS, FSU, USF); Monroe Co., *Moldenke 342* (DUKE, MO, NY, PH, US); Nassau Co., *Ward 1246* (FLAS, FSU, NCU); Okeechobee Co., *McCart 10672* (FLAS); Orange Co., *Palmer 38354* (A, MO, NY, PH, US); Osceola Co., *Cooley 12626* (MO, USF); Palm Beach Co., *Seibert 1204* (A, MO); Pasco Co., *Cooley 7280* (GH, NCU, USF); Pinellas Co., *Lakela 26857* (FLAS, GH, USF); Polk Co., 1 Apr 1965, *Conrad s.n.* (GH, NCU); Putnam Co., *Hansen 5213* (USF, WIS); Saint Johns Co., *Godfrey 74700* (FSU); Saint Lucie Co., *Baltzell 8344* (WILLI); Sarasota Co., 27 Apr 1943, *Perkins s.n.* (USF); Seminole Co., *Blanton 6499* (A, DUKE, MO, NO); Sumter Co., *Hansen 5068* (USF); Suwannee Co., 22 Aug 1955, *Kurz s.n.* (FSU); Taylor Co., *Godfrey 54716* (DUKE, NA, NCU, NY, USF, VDB); Union Co., 22 May 1942, *West s.n.* (FLAS); Volusia Co., *Rossbach 2786* (WVW); Wakulla Co., *Godfrey 53131* (FLAS, FSU,

GH, NCSC, NY, TENN, UNA, USF); Walton Co., *Godfrey 54478* (DUKE, FLAS, FSU, GH, NA, NCU, NY, USF, WIS, VDB); Washington Co., 13 Jun 1955, *Kurz s.n.* (FSU).—GEORGIA: Appling Co., 12 May 1972, *Stinson s.n.* (NCU); Baker Co., *Thorne 3750* (GEO, NA); Brantley Co., *Harper 17028* (GA); Bryan Co., *Duncan 2244* (GH, NA); Bulloch Co., 28 Mar 1966, *Park s.n.* (CM); Camden Co., *Bozeman 3907* (NCU); Charlton Co., *Harper 1499* (GH, MO, NY, US); Chatham Co., *Harbison 5* (A, PAC, US); Clinch Co., *Argus 11440* (CAN); Colquitt Co., *Demaree 49520* (VDB); Decatur Co., *Bozeman 4990* (NCU); Dougherty Co., *Bozeman 6447* (NCU); Early Co., *Thorne 2766* (GEO); Echols Co., *Argus 11446* (CAN); Effingham Co., *Baum 329* (MARY); Floyd Co., *Duncan 8443* (GA, NA); Glynn Co., *Harbison 5947* (NCU); Grady Co., *Faircloth 112* (VSC); Laurens Co., *Bozeman 56337* (NCU); Lee Co., *Duncan 2300* (GA, GH); Liberty Co., *Duncan 2200* (NA, NO); Long Co., *Bozeman 1119* (USF); Lowndes Co., *Faircloth 2221* (NCU, VSC); McIntosh Co., *Cronquist 4922* (FLAS, GH, MO, NY, PH, US); Montgomery Co., *Bozeman 5755* (NCU); Polk Co., 1 Sep 1904, *Ball s.n.* (NA); Richmond Co., *Dudley 8793* (NA); Seminole Co., 24 Oct 1973, *Gholson s.n.* (FSU); Tattnall Co., *Bozeman 4067* (NCU); Thomas Co., 21 Mar 1904, *Rohn s.n.* (PH); Wayne Co., *Duncan 4700* (GA, NCU); Wilcox Co., *Bozeman 5565* (NCU).—KENTUCKY: Campbell Co., *Meijer 51180* (KNK); Carter Co., *Braun 2476* (A, US); Grant Co., *Browne 4658* (KY, NCU); Henry Co., *Gentry 458* (KY, NCU, NY); Jefferson Co., Jan 1835, *Short s.n.* (NY); McCreary Co., *Rogers 120* (NA); Meade Co., *McInteer 970* (NA); Metcalfe Co., 19 Jul 1939, *Shacklette s.n.* (NA); Monroe Co., *Braun 4125* (A, CFR); Simpson Co., *Reed 82304* (CFR); Todd Co., *Settle 3* (APSC); Warren Co., 13 Sep 1972, *Price s.n.* (MO); Whitley Co., *McFarland 4582* (NA); Woodford Co., *Wharton 8662* (KY).—LOUISIANA: Calcascieu Par., *Small 11773A* (NY), 11 Apr 1925, *Small s.n.* (GH); Franklin Par., 7 May 1892, *Mohr s.n.* (US); Iberia Co., *Meyer 11932* (NA); Saint Mary Par., 6 Apr 189?, *Mohr s.n.* (US).—MARYLAND: Allegany Co., *Ball 3025* (MARY, NA, PH, US, VPI); Montgomery Co., *Ball 86* (MO, NA); Washington Co., *Downs 9987* (NCU); Wicomico Co., *Reed 24972* (CFR); Worcester Co., *Ball 2733* (MARY, MO, NA, NY, US).—MISSISSIPPI: Adams Co., *Meyer 11596, 11633* (NA); Hinds Co., *Tatum 25* (NLU); Pike Co., *Ray 7796* (USF); Tishomingo Co., *Harbison 13* (A), *Jacob 1448* (NCU), *Ray 7419* (USF), *Rogers 45902* (TENN).—NORTH CAROLINA: Beaufort Co., *Bloomquist 10952A* (DUKE, NA); Bertie Co., *Ahles 41381* (NCU); Bladen Co., *Rossbach 7216* (WVA, WVW); Brunswick Co., *Fox 4629* (GH, NA, NCSC, NY); Camden Co., *Reed 20715* (CFR); Carteret Co., *Argus 10922* (CAN); Chatham Co., *Radford 44616* (NCU); Chowan Co., *Ahles 39749* (FLAS, NCU); Columbus Co., 30 Apr 1935, *Matthews s.n.* (DUKE, NCU); Craven Co., *Hardin 2355* (FSU, GH, NCSC, PAC, VDB); Dare Co., *Fosberg 17867* (NCSC, TENN); Duplin Co., *Argus 10920* (CAN); Gates Co., *Fox 2282* (NA, NCSC); Halifax Co., *Ahles 11017A* (NCU); Harnett Co., *Argus 10926* (CAN); Hyde Co., *Godfrey 6872* (DUKE, GH, NCSC); Jackson Co., 17 May 1920, *Harbison s.n.* (NCU); Jones Co., *Godfrey 7071* (DUKE, GH, PH, MO, NY, US); Lee Co., *Beard 392* (NCU); Lenoir Co., *Radford 22213* (CM, NCU); Martin Co., *Radford 32173* (NCU); Nash Co., *Ahles 11747* (NCU); New Hanover Co., *Godfrey 6373* (GH, US); Onslow Co., *Boyce 523* (NA, NCSC); Orange Co., 15 Jul 1938, *Leclair s.n.* (NCU, NY); Pasquotank Co., *Ahles 39989* (NCU); Pender Co., *Weaver 2265* (DUKE); Perquimans Co., *Ahles 51146* (FSU, NCU); Pitt Co., *Helms 1183* (AUA, CLEMS, DHL, DOV, FUGR, FSU, GH, LYN, MARY, MISS, MUR, NLU, NO, NCU, NY, PAC, UNA, UNCC, USF, VDB, VSC, WILLI, WIS); Robeson Co., *Ahles 23749* (NCU); Rockingham Co., 8 Aug 1937, *Totten s.n.* (NCU, NY); Sampson Co., *Argus 10915* (CAN); Stanley Co., 20 Apr 1896, *Small s.n.* (NY); Tyrrell Co., *Radford 33768* (NCU); Wayne Co., *Brunton 191* (NCSC); Wilson Co., *Reed 115114* (CFR).—SOUTH CAROLINA: Aiken Co., 28 Apr 1914, *Sargent s.n.* (A); Anderson Co., 1886, *Earle s.n.* (VPI); Bamberg Co., *Argus 10889* (CAN); Beaufort Co., *Bell 1706* (NCU); Berkeley Co., *Godfrey 472* (DUKE, GH, MO, NY, PH, TENN, US); Charleston Co., *Ball 56* (GH, NA, US); Chesterfield Co., 25 Jul 1976, *Castrale s.n.* (WVA); Clarendon Co., *Argus 11404* (CAN); Colleton Co., *Wallace 189* (CLEMS); Darlington Co., *Coker 2* (NY); Dorchester Co., *Hunt 3297* (CLEMS, NA); Florence Co., *Ball 691* (US); Georgetown Co., *Argus 11399* (CAN); Hampton Co., *Ahles 10459* (NCU, USF); Horry Co., *Weatherby 16517* (A, GH, NY, US); Jasper Co., *Bell 2422* (NCU); Marion Co., *Radford 20* (NCU, UARK, VPI); Oconee Co., *Palmer 35414* (MO); Orangeburg Co., *Argus 10893* (CAN); Williamsburg Co., *Godfrey 380* (DUKE, GH, MO, NY, PH, TENN, US).—TENNESSEE: Bledsoe Co., *Sharp 1334* (TENN); Blount Co., *Sharp 845* (TENN); Cheatham Co., *Blum 3483* (PH, VDB); Davidson Co., 2 May 1942, *Gattinger s.n.* (PH); Dickson Co., *Sharp 397* (NCU, TENN); Franklin Co., *Svenson 9980* (GH, MUHW); Grundy Co., *Clark 11809* (NCU); Hickman Co., *Ball 2598* (GH, MO, NA, NY, TENN, UNA, US); Houston Co., *Palmer 17602* (A, MO, US); Humphreys Co., 24 May 1920, *Hall s.n.* (MUR); Lawrence Co., *Sharp 9761* (TENN); Lewis Co.,

Shanks 5782 (NCSC, TENN); Monroe Co., 22 May 1908, *Maxon s.n.* (MO); Montgomery Co., *Palmer 17660* (A, MO); Perry Co., *Temple 6116* (MISS); Robertson Co., 13 Jul 1897, *Eggert s.n.* (MO); Rutherford Co., 14 Oct 1952, *Shaver s.n.* (VDB); Stewart Co., *Stack 524* (APSC); Van Buren Co., *Caplendor 603* (VDB); Wayne Co., *Kral 55468* (VDB); White Co., 7 Jun 1963, *Rogers s.n.* (AUA); Williamson Co., *Hammer 20* (VDB).—VIRGINIA: Accomack Co., *Fernald 5286* (GH, PH, US); Arlington Co., *Tidestrom 149* (NA); Elizabeth City Co., *Robinson 479* (GH); Fairfax Co., *Maxon 21* (GH, NA, NY); Isle of Wight Co., *Fosberg 18062* (NA); James City Co., *Berans 210* (WILLI); Loudoun Co., *James 2946* (NCU); Montgomery Co., 28 Mar 1945, *Massey s.n.* (VPI); Nansemond Co., *Ecklund 133* (CM, FUGR, LYN, NCSC, NCU, NLU, USF, VSC); New Kent Co., 9 Nov 1938, *Massey s.n.* (NA, VPI); Norfolk Co., *Porter 876* (GH, MO); Northampton Co., *Fernald 5285* (CFR, GH, PH, US); Northumberland Co., *Reed 5145* (CFR); Prince George Co., *Fernald 9310* (GH, MO, PH); Princess Ann Co., *Fernald 3886* (GH, PH); Rockbridge Co., 13 Sep 1907, *Bartram s.n.* (PH); Rockingham Co., *Ball 2711* (DUKE, MARY, MUHW, NA, NCSC, TENN, UNA, US, VPI, WVA); Shenandoah Co., *Allard 9359A* (GH, NA, US, VPI, WVA); Southampton Co., *Tidestrom 4518* (A, MO); Sussex Co., *James 4173* (NCU).—WEST VIRGINIA: Fayette Co., 6 Jul 1968, *Phillips s.n.* (NLU); Grant Co., *Davis 6621* (NA); Hamphire Co., *Ball 2605* (A, DUKE, FUGR, GH, MARY, MO, MUHW, NA, NCSC, NY, TENN, UNA, US, VPI, WVA); Hardy Co., *Ball 2608* (GH, MO, NA, NY, WVA); Mineral Co., *Downs 8429* (NCU); Ohio Co., *Bartholomew 866* (WVA); Pendleton Co., 21 Sep 1976, *Tomblin s.n.* (WVA); Preston Co., 15 Aug 1953, *Bartholomew s.n.* (WVA); Raleigh Co., *Tosh 922* (US, WVA); Summers Co., *Boone 440* (WVA).

Recognition. Salix caroliniana is characterized by leaves that are lanceolate, strongly glaucous beneath, yellowish green and spreading, 4 to 7 stamens, glabrous pistils borne on long stipes, conical buds with a sharp tip, bud scales with a free-overlapping margin, and branches that are somewhat brittle at the base.

It may be distinguished from *S. nigra, S. amygdaloides,* and *S. gooddingii* by the characteristics listed in Table 1. It is sometimes confused with vegetative specimens of *S. eriocephala* and *S. alba,* but it can be easily distinguished from these by the presence of bud scales with a free-overlapping margin compared to their caplike bud scales.

Nomenclature. The Michaux herbarium, housed at the Muséum National d'Histoire Naturelle, Paris, includes two sheets of *S. caroliniana.* One consists of a single pistillate branch and bears the locality name "Carolinea," the other, which lacks a locality name, consists of two staminate flowering branches and one vegetative branch. They all are *S. caroliniana,* and Michaux's original description was based on all of them. The sheet with a single pistillate branch is selected here as the lectotype. Specimens of *S. caroliniana* were not sought in the Richard herbarium or in the main herbarium (P); other syntypes may be found there.

The name *S. subvillosa* Elliot was not validated by Nuttall (1842). Under *S. nigra* Nuttall wrote, "A variety of this tree occurs in South Carolina and Florida, in which the leaves are villous and the scales of the ament densely lanuginous. In the herbarium of Mr. Schweinitz it was marked, on the authority of Elliot, as a species *S. subvillosa.*" This incidental reference does not constitute valid publication of the name (Voss 1983, Article 34). The specimen in question could not be found in the herbarium of the Philadelphia Academy of Natural Sciences, where Schweinitz collections are located. The application of the name is uncertain, but it most likely refers to *S. caroliniana.*

The Andersson name *"venulosa"* has been placed in both *S. caroliniana* (as *"longipes"*) and *S. nigra.* It was originally described under *S. nigra* but without stated rank (Andersson 1867). In Andersson's paper the locality was given as New Mexico; an anonymous annotation on the lectotype (*Wright 1879,* GH), appar-

Table 1. Comparison of *Salix caroliniana, S. nigra, S. amygdaloides,* and *S. gooddingii.*

	S. caroliniana	*S. nigra*	*S. amygdaloides*	*S. gooddingii*
Leaves	lanceolate to very narrowly so, glaucous beneath, hypostomatous, apex acuminate to short acuminate	narrowly lanceolate to very narrowly so, non-glaucous beneath, 50% amphistomatous, apex acuminate	ovate-lanceolate to lanceolate, glaucous beneath, sometimes amphistomatous, apex acuminate to caudate.	narrowly to very narrowly lanceolate, non-glaucous beneath, often amphistomatous, apex long acuminate
Branchlets	more or less brittle glabrous to tomentose	brittle, glabrous to sparsely pilose	tenacious to more or less brittle, glabrous	tenacious or brittle, often pubescent or puberulent
Branches	dark to light brown	reddish to yellow-brown	yellowish to grayish	yellowish to yellow-brown
Petioles (length)	4.5–8–14 mm	3–6–10 mm	7–12–21 mm	3–6 (–10) mm
Stipules	usually prominent and persistent	usually small and caducous	small and caducous	small and caducous
Stipes (length)	1.25–2–5.25 mm 6 times as long as nectary	0.5–1.5 mm 3 times as long as nectary	1.6–2–3.2 mm 3 to 8 times as long as nectary	1.5–3 mm 3 to 10 times as long as nectary
Styles (length)	0.3–0.5–0.8 mm	0.2–0.3–0.4 mm	0.4–0.6 mm	0.2–0.4 mm

ently based on Wright's collection notes, corrects this as follows: "Field no. 63. '63, Salix, pebbly bars of the Nueces, 4–6 ft. tall, May 15, 1851' western Uvalde Co. Texas."

In 1893 Bebb made the combination *S. nigra* var. *venulosa.* There is no evidence that Bebb saw the type of Andersson's name. He probably used this name assuming, since the type came from New Mexico (it actually came from Texas), that it must be the same as the common Sonoran tree that he had collected in Nevada. In 1918 Schneider aligned this taxon with *S. caroliniana* by using the combination *S. longipes* var. *venulosa;* in 1919 he reaffirmed this decision and cited the Bebb combination as a synonym of *S. gooddingii.* I agree with Schneider that var. *venulosa* belongs to *S. caroliniana* and not to *S. nigra.* The type material has narrow leaves that are definitely glaucous beneath, long stipes up to 2.25 mm long, and dark brown branchlets. It is not the same as *S. gooddingii,* and the use of var. *venulosa* in this sense by Goodrich (1983) is incorrect.

No specimens were cited in Bebb's original description of *S. nigra* var. *wardii* (1881) or in his transfer of the name to specific rank (1895a). There exist, however, numerous syntypes that were collected by L. F. Ward in Washington, D.C. Some of these specimens were sent to Bebb and are now in the Field Museum (F), and others were distributed widely. A specimen at F, collected on 4 June 1881, is here selected as the lectotype. This variety is distinguished from typical specimens of *S. caroliniana* only by the presence of rounded and more persistent stipules.

The type sheet of *S. amphibia* was designated as such by Small on the label. The sheet contains a mixture of staminate and pistillate branches, and the latter are here selected as the lectotype. Other pistillate specimens, which appear to be identical to the lectotype in all respects, have been seen in eight herbaria, and they are regarded to be isolectotypes even though they lack the collection number (*2898*) that was later added to the sheet at the New York Botanical Garden.

Salix harbisonii was characterized by Schneider (1919b) as having brittle-jointed stems, glandular-dotted stipules, and amphistomatous leaves. The first two characteristics occur in both *S. caroliniana* and *S. nigra* and would not separate *S. harbisonii* from either of them. I have examined leaves of the type material for stomata in the upper leaf epidermis, but none are clearly amphistomatous although some have a few stomata along the midribs. Most of the type material is typical of *S. caroliniana* except for two paratypes that are *S. nigra* [Georgia, Stephens Co., *Harbison 3* (A), and Virginia, Smythe Co., *Small s.n.* (A, DOV, MO)].

Taxonomy. This species is a member of sect. *Humboldtiana,* which in North America also includes *S. nigra, S. amygdaloides, S. gooddingii, S. floridana, S. humboldtiana,* and *S. bonplandiana* Kunth. Some of these species are known to hybridize where their ranges overlap, but in spite of this they maintain their specific integrity. The taxonomy of the species of this section in the Southeastern States has been treated in various ways.

In DeCandolle's Prodromus Andersson (1868) regarded *S. caroliniana* (as *S. longipes*) and *S. amygdaloides* as varieties of *S. nigra* and included them in sect. *Amygdalinae.* He placed *S. humboldtiana* and *S. bonplandiana* in sect. *Austro-americanae* (or *Humboldtiana*). Schneider, in his monographic revision of *Salix* (1919b), included *S. nigra, S. gooddingii,* and *S. humboldtiana* in his sect. *Nigrae,*

S. amygdaloides in sect. *Triandrae* Dumortier, and *S. caroliniana* (as *S. longipes,* including *S. floridana* and *S. harbisonii*) and *S. bonplandiana* (including *S. laevigata* Bebb) in his sect. *Bonplandiana.* In 1949 Rehder placed *S. nigra* in series *Nigrae, S. caroliniana* (as *S. longipes* and including *S. floridana*) in series *Pentandrae,* and *S. amygdaloides* in series *Amygdalinae.* Ball (1961) also placed these species in different sections, but he included *S. nigra* in sect. *Nigrae* and *S. caroliniana* and *S. amygdaloides* in sect. *Bonplandiana.* There seems to be no justification, however, for placing these closely related species in different infrageneric taxa. Dorn (1976) discussed this problem and grouped these species, except *S. floridana,* in sect. *Humboldtiana.*

Vegetational History. Salix caroliniana may have been part of the Arcto-Tertiary Flora that occupied the Ozark uplands in Tertiary time (Fernald 1937; Steyermark 1959). With the uplift of the Ozark region at the end of the Tertiary the wet swamp flora, with which it may have been associated, was largely eliminated (Steyermark 1959) and the lowland populations of *S. caroliniana* along with it. Upland populations of the species may have survived on the Ozark Plateau and at various interior localities such as the Nashville Basin and the Blue Grass Basin.

It is also possible that *S. caroliniana* did not reach the Ozarks until the mid-Pleistocene at which time it may have moved up the Mississippi Valley along with cypress-gum forest. In either case during the last Full-Glacial Interval, when boreal spruce forest invaded the Mississippi Embayment, the Ozark population may have become isolated from the Coastal Plain populations. Cypress-gum vegetation was probably non-existent on the Gulf Coastal Plain between eastern Texas and western Florida during the Full-Glacial (Delcourt & Delcourt 1981), and Coastal Plain populations of *S. caroliniana* may have survived only on the margins of peninsular Florida. During the Hypsithermal Interval *S. caroliniana* may have spread eastward from the Ozarks and eventually came into contact with Coastal Plain populations along the East Coast. Evidently the upland populations of *S. caroliniana* have not been as successful in moving onto the lowlands of the Mississippi Embayment as has *S. nigra.* The interior and the Coastal Plain populations of *S. caroliniana* have been isolated long enough for some morphological differences to have evolved.

Taxonomic Relationship of Salix caroliniana and Salix nigra

Salix caroliniana is very closely related to *S. nigra,* and one of the objectives of this study was to evaluate their taxonomic relationship. Specimens of these taxa with well-developed leaves were sorted into two groups on the basis of the presence or absence of strong leaf glaucescence, a thick waxy deposit, on the lower leaf surface (see *S. nigra,* no. 2, for a discussion of this characteristic). Leaf glaucescence was found to be highly correlated with stipe length (Fig. 6), and to a certain extent with other characteristics including amphistomatous or hypostomatous leaves and geographical distribution. Plants with glaucous leaves occur mainly on the Atlantic Coastal Plain with major outliers in the Ozark region, the Blue Grass region, and the Nashville Basin, whereas plants with non-glaucous leaves occur throughout the study area but are absent or rare on the lower Atlantic Coastal Plain (compare Figs. 3 and 12).

The glaucous-leaved plants have been identified as *S. caroliniana,* and the plants with non-glaucous leaves as *S. nigra*. In an attempt to understand the behavior of these taxa where they are sympatric, populations from areas of overlap were studied in the field and in the herbarium.

Studies of Natural Populations. Transects were run through the region of overlap between *S. caroliniana* and *S. nigra* in North Carolina and South Carolina. Populations along a transect from Bogue Banks (Carteret Co., N.C.) west to Harnett Co. were sampled at frequent intervals. Between the coast and Black River (Sampson Co.) only *S. caroliniana* was seen. In the vicinity of Dunn (Harnett Co.) two small mixed populations were located, one 3 km east and the other 1.6 km west of Dunn [*Argus 10926–10929* (CAN) and *10930* (CAN), respectively]. Most plants in these populations were *S. nigra,* with either non-glaucous or thinly glaucous leaves, and only two plants in each sample had strongly glaucous leaves and long stipes (1.2–1.6 mm long in one plant) and were identified as *S. caroliniana.* There were no fruiting plants of *S. nigra* in these populations; the only observable difference between the species was in leaf glaucescence. About 15 km farther west of Dunn, on the Cape Fear River, and from there westward only *S. nigra* was seen.

In South Carolina two sites in the Black River Swamp (Williamsburg Co.) were sampled, one on the edge of a swamp forest [*Argus 10900–10905* (CAN)] and the other on a dike on newly cleared land [*Argus 10906–10912* (CAN)]. Three specimens collected along the dike were typical *S. caroliniana* with glaucous leaves (none was in fruit), and all others were *S. nigra* with thinly glaucous leaves; two of the latter also had the short stipes (1–1.2 mm long) typical of *S. nigra*.

In the area of overlap between *S. caroliniana* and *S. nigra* in northern Florida and southern Georgia several populations were studied by me in the field and were sampled in 1983 and 1984 by R. K. Godfrey. Populations consisting of *S. caroliniana* and some putative hybrids with *S. nigra* were sampled in Lafayette, Taylor, and Columbia counties, Florida, and populations consisting of *S. nigra* and some putative hybrids with *S. caroliniana* were sampled in Jackson and Leon counties, Florida, and Grady Co., Georgia. These populations give the impression that they consist of a single highly variable taxon, and it is often difficult to separate out the two species. The populations sampled in Madison Co., Florida, are clearly hybrid swarms.

Two Taylor Co., Florida, populations [*Godfrey 80857–80869, 80907–80926* (CAN, FSU), *81197* (FSU)] consisted mainly of *S. caroliniana,* but this species displayed great variation not only within the population but also within individuals. The leaves on a single plant varied from narrow to broad and had both cuneate and rounded leaf bases. In this population stipe length encompassed the ranges of both *S. caroliniana* and *S. nigra* (0.8–2.6–4.2 mm long), and leaves varied from strongly glaucous to non-glaucous beneath. The population consisted mainly of plants with strongly glaucous leaves and long stipes (2.0–4.2 mm long), and these were identified as *S. caroliniana;* a few plants with non-glaucous leaves and short stipes (0.8–1.6 mm long) were named *S. nigra;* and others that recombined these characteristics were identified as putative hybrids.

Populations in Columbia and Lafayette counties, Florida, [*Argus 11542* and *11463* respectively, (CAN)] consisted mainly of plants of *S. caroliniana,* but occasional plants with short stipes (only 1.4 mm long compared to 3.1 mm in typical *S.*

caroliniana in these populations) and thinly glaucous leaves were presumed to be hybrids. No specimens of *S. nigra* were seen in these populations.

A Leon Co., Florida, population [*Argus 11493* (CAN), *Godfrey 80872–80886* (CAN, FSU)] varied widely in leaf glaucescence. Most plants had non-glaucous leaves, but a few had thinly glaucous leaves and others had glaucous leaves. The glaucous-leaved plants had stipes too short to be *S. caroliniana* (0.85 mm long) and were presumed to be hybrids. The rest of the population had stipes about 1.2 mm long, well within the range of *S. nigra,* and was identified as such.

In two Jackson Co., Florida, populations [*Argus 11496, 11500* (CAN)] leaf glaucescence was found to be variable and discordant with stipe length. Some non-glaucous-leaved plants had longer stipes (ca 1.6 mm long) than did plants with glaucous leaves (1.3 mm long), but the stipe length in all of these plants was within the range of variation of *S. nigra.* Several plants in these populations with glaucous leaves and short stipes were presumed to be hybrids.

A population in Grady Co., Georgia, [*Argus 11492* (CAN)] consisted mainly of plants of *S. nigra,* but it also included plants with glaucous leaves and stipes about 1.5 mm long that were suspected to be hybrids.

Two Madison Co., Florida, populations [*Godfrey 81260, 81261* (FSU)] strongly resembled hybrid swarms of *S. caroliniana* and *S. nigra.* The leaves varied from strongly glaucous to thinly so to non-glaucous and stipes ranged in length from 0.5–1.46–2.8 mm long. Some specimens were identifiable as one species or the other, but many recombined the characteristics of both species and were suspected to be hybrids.

All of these populations were located in disturbed areas, such as wet swales between highway rights-of-way and swamp woodlands, wet areas between highways and railroads, or along causeways.

In general the easternmost populations in the zone of overlap tended toward *S. caroliniana* and the westernmost populations toward *S. nigra.* Populations from outside of the area of overlap did not show the kind of discordant variation in stipe length and leaf glaucescence that occurred in populations within the area of overlap.

These studies of natural populations show that where *S. caroliniana* and *S. nigra* come into contact on the Atlantic and Gulf Coastal Plains they intergrade into one another. In this area it is often difficult to identify individual specimens without mature leaves and fruits. It is presumed here that plants with strongly glaucous leaves and short stipes are the hybrid *S. caroliniana* × *nigra,* but plants with thinly glaucous leaves and short stipes are treated as *S. nigra.* It is possible that the thinly glaucous-leaved state represents introgression between these species, if not hybridization, but the data are equivocal (see below). Because of the wide occurrence of thinly glaucous leaves in *S. nigra* and in young *S. caroliniana,* this characteristic is regarded to be within the normal variation of both species.

Studies of Synthetic Populations (Herbarium Specimens). In areas of overlap between *S. caroliniana* and *S. nigra* on the northern Coastal Plain and in the Ozark region of Arkansas the taxonomic relationship of these species was studied in selected sets of herbarium specimens.

At the northern end of the Coastal Plain in Maryland, the geographical separation of the species is less well marked than it is further south. Herbarium

Table 2. Comparison of *Salix caroliniana* and *S. nigra* in Arkansas.

	S. caroliniana	*S. nigra*
Leaf length (Fig. 5)	longer	shorter
Leaf width (Fig. 5)	broader	narrower
Leaf length/width (Fig. 5)	broader	narrower
Petiole length/leaf width (Fig. 6)	longer relative to width	shorter relative to width
Stipe length (Fig. 7)	longer	shorter
Amphistomatous leaves	none	94%
Leaf base shape (Fig. 8)		
cuneate	41%	100%
rounded	33%	0
auriculate	26%	0
Stipule shape (Fig. 9)		
obtuse-straight	88%	30%*
acute-straight	0	10%
acute-oblique	12%	60%
Stipule presence	mostly present and conspicuous	absent or very small
Branchlet indumentum (Fig. 10)	mostly pubescent	glabrous or glabrate
Habit**	tree up to 10 m	tree 20 m or more

*Based on a small sample size; most plants were exstipulate or had very small stipules.
**Based on Glatfelter (1897).

specimens from Worcester Co., Maryland, suggest that populations there are a mixture of the two species. Many specimens from this region, however, are poorly preserved, and some seem to have been dried over excessive heat that drove off the leaf glaucescence. Well-collected specimens and population samples from the Del-Mar-Va Peninsula are needed to evaluate the possibility of hybridization and intergradation in this region.

Field and herbarium study of specimens from Arkansas revealed a marked difference in the morphology of *S. caroliniana* and *S. nigra.* (Table 2). In some characteristics, such as leaf length, leaf length/width, petiole length/leaf width, amphistomatous leaves, and shape of leaf base a shift resembling character displacement (Brown & Wilson 1956) has occurred away from the normal range of variation in one or in both species.

It is clear that these species are less alike in Arkansas, and elsewhere in the Ozark region, than they are elsewhere in their ranges; but a similar shift also appears in some characteristics in *S. nigra* in Louisiana, where the species are not sympatric, and the significance of this "character displacement" is not clear. It may be that because they occupy different habitats in this region they exhibit morphological shifts that do not occur elsewhere.

In the Ozark region of Missouri the two species are said by Steyermark (1963) and Glatfelter (1897) to occur in different habitats. They note that *S. caroliniana* occupies rocky and gravelly stream beds, and avoids "alluvial stretches as if they were poisonous" (Glatfelter 1897), whereas *S. nigra* is restricted to alluvial soils. This ecological separation was largely confirmed by my field studies in Arkansas, Missouri, and Oklahoma. *Salix caroliniana* was not seen growing on fine grained alluvial floodplains with *S. nigra,* but occasionally *S. nigra* was found growing on rocky and gravelly stream beds with *S. caroliniana.*

Taxonomic Conclusions. There is evidence that *S. caroliniana* and *S. nigra* intergrade throughout the zone of overlap on the Atlantic and Gulf coastal plains. In this area putative hybrids have been recognized. The effect of this hybridization, however, does not seem to extend far beyond the narrow zone of overlap and "pure" populations of the species may occur only a short distance away from hybrid populations. In the Ozark region, however, where *S. caroliniana* and *S. nigra* are sympatric, the species do not seem to hybridize. Here the two taxa are sharply separated by a number of characteristics (Table 2) and are morphologically and ecologically less alike than they are in regions in which they are allopatric.

The evidence suggests that *S. caroliniana* and *S. nigra* do not interact in the same way throughout their ranges. Their behavior on the Coastal Plain is compatible with subspecific rank, whereas their apparent reproductive isolation in the Ozarks supports specific rank. It is probable that the history of these two species in the two major areas of overlap is different, and that reproductive isolating mechanisms evolved in one area but in not the other. I have decided to treat these taxa as separate species, because they are clearly separable over most of their ranges and because the extent of hybridization that occurs on the Coastal Plain is limited (Fig. 10). Study of their population genetics, including hybridization experiments, and their ecological requirements is needed to help resolve their taxonomy.

Some Taxonomic Characteristics and Infraspecific Variation in Salix caroliniana

The study of infraspecific variation in *S. caroliniana* was based on specimens with conspicuous glaucescence on the lower leaf surface, as well as the presence of other characteristics that place them in sect. *Humboldtiana.* Leaf glaucescence has been shown to be highly correlated with stipe length and to a certain extent with geography. The variation within the species was analysed by grouping the specimens into five geographical regions: (1) Northern Coastal Plain Region (Virginia and North Carolina); (2) Middle Coastal Plain Region (South Carolina and Georgia); (3) Southern Coastal Plain Region (Florida); (4) Interior Region (Virginia, West Virginia, Kentucky, Tennessee, North Carolina, South Carolina, Georgia, Alabama, and Mississippi); (5) Ozark Region (Arkansas).

Within each of these regional groups morphological characteristics were measured in herbarium specimens with mature leaves. Stipe measurements were made on all specimens with mature fruits. See Materials and Methods for a description of measurement techniques and statistical analysis. The data have been grouped into five geographical regions in order to permit the analysis of geographical variation in morphological trends within the species. This grouping of the data does not, however, remove the influence of environment and, except for the relational and qualitative characteristics, the statistical differences among the synthetic populations cannot be used to imply genetic differentiation within the species.

A. *Quantitative Characters.*

1. Leaf length (Fig. 4). The mean leaf blade length of plants in the Interior and Ozark regions is greater than that on the Coastal Plain. The differences in leaf blade length seem to reflect clinal variation in this character from coastal to inland populations.

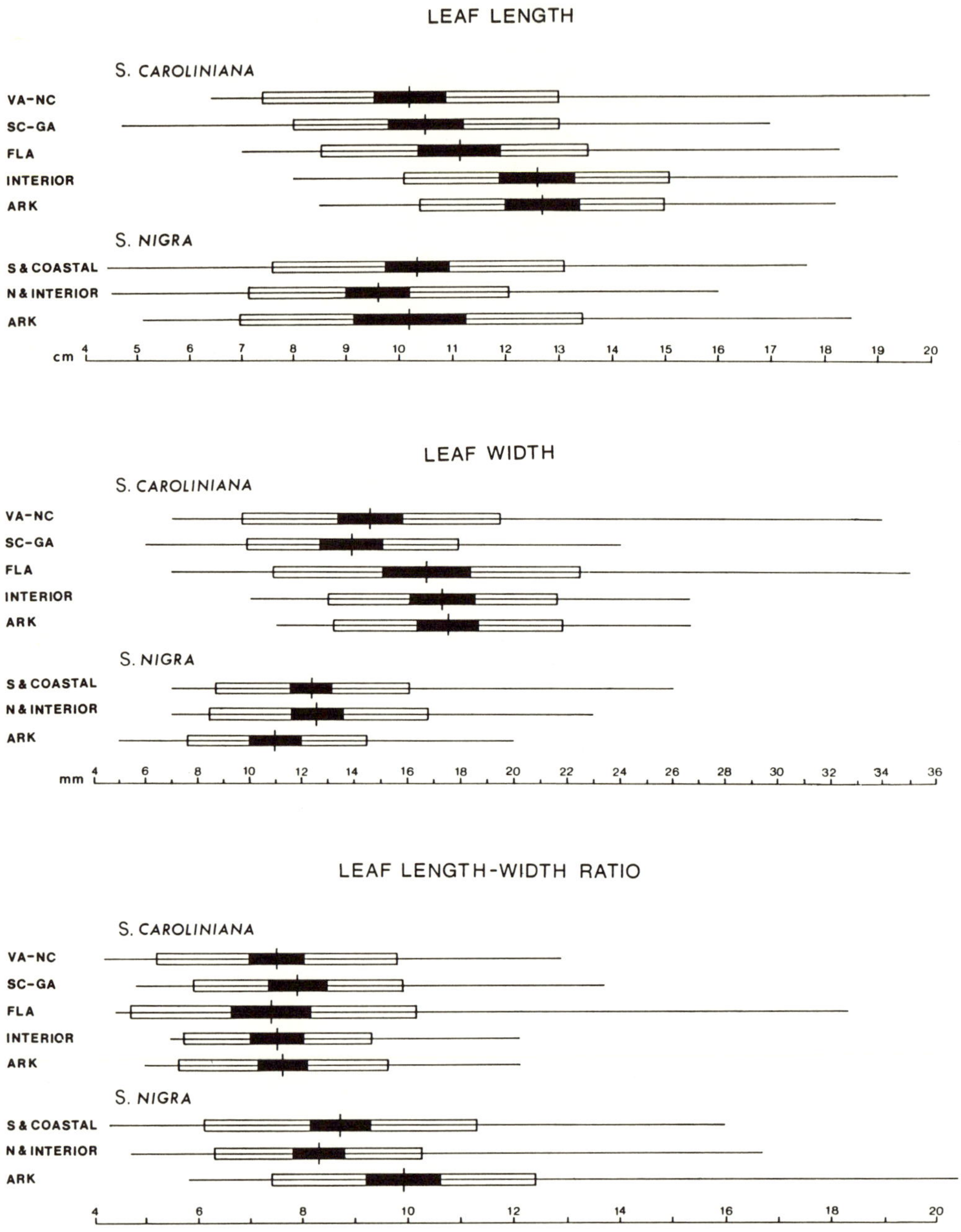

FIG. 4. Leaf length, leaf width, and leaf length-width ratio in *Salix caroliniana* and *S. nigra* in the Southeastern States. The species are subdivided into regional populations: *S. caroliniana:* VA–NC, Northern Coastal Plain Region; SC–GA, Middle Coastal Plain Region; FLA, Southern Coastal Plain Region; INTERIOR, all populations except Coastal Plain and Arkansas; ARK, Ozark Region. *S. nigra:* S & COASTAL, Southern and Coastal Plain Region from North Carolina southward; N & INTERIOR, Northern and Interior Region, north of North Carolina; ARK, Ozark Region. (See text for discussion).

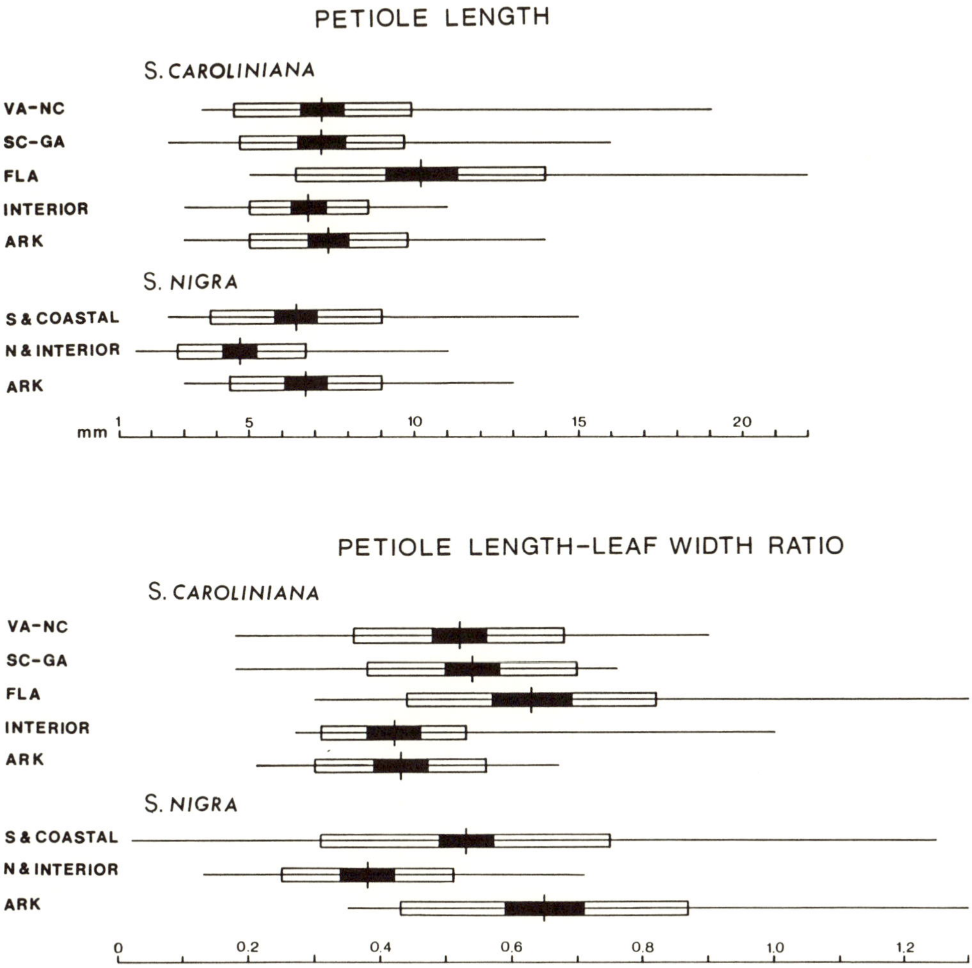

FIG. 5. Petiole length and petiole length-leaf width ratio in *Salix caroliniana* and *S. nigra* in the Southeastern States. (See Fig. 4 for definition of regional populations).

2. Leaf width (Fig. 4). The mean leaf width of plants from the Northern and Middle Coastal Plain regions is less narrow than in plants from the Interior and Ozark regions. Leaf blade length and width are correlated in that leaves that are broader are also shorter.

3. Leaf length-width ratio (Fig. 4). There is no significant difference in leaf shape, as described by the length-width ratio, within the species. The separation of plants from the Coastal Plain regions (as *S. longipes* var. *longipes*) from those in the Interior and Ozark regions (as *S. longipes* var. *venulosa*), as was done by Schneider (1919b) on the basis of the former having broader leaves than the latter, is not supported here.

4. Petiole length (Fig. 5). The petiole length is the same in all regions except the Southern Coastal Plain region in which a statistically greater mean petiole length prevails. There is considerable overlap in the values for this character. The

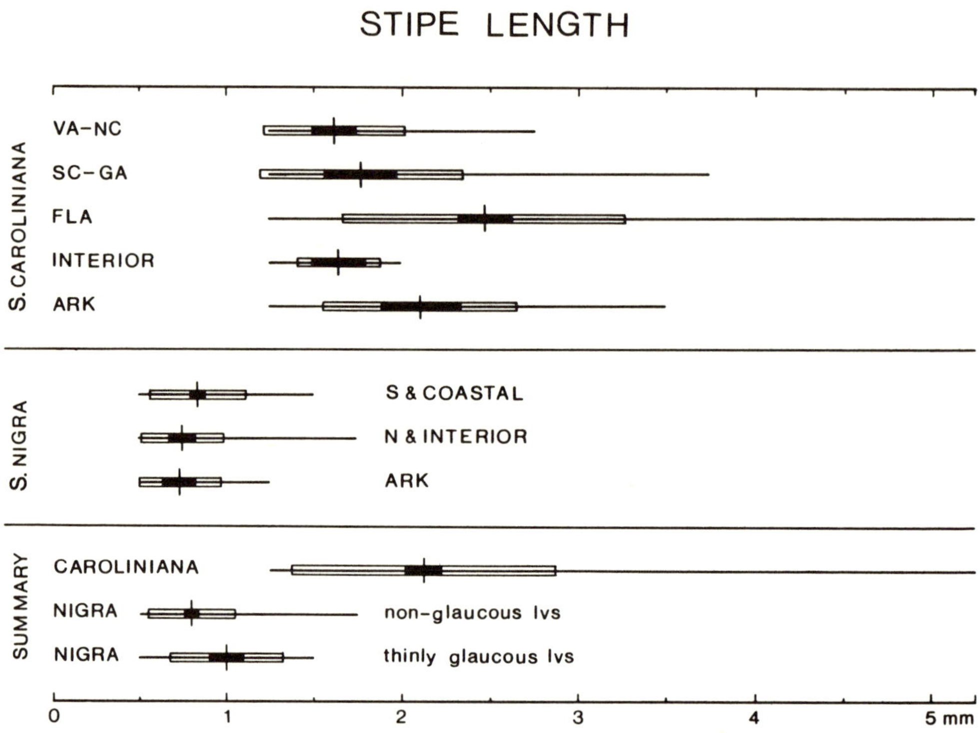

FIG. 6. Stipe length in *Salix caroliniana* and *S. nigra* in the Southeastern States. In the summary comparison of stipe length the non-glaucous and thinly glaucous-leaved specimens of *S. nigra* have been computed separately. (See Fig. 4. for definition of regional populations).

use of petiole length to separate plants from the Coastal Plain regions (as *S. longipes* var. *longipes*) from those of the Interior and Ozark regions (as var. *venulosa*), as was done by Schneider (1919b), is not supported here.

5. Petiole length-leaf width ratio (Fig. 5). The mean of this characteristic is the same throughout the Coastal Plain. In the Interior and Ozark regions plants have statistically shorter petioles relative to leaf width, but there is considerable overlap in the range of variation of this ratio.

6. Stipe length (Fig. 6). Variation in stipe length is generally clinal from north to south and from east to west, with the longest stipes occurring in the Southern Coastal Plain Region. The sample size for this characteristic in the Interior and Ozark regions is small and not statistically significant. The ranges of variation of this characteristic in the five regions completely overlap each other.

B. *Qualitative Characteristics.*

7. Amphistomatous leaves. The leaves of *S. caroliniana* are hypostomatous, i.e., stomata are present only on the lower surface. Two exceptions to this pattern have been observed. Specimens from Ohio Co., West Virginia [*Bartholomew 926, 927* (CM, NA, WVA)] have abundant stomata in the upper epidermis, and some of the syntypes of *S. harbisonii* (=*S. caroliniana*) have scattered stomata along

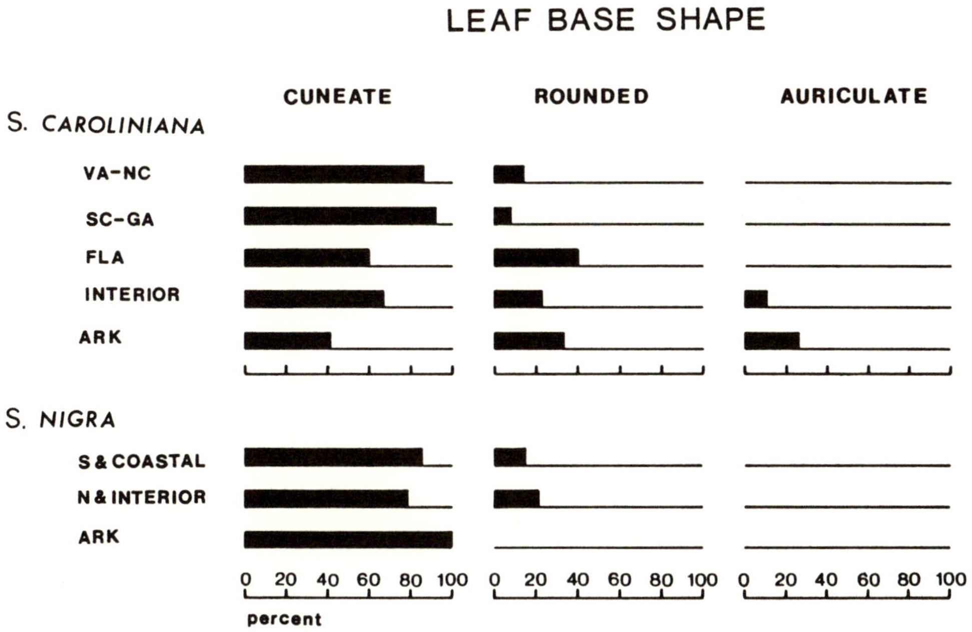

FIG. 7. Leaf base shape in *S. caroliniana* and *S. nigra* in the Southeastern States. (See Fig. 4 for definition of regional populations).

the secondary veins. The West Virginia population is so exceptional that it should be studied further.

8. Shape of leaf base (Fig. 7). The shape of the base of the leaf blade in *S. caroliniana* varies from cuneate to rounded to auriculate. In the Coastal Plain region the leaf base is mostly cuneate, although up to 40% of plants from the Southern Coastal Plain region have rounded leaf bases. The variation throughout the Coastal Plain regions is closely paralleled in the Interior and Ozark regions, but in the latter there also is a tendency toward auriculate leaf bases that is not evident elsewhere.

9. Leaf glaucescence. The mature leaves of *S. caroliniana* characteristically are strongly glaucous beneath. Immature leaves also are usually strongly glaucous, but in some populations (e.g., Georgetown Co., S.C.) young leaves have been observed to be only thinly glaucous, a condition that often occurs in *S. nigra*. The glaucescence is produced by a wax that may be driven off if the plants are dried over excessive heat. Some herbarium specimens lacking glaucescence, but identifiable as *S. nigra* by their short stipes and/or their geographical distribution, may have lost their glaucescence in drying. No geographical variation in this characteristic could be observed, because it was the characteristic by which the species was recognized.

10. Petiolar glands. Glandular dots or processes occur on the petiole near the leaf blade in 86% of the specimens, including 11% that have prominent glandular processes. There is no geographic variation in this characteristic.

11. Stipule shape (Fig. 8). Three types of stipules occur in *S. caroliniana*. The acute/oblique type has an acute apex and an axis that is curved obliquely, so that

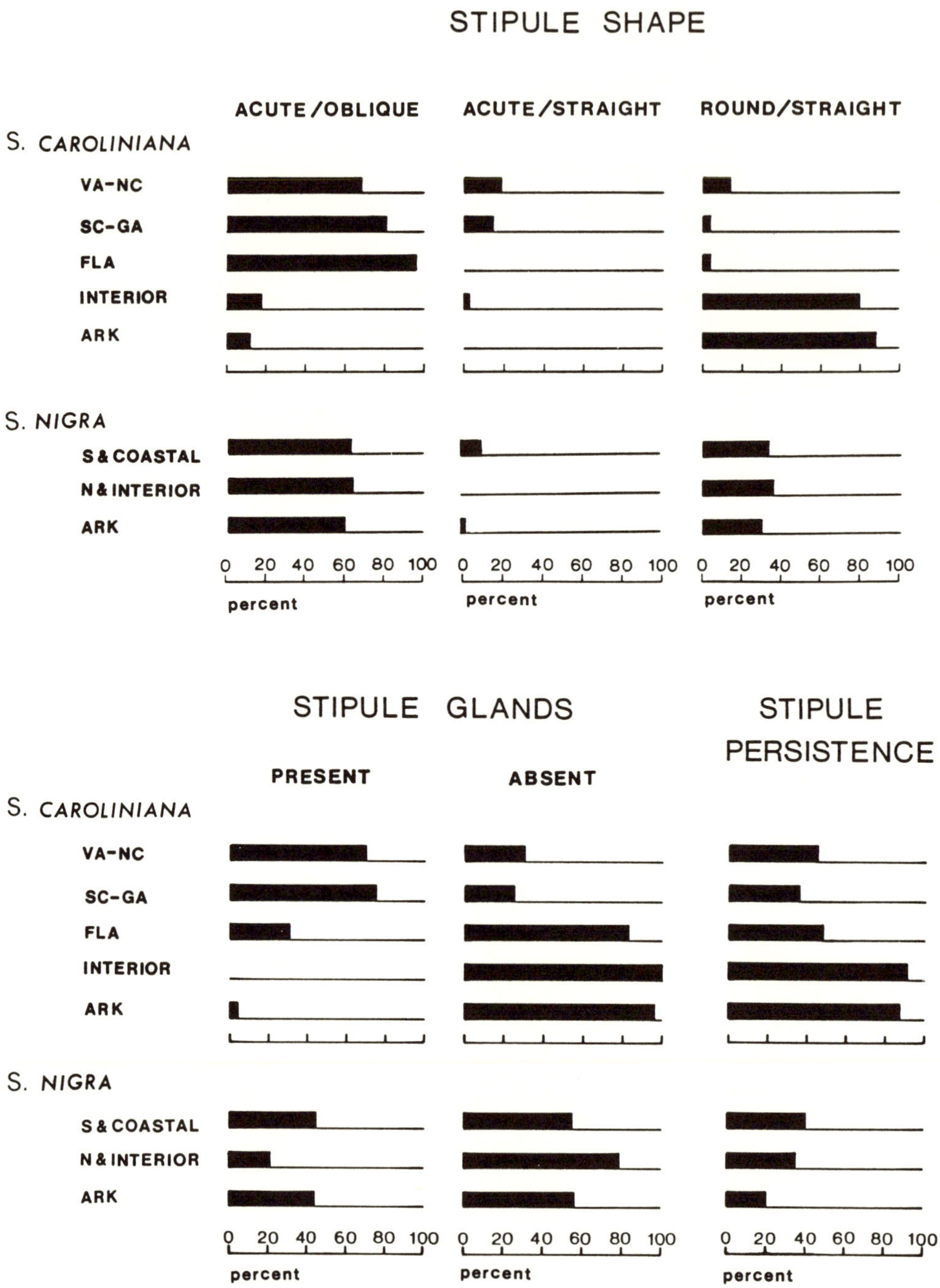

FIG. 8. Stipule shape, stipule glands, and stipule persistence in *Salix caroliniana* and *S. nigra* in the Southeastern States. Stipule shape is categorized as acute/oblique: apex acute and axis curved obliquely; acute/straight: apex acute and axis straight; round/straight: apex rounded and axis straight. Stipule shape and glandulosity data are based only on specimens with persistent stipules. In the Coastal Plain populations of *S. caroliniana* and in all populations of *S. nigra* stipules are persistent in less than half of the specimens. (See Fig. 4 for definition of regional populations).

the apex points either to the side or downward. Stipules of this shape are sometimes referred to as sickle-shaped (Skvortsov 1968). The acute/straight type has an acute apex and a straight axis, and the round/straight type has a rounded or obtuse apex and a straight axis.

The stipules in 68 to 96% of plants in the Coastal Plain regions are of the acute/oblique type. In the Interior and Ozark region 80 to 99% of the plants have round/straight stipules. Plants with acute/straight stipules occur in low frequency or are absent in all populations. The occurrence of rounded stipules, which was used by Schneider (1919b) to characterize *S. caroliniana* (as *S. longipes* var. *wardii*), usually distinguishes plants from the Interior and Ozark regions from those in the Coastal Plain regions.

12. Stipule glandulosity (Fig. 8). Stipules of *S. caroliniana* may or may not have glandular dots or processes on the adaxial surface. Populations in the Northern and Middle Coastal Plain regions mainly have glandular stipules with about 20 to 25% eglandular. In the Southern Coastal Plain Region this pattern is reversed and the percentage of eglandular stipules rises to 83%; this reversal is paralleled in plants in the Interior and Ozark regions in which almost all plants have eglandular stipules.

13. Stipule persistence (Fig. 8). Stipule persistence varies widely in the species. In general stipules are persistent in less than 50% of plants in the Coastal Plain regions, but almost 90% of plants in the Interior and Ozark regions have persistent stipules. Persistent stipules and straight stipules with a rounded apex are correlated.

14. Branchlet brittleness. Differences in the ease with which branchlets break from the stem when flexed at the base are not easily quantified; but striking differences were observed in the field between plants with tenacious or flexible branchlets that resist breaking and those with branchlets that readily break off at the base. Field observations showed that the branchlets of *S. caroliniana* are neither as highly brittle as those of *S. nigra* and other species, nor are they as tenacious as those of *S. discolor* and *S. humilis*. They seem to vary from brittle to only slightly so. The use of brittle branchlets to separate *S. harbisonii* from the supposedly non-brittle *S. caroliniana* (as *S. longipes*) (Schneider 1919b; Coker & Totten 1937) is not supported by these observations.

15. Branchlet indumentum (Fig. 9). The pubescence of branchlets ranges from glabrous to very sparsely villous (and soon glabrate) to persistently villous or tomentose, sometimes very densely so. Branchlets of plants from the Northern Coastal Plain Region are mostly glabrous or glabrate (75%). Southward this condition is reversed, and on the Southern Coastal Plain 70% of plants have villous or tomentose branchlets. Similarly, to the west the percentage of plants with villous or tomentose branchlets rises from 60% in the Interior Region to 92% in the Ozark Region, where most plants have branchlets with persistent, densely villous indumentum. Schneider's characterization of the branchlets of *S. longipes* var. *wardii* (Interior and Ozark regions) as glabrous or glabrate and the branchlets of *S. longipes* var. *longipes* (Coastal Plain) as tomentose or villous is contrary to these data.

16. Pistil indumentum. The pistils of *S. caroliniana* are characteristically glabrous. The report of thinly pubescent pistils that become glabrate in age (Ball 1938, 1961) has not been confirmed in this study.

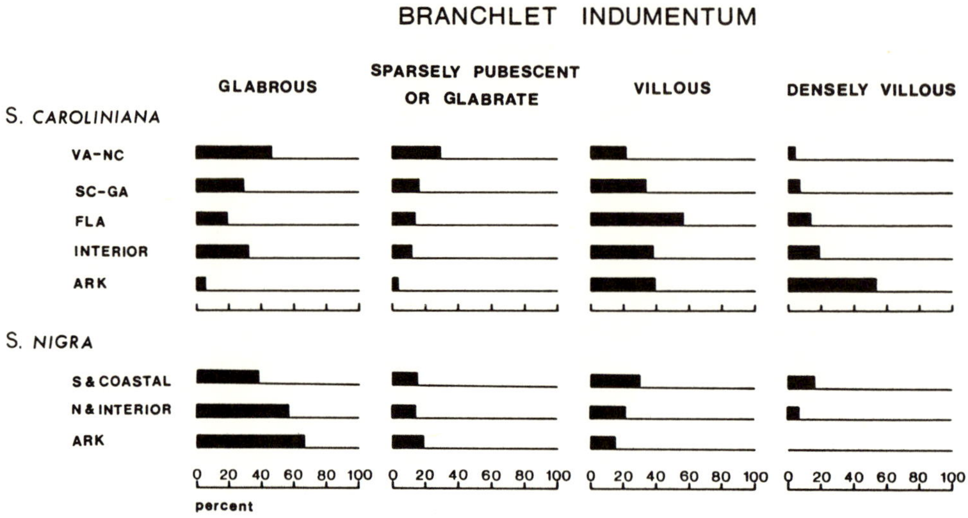

FIG. 9. Branchlet indumentum in *Salix caroliniana* and *S. nigra* in the Southeastern States. (See Fig. 4 for definition of regional populations).

17. Ament position. The aments of *Salix,* when borne on a leafy flowering branchlet, are terminal. It has been noted by some authors that the aments of *S. caroliniana* sometimes appear to be lateral or borne on the side of the flowering branchlet. In *S. caroliniana* and *S. nigra* the bud in the axil of the distal leaf or the leaf just below the ament itself may develop in the year it was formed and produces a secondary shoot that grows parallel to the ament which then appears to be lateral. The occurrence of such pseudolateral aments is not common in *Salix,* and it may prove to be confined to members of sect. *Humboldtiana.* Geographical variation of the characteristic in *S. caroliniana* could not be detected.

Taxonomic Significance of Infraspecific Variation in Salix caroliniana

The study of some morphological characteristics in *S. caroliniana* reveals several tendencies within the species. In the Interior and Ozark regions the species tends to have longer mean leaf length, shorter petioles in relation to leaf width, rounded and auriculate leaf bases, obtuse/straight stipules, more persistent, eglandular stipules, and higher frequency of pubescent branchlets. In the Southern Coastal Plain region the species shows tendencies toward longer mean petiole length, longer mean stipe length, higher frequency of rounded leaf bases, eglandular stipules, and higher frequency of pubescent branchlets.

It would seem that such an array of characteristics would be sufficient for the recognition of infraspecific taxa in *S. caroliniana.* These variations, however, are only tendencies and do not seem to represent significant character differentiation. The quantitative differences, such as leaf length, petiole length, and stipe length, and the ratio of petiole length and leaf width in some populations have statistically different means; but there is so much overlap in the measurements that on the basis of them many specimens could not be placed confidently into one group or another.

Similarly, the qualitative characters are also highly variable. For example, the auriculate leaf bases that are prominent in plants in the Ozark region represent only 26% of the total, and cuneate (41%) and rounded (33%) leaf bases still predominate. Pubescent branchlets seem to vary clinally southward along the Coastal Plain and westward through the Interior and Ozark regions; they reach a level of 92% in the Ozarks. All of the other qualitative characters are also clinal and do not permit the unequivocal recognition of infraspecific taxa.

The only qualitative characteristic that could be useful in itself for the recognition of an infraspecific taxon is stipule shape. The occurrence of plants with stipules with a rounded apex and a straight axis was the main basis for Bebb's recognition of *S. nigra* var. *wardii,* later treated by him as a species and then as *S. longipes* var. *wardii* by Schneider. In 1919 Schneider wrote, "The best way to distinguish var. *Wardii* is by the obtuse stipules, all the other characters are apparently too unstable . . . " (p. 29). Obtuse or rounded stipules are conspicuous in plants from the Interior and Ozark regions, but even in these populations 12 to 21% of specimens have acute stipules; in the Coastal Plain regions rounded stipules occur in 4 to 14% of specimens. It must be concluded that, although there is a tendency for some characters to be correlated with geography, this correlation is not strong enough for the recognition of geographical races.

Salix caroliniana is a highly variable species. Several factors may account for this variation including hybridization with *S. nigra,* character displacement where it is sympatric with *S. nigra,* the adaptation of populations to different environments, or to the genetic divergence of isolated populations. It is probable that all of these factors, and others unrecognized here, are influencing the variation displayed by *S. caroliniana.*

Hybridization

Salix caroliniana has been reported to hybridize with *S. nigra, S. sericea, S. eriocephala* (as *S. rigida* Muhlenberg) by Glatfelter (1897), Hitchcock and Standley (1919), and Steyermark (1963), and with *S. amygdaloides* (Glatfelter 1897). In the southeastern species of *Salix* I have recognized only *S. caroliniana* × *nigra.* This putative hybrid is characterized by it glaucous leaves and very short stipes. Plants of this type occur mainly in northern Florida and adjacent Georgia, on the Atlantic Coastal Plain, and at a few Interior localities where the distributions of the two species overlap (Fig. 10). Plants with leaves very thinly glaucous beneath and stipes that have a tendency to be slightly longer than usual in *S. nigra* may also be this hybrid; but they are included here in *S. nigra* (see Fig. 6 and *S. nigra,* no. 2, for a discussion of these characters). Glatfelter (1897) recognized hybrids between *S. caroliniana* and *S. nigra* in the vicinity of St. Louis, Missouri, by the varying degrees of leaf glaucescence that he thought were recombined in the hybrids. I have studied specimens of *S. nigra* that have a thin glaucescence on the lower leaf surface, but I have not been able to recognize any other strongly correlated indications of hybridity.

The area of intergradation between the Coastal Plain *S. caroliniana* and the continental *S. nigra* in northern Florida and southern Georgia lies in a "suture-zone," described by Remington (1968) as a zone of hybridization between a number of peninsular Florida animal species and subspecies and their continental

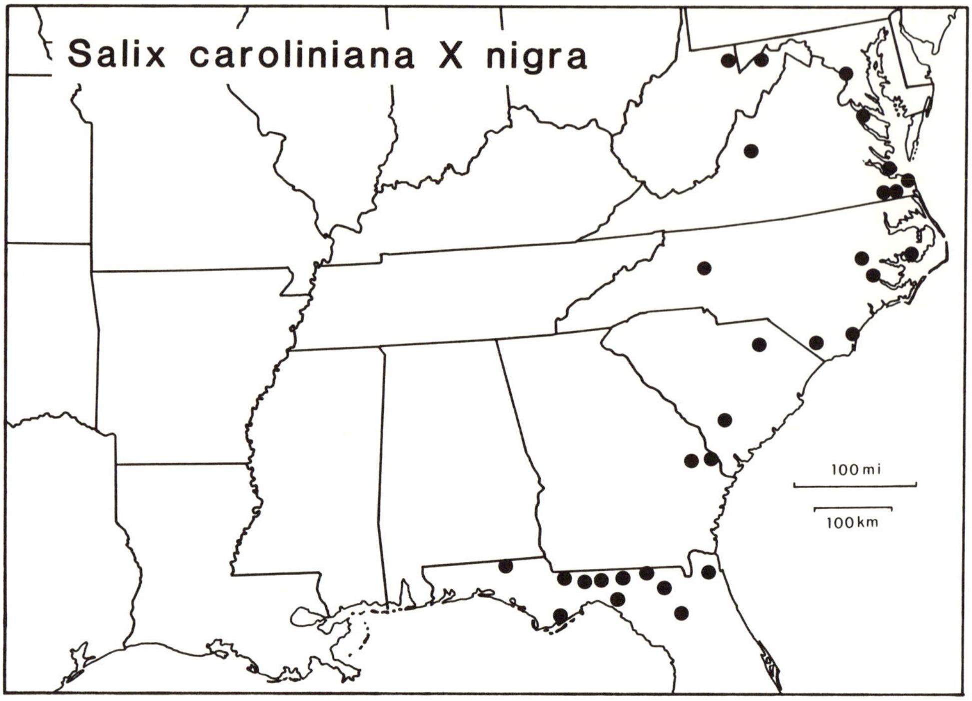

FIG. 10. Distribution of *Salix caroliniana* × *nigra* in the Southeastern States. Solid dots, herbarium specimens.

near relatives. Remington suggested that the recent break-up of forest cover in this region, due to human activity, may have allowed species of semi-wooded habitats to invade and overlap. The zone of hybridization between *S. caroliniana* and *S. nigra* is very narrow, and it is likely that they have only recently come into contact in a way similar to that suggested by Remington. All known hybrid populations of these species occur in man-made habitats (see discussion of natural populations above), but a study of the structure of naturally disturbed habitats may help our understanding of the initiation of hybridization between these species.

REPRESENTATIVE SPECIMENS. U.S.A. DISTRICT OF COLUMBIA: 10 May 1892, *Blanchard s.n.* (A).—FLORIDA: Alachua Co., 16 Apr. 1941, *West s.n.* (FLAS); Columbia Co., 8 May 1901, *Hume s.n.* (FLAS); Duval Co., *Harbison 2* (A, US); Franklin Co., *Godfrey 54482* (FLAS, USF); Gadsden Co., *Godfrey 54683* (USF); Hamilton Co., 11 May 1901, *Hume s.n.* (FLAS); Jackson Co., *Argus 11500A* (CAN); Jefferson Co., *Godfrey 54517;* Leon Co., *Argus 11493A* (CAN); Madison Co., *Meyer 10689A* (NA); Taylor Co., 10 Apr 1956, *Kurz s.n.* (FLAS, FSU, GH, NA, NCU, NY).—GEORGIA: Bullock Co., 8 Apr 1963, *Youmans s.n.* (FLAS); Grady Co., *Argus 11492A* (CAN).—NORTH CAROLINA: Bladen Co., *Argus 11389* (CAN); Columbus Co., 25 Apr 1958, *Radford s.n.* (WVA); Iredell Co., *Hardin 2369* (FSU); New Hanover Co., *Boufford 9002* (CM); Pitt Co., *Radford 32404* (NCU).—SOUTH CAROLINA: Bamberg Co., *Argus 11415* (CAN); Darlington Co., *Harbison 14158* (NCU).—VIRGINIA: Elizabeth City Co., *Robinson 479* (A); Nansemond Co., *Huff 70* (UNA); Norfolk Co., *Fernald 15082* (GH, PH, WIS); Princess Ann Co., *Fernald 4625* (GH, PH, US); Rockbridge Co., 29 May 1909, *Bartram s.n.* (GH); Westmoreland Co., *Edwin 352* (NA); York Co., *Kirkman 133* (LYN,

WILLI).—West Virginia: Mineral Co., 14 Jun 1933, *Strausbaugh s.n.* (WVA); Preston Co., 4 Jul 1909, *Jennings s.n.* (CM).

Salix caroliniana may also hybridize with *S. amygdaloides,* but the latter species is very uncommon in the study area, and the hybrid has not yet been recognized there. Specimens recognizable as hybrids between *S. caroliniana* and *S. sericea* or *S. eriocephala* have not been seen, even from the northern part of the study area where the species overlap, and it is doubtful that these hybrids occur naturally.

The recognition of hybrids is difficult, especially on the basis of herbarium specimens, and it may be questioned if even the few putative hybrids recognized here are in fact hybrids. Experimental evidence is needed to confirm the ability of these species to cross and to recognize the character combinations that may result.

2. Salix nigra Marshall, *Arbust. amer.* 139. 1785.—Type: unknown.—Black willow.

Salix flavo-virens Hornemann, *Enum. pl. hort. hafn.* Supplementum 2: 11. 1809.—Type: unknown.

Salix ligustrina Michaux f., *Hist. arbr. forest.* 3: 226. 1812. *Salix nigra* [var.] β *longifolia* Andersson, *Monogr. Salicum* 20. 1867.—Type: New York. "sur les bords du lac Champlain, et notament à sa naissance près de Skeenborough," *Michaux s.n.* (lectotype, here designated: P!, photo: CAN).

? *Salix falcata* Pursh, *Fl. Amer. sept.* 2: 614. 1814. *Salix purshiana* Sprengel, Syst. 5: 608. 1828, nom. superfl. *Salix nigra* var. *falcata* (Pursh) Torrey, *Fl. New York* 2: 209. 1847. *Salix nigra* f. *falcata* (Pursh) Rehder, *Bibl. cult. trees* 73. 1949.—Type: unknown.

? *Salix ambigua* Pursh, *Fl. Amer. sept.* 2: 617. 1814, non Ehrhart, 1791. *Salix dubia* Trautvetter, Bull. Acad. Sci. Petersb. 1: 132. 1867.—Type: unknown.

? *Salix denudata* Rafinesque, *Fl. ludov.* 116. 1817.—Type: unknown. (Probably no specimens preserved, fide Merrill, 1949).

? *Salix ludoviciana* Rafinesque, *Fl. ludov.* 116. 1817, nomen nudum.

Salix nigra [var.] **S. marginata* Wimmer ex Andersson, *Monogr. Salicum* 21. 1867.—Type: Louisiana. New Orleans, 1832, *Drummond 303* (lectotype, here designated: GH!; isotypes: K, photo: A! NY!).

Salix nigra var. *altissima* Sargent, *Trees and shrubs* 2: 216. 1913.—Type: Arkansas. Fulton, 20 May 1909, *Bush 5654* (lectotype, here designated: A!).

Trees up to 20 m or more tall; bark rough, furrowed, blackish brown with wide, thick-scaled ridges and narrower connecting ridges (Sudworth 1934). Branches reddish brown to yellowish brown, glabrous; branchlets brown to reddish brown, glabrous, sometimes sparsely pilose, rarely tomentose or densely villous with white and/or ferruginous hairs, very brittle at base; buds tawny, glabrous, apex sharply pointed, bud scale margins free and overlapping. Leaf blades narrowly lanceolate to very narrowly so, the larger ones (5–) 7–10–13 (–19) cm long, (0.6–) 0.75–1.2–1.7 (–2.3) cm wide, length/width 6–9–13, apex acuminate, base acute, sometimes rounded, margin glandular-serrulate; immature leaf blades reddish to yellowish green, translucent, glabrous, sometimes sparsely pilose to densely sericeous, midrib pilose or sericeous with white hairs sometimes intermixed with ferruginous hairs; mature leaf blades glabrous to sparsely pilose above, lower surface paler green but non-glaucous, rarely with thin transparent wax, glabrous or sparsely pilose, midrib often pilose on both sides, often amphistomatous; petioles yellowish brown, (2–) 3–6–10 (–15) mm long, sparsely pilose adaxially or all over or glabrate, glands usually present at distal end; stipules usually small and caducous, apex acute, usually eglandular adaxially, maximum length about 12 mm. Aments coetaneous, borne on leafy flowering branchlets, appearing lateral if bud in axil of distal leaf emerges (syllepsis), leaves deciduous after flowering, aments finally deciduous, abscissing above distal leaf. Staminate aments 1.7–4.7–7.5 cm long, borne on flowering branchlets 0.4–0.8–1.5 cm long; stamens usually 6, rarely 4 or 5, filaments pilose on proximal half, dry anthers 0.4–0.6 mm long; nectaries usually 2 or more, one short stout adaxial lobe and 2–3 slender abaxial lobes, sometimes only one and then adaxial; bracts tawny, 1–3 mm long, obovate, apex round or acute, crisp-villous on both sides or glabrate abaxially. Pistillate aments 2.3–4.8–7 cm long, usually densely flowered, borne on flowering branchlets 0.5–1.5–2.3 cm long, pistils greenish brown, glabrous, white-punctate with stomata, pyriform to obclavate, short-beaked, capsules brownish, 3–5 mm long, fruits often deciduous, styles (including the 2-branched stigmas) 0.2–0.4 mm long, stipes 0.5–1.5 mm long; nectary 1, adaxial, short and blunt, 0.2–0.4 mm long, ⅓ length of stipe; ovules 6–9 per carpel; bracts as in staminate flowers but often oblong and with acute or irregularly toothed apex, deciduous after flowering. Chromosome number 2n = 38. Fig. 11

Phenology. Flowering in the South (Georgia to Florida and Louisiana) from early February to early May with a mean in early April, and to the north and west from late March to the end of May with a mean in late April; based on 426 records.

Distribution. (Figs. 11-inset, 12). *Salix nigra* is a very wide-ranging North American species that occurs from southern Ontario, the eastern townships of Quebec, and New Brunswick westward to southeastern Minnesota and southward to the panhandle of Florida and central Texas. South of Virginia it is very rarely encountered on the outer or lower Atlantic Coastal Plain. The distribution map by Fowles (1965) includes more of the Coastal Plain than it should, but otherwise it is accurate. The map in Little (1971, 1976a) includes also *S. gooddingii,* accounting for that portion of the range from western Texas to California and south to Mexico.

Salix nigra is primarily a tree of river margins and floodplains. As a pioneer species on wet, fine-textured alluvial soils (Harlow et al. 1979) it may form large pure stands. Often its habitat is a very narrow band along streams, and as it is shaded by taller trees on the shoreward side and undermined by water erosion its stems characteristically lean toward the river (Lindsey et al. 1961). In the Ozark region it usually does not occur on gravel beds next to the streams themselves, but rather on the deeper alluvial soils on the river banks (Steyermark 1963). It is not, however, highly restricted in its habitat and grows wherever light and moisture conditions are favorable (Glatfelter 1897; Fowles 1965). Specimens have been seen from a wide variety of wet to mesic habitats, including the edges of ponds

FIG. 11. *Salix nigra.* a. Branch bearing young vegetative branchlets and staminate aments (×0.75). b. Mature leaf (×0.75). c. Staminate flower (×10). d. Pistillate ament in fruit with sylleptic side branch, arrow (×0.75). e. Young pistillate flowers (×10). Inset: map of total distribution, based on Argus (1964), Fowles (1965), Little (1976a), and herbarium specimens.

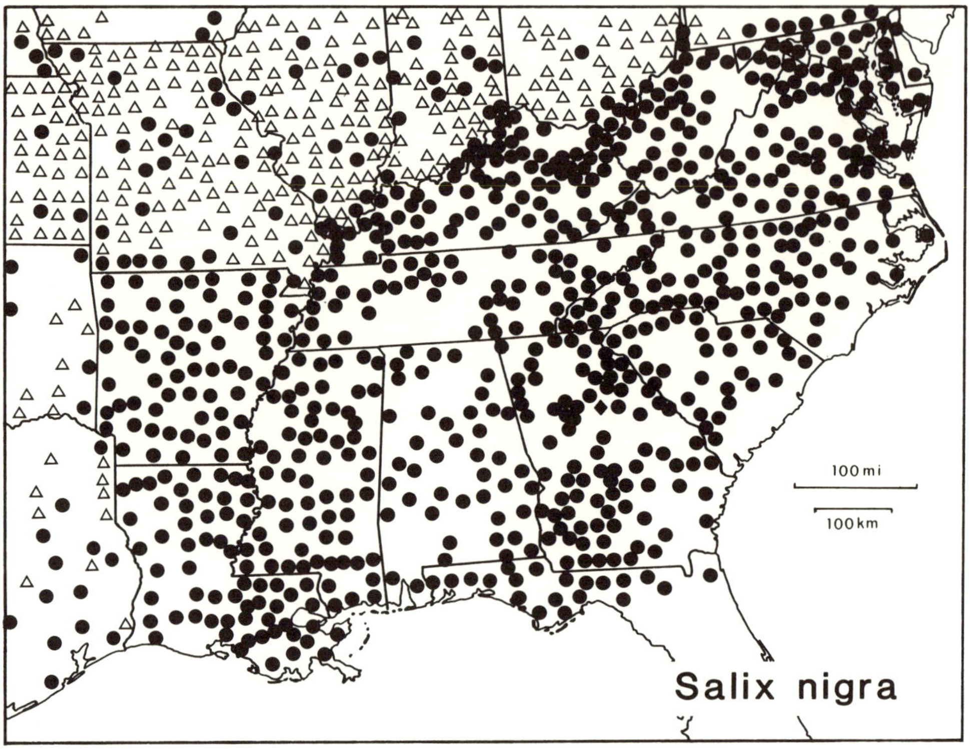

FIG. 12. Distribution of *Salix nigra* in the Southeastern and peripheral states. Solid dots, herbarium specimens; diamonds, W. Duncan (pers. comm.); open triangles, literature reports.

and lakes, swamps, marshes, white cedar bogs, wet meadows, open fields, and roadside ditches, as well as in mixed upland deciduous woods along rivers. A comparison of the habitats of *S. nigra* and *S. caroliniana* is given under the latter species (no. 1).

REPRESENTATIVE SPECIMENS. U.S.A. ALABAMA: Baldwin Co., *Kral 41800* (VDB); Barbour Co., 25 Apr 1967, *Drake s.n.* (AUA); Bibb Co., *Argus 11515* (CAN); Calhoun Co., *Haynes 7133* (UNA); Choctaw Co., 3 Apr 1974, *Gwaltney s.n.* (AUA); Cleburne Co., *Clark 2661* (NCU); Coffee Co., *Argus 11508* (CAN); Colbert Co., 30 Mar 1967, *Wisenant s.n.* (NCU); Conecuh Co., *Clark 10279* (NCU); Coosa Co., *Argus 11510* (CAN); Dale Co., 8 May 1971, *Gwaltney s.n.* (AUA); Dallas Co., *Harbison 38* (GH, NCU); DeKalb Co., *Thomas 52660* (NLU); Elmore Co., *Kral 29922* (PH, TENN, VDB); Escambia Co., *Gaines 192* (GH, NCSC, NY); Etowah Co., *Whetstone 4094* (NCU); Franklin Co., 6 Apr 1967, *Harima s.n.* (GH); Geneva Co., *McDaniel 8649* (UNA, VDB); Greene Co., *Wiersema 1633* (UNA); Hale Co., *Thomas 811* (GH, UNA); Houston Co., *Argus 11505* (CAN); Jefferson Co., 15 Apr 1965, *Wright s.n.* (AUA); Lamar Co., *Clark 10605* (NCU); Lauderdale Co., 15 Apr 1970, *Flurry s.n.* (AUA); Lawrence Co., *Clark 12895* (NCU); Lee Co., *Pollard 23* (GH, NA, NY, US); Limestone Co., 12 May 1968, *White s.n.* (NCU); Lowndes Co., *Kral 55040* (VDB); Macon Co., *Sessler 21* (AUA); Madison Co., *Bryson 499* (VDB); Marengo Co., *Argus 11517* (CAN); Marshall Co., *Brown 71* (NCU); Mobile Co., *Boufford 18595* (MO, NCU); Montgomery Co., *Kral 55083* (VDB); Pike Co., *Palmer 27226* (A, MO, NA); Russell Co., *Haynes 7680* (UNA); Saint Clair Co., *Haynes 7156* (UNA); Shelby Co., *Clark 13229* (NCU); Sumter Co., *Jones 15569* (VDB); Talladega Co., *Ward 126* (AUA); Tallapoosa Co., *Brown 21* (AUA); Tuscaloosa Co., *Hoener 1234* (UNA); Walker Co., *Haynes 6096* (NCU, UNA); Wilcox Co., 28 Jun 1958, *Matschke s.n.* (AUA).—

ARKANSAS: Arkansas Co., *Demaree 21096* (MO, NA, NY); Ashley Co., *Demaree 14727* (NA, NY, UARK); Baxter Co., *Palmer 6044* (A, MO, US); Benton Co., 6 Sep 1945, *Larimore s.n.* (UARK); Bradley Co., *Demaree 14390* (A, MO, NA, NY); Calhoun Co., *Demaree 14440* (A, NA, NO, NY); Carroll Co., *Palmer 5627* (A, MO, US); Chicot Co., *Demaree 20934* (MO, NA, NY); Clark Co., *Demaree 17196* (DAO, MO, NA, NY); Clay Co., *Demaree 28797* (DUKE, NA, UNA); Cleburne Co., *Thomas 10218* (NLU); Cleveland Co., *Demaree 13572* (MO, NA, NY); Columbia Co., *Demaree 20789A* (MO, NA, NY); Conway Co., *Demaree 20915* (MO, NA); Craighead Co., *Demaree 28785* (NA, NCU, TENN, WILLI); Crawford Co., *Demaree 20868* (MO, NA, NY); Crittenden Co., *Demaree 14885* (A, GH, MO, NA, NY, TENN); Cross Co., *Demaree 27608* (NA); Desha Co., *Demaree 13770* (NA, NY, UARK); Drew Co., *Demaree 14352* (A, NA, NO, NY); Faulkner Co., *Demaree 54* (UARK); Franklin Co., *Barber 911* (UARK); Fulton Co., *Lipscomb 292* (UARK); Garland Co., *Demaree 15948* (MO, NA, NY); Grant Co., *Locke 2391* (UARK); Greene Co., *Demaree 27620* (NA, NCSC); Hempstead Co., *Demaree 63423* (NLU, WILLI); Hot Springs Co., *Demaree 17890* (DAO, NA); Howard Co., *Lawson 2142* (NLU); Independence Co., *Demaree 17041* (MO, NA, NY); Izard Co., 20 Aug 1958, *Robinson s.n.* (UARK); Jackson Co., *Demaree 16972* (MO, NA, NY, UARK); Jefferson Co., *Demaree 14455* (MO, NA, NY); Johnson Co., *Demaree 19909* (MO, NA); Lawrence Co., *Demaree 28899* (FSU, NA, NCU, TENN); Lincoln Co., *Locke 2953* (UARK); Little River Co., *Demaree 58164* (NCU, TENN); Logan Co., *Britton 68* (UARK); Lonoke Co., *Demaree 22965* (MO, NA, NY); Marion Co., *Smith 65* (UARK); Miller Co., *Demaree 13480* (MO, NA, NY, TENN); Mississippi Co., *Demaree 28738* (DUKE, TENN, UNA, WILLI); Monroe Co., *Demaree 14937* (A, GH, MO, NA, NY); Nevada Co., *Redfearn 30566* (NCU); Newton Co., *Thompson 266* (NLU, UARK); Ouachita Co., *Demaree 16826* (MO, NA, NY, UARK); Perry Co., *Demaree 20147* (NA); Phillips Co., *Demaree 37448* (NA); Pike Co., *Demaree 36787* (FSU, GH, NCU, NO); Poinsett Co., *Demaree 27601* (NA, NCSC, NY); Polk Co., *Haynes 5725* (LSUS); Pope Co., *Demaree 55641* (AUA, NCU, TENN); Prairie Co., *Demaree 57628* (APCR, NCU, NLU); Pulaski Co., *Demaree 17300* (MO, NA, NY); Randolph Co., *Demaree 30961* (GH, USF); Saint Francis Co., *Howton 65* (UARK); Saline Co., *Demaree 21741* (MO, NA, NY); Scott Co., 1 Aug 1974, *Lynn s.n.* (UARK); Searcy Co., *Thomas 9493* (NLU); Sebastian Co., *Meeks 154* (UARK); Sevier Co., *Demaree 10034* (GH, NA); Stone Co., *Lawson 841* (NLU); Union Co., *Demaree 22047* (MO, NA, NY); Van Buren Co., *Argus 11655* (CAN); Washington Co., *Palmer 23256* (A, MO); White Co., *Thomas 36857* (NLU); Yell Co., *Demaree 20123A* (MO, NA).—DELAWARE: New Castle Co., *Tatnall 1757* (DOV); Sussex Co., *Tatnall 2510* (DOV, PH).—DISTRICT OF COLUMBIA: *Ball 2592* (DUKE, GH, NCSC, NY, UNA).—FLORIDA: Bay Co., *Martin 1632* (DUKE, NA); Columbia Co., *Argus 11449* (CAN); Duval Co., *Curtiss 4524B* (A, MO, NY); Escambia Co., *Godfrey 54589* (DUKE, FSU, GH, NA, NCU, NY, USF); Franklin Co., *Harbison 13892* (NCU); Gadsden Co., *Godfrey 54451* (DUKE, FSU, GH, NCU, NY, USF, VDB); Gulf Co., 15 Apr 1948, *Haynie s.n.* (FLAS); Hamilton Co., *Faircloth 7489* (NCU, VSC); Holmes Co., *Godfrey 54476* (DUKE, FSU, GH, NA, NCU, USF, VDB); Jackson Co., *Godfrey 54691* (DUKE, FSU, GH, NY, USF); Jefferson Co., *Godfrey 80849* (CAN); Leon Co., *Godfrey 54447* (DUKE, FLAS, FSU, NA, NCU, NY, USF, WIS); Liberty Co., *Thorne 16576* (FLAS); Madison Co., *Dean 225* (VSC); Okaloosa Co., *Godfrey 54621* (DUKE, FSU, GH, NA, NCU, NY, USF); Santa Rosa Co., *Martin 64* (NCU); Taylor Co., *Godfrey 81197P* (FSU); Wakulla Co., *Kreager 1* (NA, US); Walton Co., *Godfrey 55242* (FSU, NA); Washington Co., 11 Jun 1955, *Kurz s.n.* (FSU).—GEORGIA: Baker Co., *Thorne 4115* (GA); Bartow Co., *Greear 66324* (UNA); Ben Hill Co., *Faircloth 4615* (GA, MO, NCU, VSC); Bibb Co., *Duncan 1920* (GA); Bleckley Co., *Argus 11425* (CAN); Brooks Co., *Duncan 1315* (GA, PH); Bulloch Co., *Argus 11420* (CAN); Calhoun Co., *Bozeman 6533* (NCU); Candler Co., *Bozeman 4216* (NCU); Carroll Co., *Carnes 21* (GA); Clarke Co., *Duncan 27* (NO); Clayton Co., *Harbison 1* (MO); Clinch Co., *Ernest 14* (GA, VSC); Coffee Co., *Argus 11433* (CAN); Colquitt Co., *Demaree 29521* (NCU, NO, VDB); Columbia Co., *Duncan 9530* (GA, MISS); Crisp Co., *Lane 180* (GA); Dade Co., *Whetstone 37* (GA); Decatur Co., *Thorne 2416* (GEO, NA); DeKalb Co., *Duncan 2262* (GA); Dodge Co., *Bozeman 5362* (NCU); Dougherty Co., *Watson 88* (GA, VSC); Early Co., *Thorne 3621* (GA, GEO); Effingham Co., *Duncan 2184* (GA, GH, NA); Elbert Co., *Coile 1683* (FLAS, GA, NCU); Emanuel Co., *Argus 11422* (CAN); Fannin Co., *Smith 2572* (WIS); Floyd Co., *Duncan 15257* (LTU); Franklin Co., *Argus 10883* (CAN); Fulton Co., *McDowell 322* (GEO); Glynn Co., 4 May 1969, *Music s.n.* (VSC); Gordon Co., *Duncan 2459* (GA, NA); Grady Co., *Faircloth 748* (GA, NCU, VSC); Greene Co., *Duncan 2284* (GA, NA); Hall Co., *Duncan 18602* (GA, MISS); Haralson Co., 20 Jul 1966, *Ezzell s.n.* (GA); Hart Co., *Argus 10884* (CAN); Henry Co., 9 Jul 1966, *Lewis s.n.* (GA); Houston Co., *Dixon 1450* (NLU); Jackson Co.,

Underwood 44 (GA); Johnson Co., *Argus 11424* (CAN); Jones Co., *Duncan 1890* (GA); Lanier Co., *Bozeman 4710* (NCU); Laurens Co., *Bracewell 52* (NCU, VSC); Liberty Co., *Duncan 2197* (GA, NA); Lincoln Co., *Pyron 82* (GA, PH); Lowndes Co., *Faircloth 2220* (GA, NCU, VSC); Macon Co., 1 Jul 1966, *Averill s.n.* (GA); Madison Co., *Coile 299* (GA); Meriwether Co., *Duncan 5166* (GA); Mitchell Co., *Faircloth 6599* (VSC); Murray Co., *Duncan 16455* (GA); Muscogee Co., 16 May 1965, *Merritt s.n.* (AUA); Oglethorpe Co., *Duncan 22493* (DUKE, FSU, FUGR, GA, LTU, MO, NCSC, NCU, NLU, NO, NY, PH, TENN, UNA, USF, WVA); Polk Co., *Pyron 2621* (GA); Pulaski Co., *Argus 11426* (CAN); Quitman Co., *Thorne 3146* (GA, GEO); Rabun Co., *Duncan 16477* (GA, NCSC); Randolph Co., *Harbison 13926* (NCU, NY); Richmond Co., *Harbison 3999* (NCU); Rockdale Co., *Eyles 7158* (DUKE, NA); Schley Co., *Henry 2053* (PH); Screven Co., *Argus 11418* (CAN); Spalding Co., 31 Mar 1935, *Drury s.n.* (GA); Stephens Co., 22 Jul 1972, *Doornbos s.n.* (FUGR); Sumpter Co., *Demaree 49551* (FSU, GA, NCSC); Tattnall Co., *Padgett 194* (GA); Taylor Co., *Faircloth 7569* (NCU, VSC); Terrell Co., *Dean 328* (VSC); Thomas Co., *Faircloth 3091* (GA, NCU, VSC); Tift Co., *Bozeman 5303* (NCU); Turner Co., *Faircloth 4359* (GA, MO, NCU, VSC); Union Co., *Duncan 16325* (GA); Walton Co., Jul 1970, *Harcourt s.n.* (GA); Ware Co., *Argus 11438* (CAN); Warren Co., *Duncan 2295* (GA, NA); Webster Co., *Thorne 9351* (GA); Wheeler Co., *Duncan 4718* (GA, MO); White Co., *Duncan 22678A* (GA); Wilcox Co., *Argus 11428* (CAN); Worth Co., *Faircloth 3878* (GA, NCU, VSC).—KENTUCKY: Adair Co., *Gibson 670* (F); Ballard Co., *Argus 11541* (CAN); Bath Co., *Warden 205* (USF); Bell Co., *Horsey 1971* (A); Boone Co., *Reed 44763* (CFR); Bourbon Co., *Guhardja 129* (KY); Boyd Co., *Demaree 1320* (NA, NY); Bracken Co., *Braun 4468* (US); Breathitt Co., *Horsey 1130* (A); Breckenridge Co., *McFarland 16* (NA); Bullitt Co., *McFarland 9* (NA); Caldwell Co., *Reed 18873* (CFR); Calloway Co., *McLemore 26* (MUR); Carlisle Co., *Wilson 1462* (MUR); Carroll Co., *McInteer 379* (NA); Carter Co., *Warden 168* (KY); Casey Co., *Wharton 4280* (KY); Christian Co., *Labisky 311* (WIS); Clark Co., *Beckett 661* (KY, NA); Crittenden Co., *McFarland 29* (NA); Daviess Co., *Brown 7290* (KY); Edmonson Co., *Elmore 791* (WKU); Elliott Co., *Reed 15372* (CFR); Estill Co., *Argus 10842* (CAN); Fayette Co., *Adams 60* (KY); Franklin Co., *Wharton 10281* (KY); Fulton Co., *McInteer 1755* (KY); Grant Co., 29 Jul 1961, *Browne s.n.* (KY); Greenup Co., *Braun 1797* (US); Hardin Co., 10 Jun 1950, *Davies s.n.* (NCU); Harlan Co., *McInteer 1553* (KY); Hart Co., *McFarland 65* (NA); Henderson Co., *Marx 712* (NLU); Henry Co., *Gentry 296* (KY, NCU, NY); Hickman Co., *Kerrick 2* (MUR); Hopkins Co., *Sullivan 17* (WKU); Jefferson Co., *Wharton 4479* (NA, NY); Johnson Co., *Horsey 931* (A); Knox Co., *Evans 318* (KY); Laurel Co., *McInteer 353* (NA); Lawrence Co., *McKenzie 22* (MUHW); Leslie Co., *McInteer 1539* (KY); Letcher Co., *Sole 744* (EKY); Lincoln Co., *Wharton 1430* (NA); Livingston Co., 2 May 1972, *Wilson s.n.* (MUR); Logan Co., 27 May 1969, *Meijer s.n.* (KY); Lyon Co., 15 May 1971, *Yates s.n.* (MUR); Madison Co., *Wharton 5686* (KY, NA); Magoffin Co., *McFarland 3428* (NA); Marion Co., *Wharton 4207* (KY); McCracken Co., *McFarland 35* (NA); McLean Co., *Conrad 751* (KY); Meade Co., *Braun 3863* (A, US); Menifee Co., *Netherly 86* (CFR); Mercer Co., *Varner 11571* (EKY); Metcalfe Co., *Shacklette 25* (NA); Montgomery Co., *Wharton 1878* (MO, NA, NY); Morgan Co., 6 May 1950, *Perkins s.n.* (CFR); Muhlenberg Co., *Conrad 136* (KY, NCU, TENN); Nelson Co., *Braun 2988* (A, US); Ohio Co., *Browne 5503* (KY, NCU); Oldham Co., *Woodward 2726* (DHL); Owen Co., *Lovell 850* (NLU); Pike Co., *Horsey 953* (A); Powell Co., *Wharton 6123* (NA, NY); Pulaski Co., *Reed 7931* (CFR); Rockcastle Co., *Hanna 14* (EKY); Rowan Co., *Argus 10841* (CAN); Scott Co., *Wharton 10981* (KY); Shelby Co., *McFarland 1* (NA); Todd Co., *McFarland 63* (MO, NA); Trigg Co., *Woford 1688* (NLU); Union Co., *McFarland 26* (NA); Warren Co., *Nicely 1799* (WKU); Washington Co., *Reed 25207* (CFR); Whitley Co., *McInteer 370* (NA); Wolfe Co., *McFarland 3316* (NA).—LOUISIANA: Acadia Par., *McWilliams 218* (LTU); Ascension Par., *Watters 1171* (LTU); Assumption Par., 6 Apr 1918, *Marionneaux s.n.* (LSU); Beauregard Par., *Ewan 20066* (NO); Bienville Par., *Thomas 5212* (NLU); Boissier Par., *Barbour 1142* (LSUS, NO); Caddo Par., *Allen 8014* (DUKE, FLAS, NCU, USF, WVA); Calcasieu Par., 11 Apr 1925, *Small s.n.* (GH, NY, US); Caldwell Par., *Hendershot 15* (LTU); Cameron Par., 26 Mar 1977, *Shapii s.n.* (LSU); Catahoula Par., *French 109* (NLU); Claiborne Par., *Lewis 390* (NLU); Concordia Par., *Rhodes 15428* (LTU); DeSoto Par., *Thomas 68736* (NLU); East Baton Rouge Par., *Sijam 98* (FLAS, LTU); East Carroll Par., *Thomas 3306* (LYN, NLU); East Feliciana Par., *Smith 155* (LTU); Evangeline Par., *Cormier 606* (NLU); Franklin Par., *Thomas 2308* (NCU, NLU); Grant Par., *Thomas 28905* (NLU); Iberia Par., 14 Oct 1975, *Naquin s.n.* (LSU); Iberville Par., *Lewis 1111* (NLU); Jackson Par., *Reid 1295* (AUA, NLU); Jefferson Par., *Stone 175* (GH, NO); Jefferson Davis Par., *Palmer 7670* (A, MO); Lafayette Par., *Thieret 22836* (DUKE, US); Lafourche Par., *Lasseigne 5610* (KNK); LaSalle Par., *Thieret 26263* (DUKE, FSU); Lincoln Par.,

Watters 76 (LTU); Livingston Par., *Montz 1996* (LTU); Madison Par., *Thomas 2354* (LYN, NLU); Morehouse Par., *Lynch 36* (LTU); Natchitoches Par., *Thomas 72520* (NLU); Orleans Par., *Ball 377* (GH, NA, NY, US); Ouachita Par., *Thomas 22589* (AUA, CAN, CFR, CM, DHL, FLAS, LSU, LYN, MISS, NO, TENN, UARK, UNA, VSC); Plaquemines Par., *Rich 444* (NLU); Pointe Coupee Par., *Lieux 16* (NCU, NLU); Rapides Par., *Ball 477* (GH, NA, NY); Red River Par., *Thomas 45041* (NLU); Richland Par., *Jones 187* (NCU, NLU); Sabine Par., *Demaree 48248* (GH, NO); Saint Bernard Par., *Demaree 13501* (NA, NY); Saint Charles Par., *Carter 101* (LTU); Saint Helena Par., *Allen 6526* (NCU, USF); Saint James Par., 12 Sep 1972, *Guidroz s.n.* (NLU); Saint John The Baptist Par., *Breeland 246* (LTU); Saint Landry Par., *Thieret 10500* (DAO, DUKE, FSU) Saint Martin Par., *Thieret 17043* (DUKE, FSU, USF, WIS); Saint Mary Par., *Montz 3920* (LSU, NO); Saint Tammany Par., *Arsene 12190* (GH, US); Tangipahoa Par., *Kirby 153* (LSUS); Tensas Par., *Iltis 25274* (NCU, WIS); Terrebonne Par., *Guidroz 19* (NLU); Union Par., *Thomas 13061* (NCU, NLU); Vernon Par., *Cooley 3887* (NO, USF); Washington Par., *Thomas 63422* (NLU); Webster Par., *Thomas 72073* (NLU); West Baton Rouge Par., *Demaree 13486* (NA); West Carroll Par., *Gough 248* (NLU); West Filiciana Par., *Thieret 19449* (DUKE, FSU); Winn Par., *Hutchinson 81* (NLU).—MARYLAND: Allegany Co., *Ball 3023* (NA); Anne Arundel Co., 27 Jul 1940, *Morton s.n.* (MARY); Baltimore Co., *Reed 44390* (CFR); Calvert Co., *Seymour 16687* (WIS); Caroline Co., *Quidas 42* (MOAR); Carroll Co., *Adams 4757* (MOAR); Cecil Co., *Benner 7104* (GH, PH); Charles Co., *Walker 5564* (MARY); Dorchester Co., *Reed 1669* (CFR); Frederick Co., *Hickey 272* (NCU); Garrett Co., *Ball 3014* (MARY, NA); Harford Co., *Reed 23354* (CFR); Howard Co., *Solomon 429* (MARY); Kent Co., 13 May 1965, *Gladu s.n.* (MARY); Montgomery Co., *Ball 2592* (NA, VPI); Prince Georges Co., *Hill 8996* (NY); Queen Annes Co., *Tidestrom 8089* (A); Queen Marys Co., *Reed 4919* (CFR); Somerset Co., *Wolfe 41* (CFR); Talbot Co., *Ball 2731* (DUKE, MUHW, NA, NCSC, TENN, UNA, US, VPI, WVA); Washington Co., *Argus 10819* (CAN); Worcester Co., *Ball 2737* (MARY, MO, NY, US). —MISSISSIPPI: Adams Co., *Jones 11604* (MISS); Alcorn Co., *Temple 4703* (MISS, NCU); Amite Co., *Jones 7266* (MISS); Attala Co., *Jones 16990* (MISS); Benton Co., *Pullen 67525* (MISS); Calhoun Co., *Pullen 66793* (MISS); Carroll Co., *Pullen 70634* (MISS); Chickasaw Co., *Pullen 66868* (MISS); Choctaw Co., 22 Jun 1905, *Jensen s.n.* (MO); Clay Co., *Pullen 69852* (MISS); Coahoma Co., *Argus 11528* (CAN); Copiah Co., *Pullen 8895* (MISS); Covington Co., *Jones 12629* (MISS); Desoto Co., *Argus 11530* (CAN); Forrest Co., *Rogers 4348* (NCU); Franklin Co., *Jones 7439* (MISS); George Co., *Jones 6970* (MISS); Greene Co., *Jones 8490* (MISS); Grenada Co., *Pullen 64210* (MISS); Hancock Co., *Jones 9535* (MISS); Harrison Co., *Darwin 1340* (ASU, FLAS, LSUS, NLU, NO, UNA, USF); Hinds Co., *Temple 8219* (MISS); Holmes Co., *Pullen 70381* (MISS); Humphries Co., *Pullen 66974* (MISS); Itawamba Co., *Pullen 661046* (MISS); Jackson Co., *Jones 8116* (MISS); Jasper Co., *Dearing 45* (MISS); Jefferson Co., *Jones 6236* (MISS); Jones Co., *Jones 9012* (MISS); Kemper Co., *Temple 8717* (MISS); Lafayette Co., *Palko 23* (APCR); Lamar Co., *Hillman 132* (AUA); Lauderdale Co., *Jones 7035* (MISS); Lawrence Co., *Jones 7766* (MISS); Lee Co., 22 Sep 1891, *Seymour s.n.* (DUKE); Leflore Co., *Argus 11524* (CAN); Madison Co., *McDougall 1531* (US); Marion Co., *Jones 6591* (MISS); Marshall Co., *Lassetter 610* (MISS); Montgomery Co., *Pullen 66724* (MISS); Newton Co., *Jones 15647* (MISS); Noxubee Co., *Argus 11521* (CAN); Oktibbeha Co., *Harbison 26* (A, US); Penola Co., 16 Sep 1964, *Sartor s.n.* (MISS); Pearl River Co., *Jones 8559* (MISS); Perry Co., *McDaniel 21192* (FLAS, FSU); Pike Co., *Jones 19723* (MISS); Quitman Co., *Pullen 67891* (MISS); Rankin Co., *Jones 19015* (FSU); Scott Co., *Jones 16491* (FLAS, MISS); Sharkey Co., *Ray 4932* (GH, NCU, NY, USF); Simpson Co., *Thomas 7042* (LYN, NLU); Smith Co., *Jones 5841* (MISS); Tallahatchie Co., *Argus 11525* (CAN); Tishomingo Co., *Searcy 639* (MISS); Tunica Co., *Temple 2953* (MISS, NCU); Walthall Co., *Jones 8762* (MISS); Warren Co., *Demaree 14106* (MO, NA, NY); Washington Co., *Muntz 8682* (A, DUKE, FLAS, GH, NY, US); Wilkinson Co., *Jones 7536* (MISS); Winston Co., *Argus 11522* (CAN); Yalobusha Co., *Pullen 70713* (MISS); Yazoo Co., *Cooley 3339* (GH, USF).—NORTH CAROLINA: Alleghany Co., *Correll 14633* (DUKE, NA); Anson Co., *Radford 43652* (NCU); Bladen Co., *Argus 11386* (CAN); Buncombe Co., *Biltmore Herb. 1230* (MO, NY, US); Burke Co., *Bell 6546* (NCU); Cabarrus Co., *Bell 2242* (NCU); Caswell Co., *Ahles 54012* (NCU); Catawba Co., *Bell 6825* (NCU); Chatham Co., *Crutchfield 1314* (NCU); Cherokee Co., 17 Apr 1976, *Schutz s.n.* (NLU); Clay Co., *Pittillo 3366* (NCU); Cleveland Co., *Ahles 15298* (NCU); Columbus Co., *Argus 11387* (CAN); Craven Co., *Harbison 14076* (NCU); Cumberland Co., *Ahles 24313* (NCU); Dare Co., *Radford 7590* (NCU); Davie Co., *Radford 10860* (NCU); Duplin Co., *Ahles 23992* (NCU); Durham Co., *Ahles 53224* (AUA, CM, DAO, FLAS, FSU, NCU, NY, PAC, TENN, UARK, UNA, USF, WIS, WVA); Edgecombe Co., *Smith 90* (NCU); Forsyth Co., *Ahles 40866* (MUR, NCU); Franklin

Co., *Ahles 11390* (NCU); Gaston Co., *Ahles 14976* (NCU); Gates Co., *Wilbur 19666* (DUKE); Graham Co., *Ahles 13251* (NCU); Granville Co., *Ahles 11484* (NCU); Greene Co., *Radford 32587* (NCU); Guilford Co., *Bell 11658* (NCU); Halifax Co., *Ahles 14857* (NCU, TENN); Harnett Co., *Argus 10929* (CAN); Haywood Co., *Wood 1525* (NCU, UNCC); Henderson Co., 24 Apr 1887, *Memminger s.n.* (NCU); Hoke Co., *Ahles 25152* (NCU); Iredell Co., *Hardin 2369* (NCSC); Jackson Co., *Pittillo 2974* (NCU); Johnston Co., *Fosberg 17715* (NA, NCSC); Lee Co., *Argus 11381* (CAN); Macon Co., *Keener 922* (DUKE); Madison Co., *Boufford 18123* (MO); Martin Co., *Radford 32245* (NCU); McDowell Co., *Bell 3598* (GH, NCU); Mecklenburg Co., 20 Jun 1974, *James s.n.* (UNCC): Mitchell Co., *Harbison 5777* (NCU); Montgomery Co., *Wood 120* (NCU); Moore Co., *Carter 481* (NCU); Northampton Co., *Carroll 305* (NCU); Orange Co., *Ahles 53228* (AUA, PAC, UARK, UNA, US, WIS); Pamlico Co., *Radford 31983* (NCU); Person Co., *Batson 756* (DUKE); Polk Co., *Fosberg 18855* (NA, NO); Randolph Co., *Radford 42829* (NCU); Richmond Co., 18 Apr 1965, *Mullen s.n.* (NCSC); Rockingham Co., *Radford 10753* (NCU); Rowan Co., 29 Jul 1979, *Halshauser s.n.* (UNCC); Rutherford Co., *Smith 60* (NCU); Sampson Co., *Argus 10917B* (CAN); Scotland Co., *Sharp 1958* (NCU); Stanley Co., *Morgan 966* (NCU); Stokes Co., *McCurdy 202* (NCU); Surry Co., *Williams 522* (DUKE); Swain Co., *Argus 10861* (CAN); Vance Co., *Lee 274* (DUKE); Wake Co., *Godfrey 3584* (GH, NCSC); Warren Co., *Bell 2858* (FLAS, NCU); Washington Co., *Wilbur 19676* (DUKE); Wilks Co., *Keever 234* (DUKE); Wilson Co., *Hardin 2359* (NCSC).—South Carolina: Abbeville Co., *Argus 10887* (CAN); Aiken Co., *Faircloth 4218* (NCU, VSC); Allendale Co., *Argus 11417* (CAN); Anderson Co., *Anderson 1396* (US); Bamberg Co., *Argus 10890* (CAN); Barnwell Co., *Ahles 56479* (NCU); Calhoun Co., *Ahles 21793* (NCU); Cherokee Co., *Ahles 22731* (NCU); Chester Co., *Rodgers 76129* (CLEMS); Chesterfield Co., 20 Sep 1932, *Coker s.n.* (NCU); Clarendon Co., *Argus 11410* (CAN); Darlington Co., *Mellichamp 667* (UNCC); Dillon Co., *Ahles 23205* (NCU); Fairfield Co., *Radford 43479* (MUR, NCU); Greenville Co., *Rogers 76211* (FUGR, MO); Greenwood Co., *Radford 20139* (NCU); Hampton Co., *Ahles 57881* (NCU); Horry Co., *Argus 11390* (CAN); Kershaw Co., *Radford 20791* (NCU); Lancaster Co., *Bozeman 8829* (DHL, MISS, NCU, UNCC, WILLI); Lexington Co., *Radford 20471* (NCU, USF); Marion Co., *Bell 6291* (NCU); Marlboro Co., *Argus 10913B* (CAN); McCormick Co., *Argus 10888* (CAN); Oconee Co., *Rodgers 73260* (CITA, NLU, WILLI); Orangebury Co., *Ahles 21597* (NCU); Pickens Co., *Hendricks 92* (CLEMS); Spartanburg Co., *Walker 959* (FUGR); Sumter Co., *Radford 21079* (NCU); Williamsburg Co., *Argus 10905* (CAN); York Co., *Boufford 12611* (CM).—Tennessee: Anderson Co., *Ellis 28744* (TENN); Benton Co., *Thomas 13417* (NLU); Bledsoe Co., *Iltis 3418* (TENN); Blount Co., 24 Jul 1965, *Thomas s.n.* (TENN); Campbell Co., *Marx 904* (NLU); Carroll Co., *Sharp 13041* (TENN, VDB); Carter Co., *Palmer 42558* (MO, NY); Cheatham Co., *Morton 43424* (VDB, VPI); Cocke Co., 19 May 1977, *Genelle s.n.* (USF); Crocket Co., *Gonsoulin 2002* (VDB); Cumberland Co., *Shanks 14174* (TENN); Davidson Co., *Blum 3488* (FUGR, PH); Dickson Co., *Kral 55377* (VDB); Dyer Co., *Argus 11536* (CAN); Fayette Co., *Hebb 409* (TENN); Franklin Co., *Sasnett 44* (NCU); Gibson Co., *Sharp 12363* (TENN, VDB); Grundy Co., *Patrick 210* (TENN, VDB); Hamilton Co., *Argus 10960* (CAN); Hardeman Co., *Sharp 10283* (TENN, VDB); Hawkins Co., *Reed 12725* (CFR); Henderson Co., *Godfrey 78917* (FSU); Humphreys Co., *Kral 45774* (VDB); Jefferson Co., May 1842, *Rugel s.n.* (DAO); Johnson Co., *Shanks 18567* (NCU, TENN); Knox Co., *Ellis 28980* (NCU, TENN); Lake Co., *Shanks 15874* (NCU, TENN); Lauderdale Co., *Sharp 12260* (TENN); Marion Co., *Shanks 15459* (CM, DAO, DHL, DUKE, FSU, GH, NCSC, NCU, NY, TENN, US, WVA, WVW); Monroe Co., *Malter 53935* (TENN); Montgomery Co., *Chester 3292* (NCU); Obion Co., *Webb 897* (VDB); Pickett Co., *Gonsoulin 1735* (KNK); Polk Co., *Wofford 80* (TENN); Roan Co., *Nease 240* (DUKE, US, WVA); Robertson Co., *Quarterman 1141* (DUKE, VDB); Scott Co., *Argus 10844* (CAN); Sevier Co., *Ball 2714* (NA, TENN, US); Shelby Co., *Demaree 20918* (MO, NA, NY, PH); Stewart Co., *Phillips 2569* (NCU, NLU); Sumner Co., *Blum 3308* (VDB, WILLI); Tipton Co., *Palmer 17327* (A, MO, NY); Unicol Co., *Lyle 19684* (TENN); Union Co., 23 Jun 1936, *Kalter s.n.* (TENN); Van Buren Co., *Iltis 3410* (TENN); Warren Co., *Shanks 15791* (TENN); White Co., *Shanks 15550* (TENN, VDB); Williamson Co., *Catignan 24* (VDB); Wilson Co., *Blum 3420* (AUA, TENN, VDB).—Virginia: Albemarle Co., *Ball 2743* (NA, VPI); Alleghany Co., *Ball 3023* (MARY, VPI); Amelia Co., *Lewis 412* (VPI); Appomattox Co., *James 7340* (NCU); Arlington Co., *Ball 2593* (GH, NA, NY); Augusta Co., 4 Jul 1936, *Carl s.n.* (VPI); Bedford Co., *Moore 43* (LYN); Botetourt Co., *Freer 1628* (GH, LYN, NCU); Brunswick Co., *James 13331* (NCU); Buchanan Co., *James 9972* (NCU); Buckingham Co., *Little 14629* (US, VPI); Campbell Co., *Wood 4453* (CM, PH); Charles City Co., *Argus 10966* (CAN); Charlotte Co., *James 2735* (NCU); Clarke Co., *Meyer 9168* (DAO, NA, WVA); Craig Co., *Fogg*

13147 (MO, PH); Cumberland Co., *James 7176* (NCU); Dickenson Co., *James 9845* (NCU); Dinwiddie Co., *Reed 99167* (CFR); Fairfax Co., *Bradley 10457* (WILLI); Fauquier Co., *Allard 9313* (CM, GH, NY, VPI); Fluvanna Co., *Diggs 1496* (WILLI); Franklin Co., *James 10682* (NCU); Frederick Co., 8 Apr 1978, *Deane s.n.* (LFCC); Giles Co., *Argus 10951* (CAN): Gloucester Co., *Greaves 1258A* (WILLI); Greensville Co., *Fernald 14815* (GH, MO, PH, TENN, US, WVA); Halifax Co., *Kral 13622* (FSU, VPI); Hanover Co., *James 13431* (NCU); Henrico Co., 26 Apr 1968, *DeChalmot s.n.* (US); James City Co., *Barans 196* (LYN, NCU, WILLI); King and Queen Co., *James 5307* (NCU); King George Co., *James 13698* (NCU), Lancaster Co., *James 6122* (NCU); Loudoun Co., *Ball 3010* (NA, VPI); Louisa Co., *James 13592* (NCU); Lunenburg Co., *James 3072* (NCU); Mathews Co., *Montfrans 317* (WILLI); Mecklenburg Co., *Seaman 3600* (NCU); Middlesex Co., *Reed 43889* (CFR); Montgomery Co., *Smythe 4736* (VPI); Nansemond Co., *Argus 10964A* (CAN); Nelson Co., *Bidwell 30* (LYN); New Kent Co., *Gillespie 34* (NCU, WILLI); Northampton Co., *James 7064* (NCU); Page Co., *Ball 3006* (MO, NA, NY, PH, US, VPI); Patrick Co., *James 10298* (NCU); Pittsylvania Co., *Ramsey 4890* (LYN); Powhatan Co., *Corcoran 1635* (WILLI); Prince George Co., 22 Aug 1908, *Rehder s.n.* (A); Prince William Co., *Allard 8518* (CM, MO, VPI); Princess Anne Co., *Gates 35* (UNA); Pulaski Co., *James 8452* (NCU); Rappahannock Co., *Walker 2421* (US); Richmond Co., *Tidestrom 7727* (NA, US); Roanoke Co., *Wood 3292* (PH); Rockbridge Co., 29 Apr 1915, *Churchill s.n.* (GH, MO, WVA); Rockingham Co., *Ball 2709* (DUKE, NA, NCSC, TENN, UNA, VPI); Russell Co., *James 9336* (NCU); Scott Co., *Peake 792* (WILLI); Shenandoah Co., *Allard 9361* (US, VPI); Smythe Co., *James 9036* (NCU); Southampton Co., *Fernald 15083* (GH, MO, PAC, PH, US); Surry Co., *Ware 5347* (WILLI); Warren Co., *Appler 443* (NLU, WILLI); Washington Co., *James 9045* (NCU); Westmoreland Co., *Reed 87034* (CFR); Wise Co., *James 9456* (NCU); York Co., *Salle 148* (NCU, VPI, WILLI).—West Virginia: Berkeley Co., *Downs 3648* (NCU); Boone Co., 10 May 1972, *Dunlap s.n.* (WVA); Braxton Co., *Gilbert 1002* (MUHW, NA, WVA); Cabell Co., *Gilbert 618* (A, CM, DUKE, GH, MO, MUHW, NY, PH, TENN, WIS, WVA); Calhoun Co., 7 May 1933, *Harris s.n.* (WVA); Clay Co., 6 Oct 1955, *Sizemore s.n.* (CM); Doddridge Co., 21 Sep 1969, *Bartholomew s.n.* (WVA); Fayette Co., 6 Jul 1968, *Phillips s.n.* (LSU); Gilmer Co., 11 Jul 1959, *Woofter s.n.* (WVA); Grant Co., *Davis 6620* (NA, WVA); Greenbrier Co., 9 Jun 1954, *Botany Exp. s.n.* (CM); Hampshire Co., *Ball 2606* (GH, MO, NA, NY, US, WVA); Hardy Co., *Argus 10824* (CAN); Jackson Co., *Bartholomew 7* (WVA); Kanawha Co., 12 May 1969, *Denison s.n.* (WVA); Logan Co., *Rossbach 2575* (WVW); Marion Co., 7 May 1961, *Steele s.n.* (WVA); Marshall Co., *Bartholomew 27* (WVA); Mason Co., *Argus 10839* (CAN); Mercer Co., 1 Sep 1940, *Fox s.n.* (WVA); Mineral Co., Aug 1956, *Brown s.n.* (NA); Mingo Co., 30 Jul 1972, *White s.n.* (NLU); Monongalia Co., *Millspaugh 113* (NY, WVA); Morgan Co., *Downs 10123* (NCU); Nicholas Co., 9 Aug 1940, *Fox s.n.* (WVA); Ohio Co., *Bartholomew 950* (GH, US); Preston Co., 1931, *Zucchero s.n.* (WVA); Putnam Co., *Argus 10837* (CAN); Raleigh Co., 1 Oct 1972, *Grafton s.n.* (WVA); Ritchie Co., *Bartholomew 72* (CM, WVA); Roane Co., *Bartholomew 8* (WVA); Summers Co., 27 May 1967, *Cain s.n.* (NLU); Tyler Co., *Berkley 587* (MO); Upshur Co., 17 May 1954, *Rossbach s.n.* (MO, WVW); Wayne Co., *Williams 685* (A, CLEMS, CM, DUKE, GH, MO, MUHW, NA, NY, PH, TENN, US, WIS, WVA); Webster Co., *Argus 10835* (CAN); Wetzel Co., *Bartholomew 60* (CM, NCU); Wirt Co., *Bartholomew 4113* (NCU); Wood Co., *Bartholomew 31* (WVA); Wyoming Co., *Lafferty 1144* (MUHW).

Recognition. Salix nigra is a tree willow that is characterized by leaves that are narrowly lanceolate, dark green on both sides and often pendulous, buds with a single bud scale with a free, overlapping margin, branchlets that are very brittle at the base, staminate aments with six stamens, glabrous pistils borne on short stipes, and by caducous pistillate floral bracts.

Nomenclature. There seems to be no type material of *S. nigra* in existence. According to Ewan (1967) Marshall did not preserve specimens, but the suggestion has been made that herbarium specimens may have been preserved by his nephew Moses Marshall (Pennell 1942) and that collections may have been sent to England. A search of the Darlington Herbarium (DWC), where some specimens of plants from Marshall's garden are preserved, and at the Philadelphia Academy of Natural Sciences did not turn up any Marshall specimens of *Salix*. The description of *S. nigra,* provided by Marshall (1785: 139–140), reads as follows: " . . . rises often with a leaning or crooked trunk to the height of about twenty feet, covered with a dark coloured rough bark. The leaves are smooth and of equal colour on both sides; narrow, lance-shaped, and very slightly serrated. The catkins are long and slender." This description perfectly fits *S. nigra,* as it is presently understood, and it has been used to fix the name of this species.

Type material of *S. flavo-virens* Hornemann could not be located at the Botanical Museum and Herbarium, Copenhagen; it is probable that Hornemann did not preserve herbarium specimens (A. Fox Maule, pers. comm.). The original description seems to fit *S. nigra.*

A vegetative specimen annotated *S. ligustrina* Michaux f., located in the Muséum National d'Histoire Naturelle, Paris, has been selected as the lectotype of this name. It bears the locality "Am Sept" and the collector's name "A. Michaux." A vegetative branch on a sheet in the Michaux herbarium that bears the locality name "New York" may be an isolectotype or a syntype.

The type of *S. falcata* Pursh could not be found in the Philadelphia Academy of Natural Sciences where many Pursh specimens are housed. This species traditionally has been placed with *S. nigra,* where it probably belongs; but its description also fits *S. caroliniana.* Pursh (1814) seems to have included members of sect. *Humboldtiana* under three names: 24. *S. houstoniana* Pursh (the type is *S. humboldtiana* Willdenow, holotype: BM, photo: CAN!), 25. *S. falcata* Pursh, and 26. *S. nigra* Marshall (including *S. caroliniana* Michaux as a synonym). It is possible, but not certain, that Pursh used the name *S. falcata* for *S. nigra* and the name *S. nigra* for *S. caroliniana.*

Andersson published the Wimmer name "*marginata*" as a variety of *S. nigra* in 1868. The holotype, cited by Andersson as in "herb. Vindob," later housed in the Naturhistorisches Museum, Wien, Austria, was lost to fire during World War II. Schneider (1919b) referred var. *marginata* to his *S. longipes* var. *venulosa* and commented that the "ovaries of the New York specimens are partly pilose and Wimmer's form needs further observation" (p. 26). Sargent (1913) thought that it was possible that this variety was the same as his var. *altissima.* Specimens of this taxon that I have seen in the Gray Herbarium and at the New York Botanical Garden have short stipes (about 1 mm long) and leaves that are thinly glaucous beneath. These plants fit well into *S. nigra,* but the partly pilose pistils are unusual and may reflect a link or perhaps intergradation with *S. gooddingii* (see below).

Sargent (1913) described *S. nigra* var. *altissima* to include the very tall trees that occur in southern Arkansas and Louisiana. He noted that these trees are "sometimes 40 metres high, with a trunk from 1 to 1.5 metres in diameter" (p. 216). Where they grow with the typical variety they flower two to three weeks later. Schneider (1919b) further separated this group from typical *S. nigra* by its more pubescent young shoots, leaves, and petioles, and by its longer and more acutely based leaves. These characteristics, however, are variable and probably not taxonomically important. I cannot separate the series of syntypes of this variety in the Arnold Arboretum from the species. I am not familiar with the described differences in habit and flowering time, and I do not know if these differences warrant varietal status.

Taxonomy. This species is closely related to *S. caroliniana, S. amygdaloides,*

S. gooddingii, S. floridana, S. humboldtiana, and *S. bonplandiana.* In the Southeastern States it may be distinguished from the first three species by the characteristics in Table 1. See also *S. caroliniana,* no. 1, for a discussion of taxonomic relationships.

Some Morphological Characteristics and Infraspecific Variation in Salix nigra

In this study of infraspecific variation specimens of *S. nigra* were recognized by having leaves non-glaucous beneath (green on both sides) and very short stipes. Infraspecific morphological variation was analysed by grouping the specimens into three geographical regions: (1) Southern and Coastal Plain Region (North Carolina, South Carolina, Georgia, Alabama, Mississippi, Florida, and Louisiana); (2) Northern and Interior Region (Maryland, Virginia, West Virginia, Kentucky, and Tennessee); and (3) Ozark Region (Arkansas).

Within each of these regional populations vegetative morphological characteristics were measured in individuals with mature leaves. Stipe measurements were made on all plants with mature fruits. See Materials and Methods for discussion of method of measurement and statistical analysis.

A. *Quantitative Characters.*

1. Leaf length and leaf width (Fig. 4). There are no significant infraspecific differences in these characters in *S. nigra.*

2. Leaf length-width ratio (Fig. 4). Leaf shape, as expressed by the length-width ratio, shows clinal variation from narrower leaves in the Ozark Region to broader leaves in the Northern and Interior Region. The differences in the means is statistically significant, but the overlap in values is very great.

3. Petiole length (Fig. 5). Plants in the Northern and Interior Region have a statistically shorter mean petiole length than do other populations, but the overlap in the values is very great.

4. Petiole length-leaf width ratio (Fig. 5). The means of this ratio in the three regional groups all are statistically significant. In the Northern and Interior Region the leaves tend to have shorter petioles in relation to leaf width, and in the Ozark Region the petioles tend to be longer in relation to leaf width. The ratio in the Southern and Coastal Plain Region falls between the other two.

6. Stipe length (Fig. 6). There is no significant infraspecific variation in stipe length in *S. nigra.*

B. *Qualitative Characters.*

7. Amphistomatous leaves. Most plants of *S. nigra* (94 to 100%) in Arkansas, Louisiana, and Mississippi have stomata on both leaf surfaces. In the other regions there is a lower incidence, varying from 23 to 26%, of amphistomatous leaves.

8. Shape of leaf base (Fig. 7). In both the northern and southern populations of *S. nigra* cuneate leaf bases occur in 80 to 85% of the plants. In the Ozark Region 100% of plants have cuneate leaf bases.

9. Stipule shape (Fig. 8). Stipules in *S. nigra* are usually very small or caducous, and in only a small percentage of specimens could stipule shape be ob-

served. In plants with stipules the most common shape is the acute/oblique type (apex acute and axis recurved), but up to 30 to 40% of stipulate plants have stipules with a rounded apex and a straight axis. This variation is not geographically correlated.

10. Stipule glandulosity (Fig. 8). In *S. nigra* about as many plants bear stipules that are glandular-dotted adaxially as have eglandular stipules. There is an apparent increase in eglandular-stipuled plants from the Southern and Coastal Plain Region, but because of the small numbers of plants with stipules these data are not statistically significant.

11. Stipule persistence (Fig. 8). Only 20 to 40% of plants have persistent stipules. This characteristic is not geographically correlated.

12. Petiolar glands. Glandular dots or processes are present on the petiole near the base of the leaf blade in 74% of specimens, including 4% with prominent glandular processes.

13. Leaf glaucescence. The leaves of *S. nigra* are typically green and non-glaucous beneath. In some plants, however, a very thin, transparent wax appears on the lower surface. The presence of this wax, which is not thick enough to be scratched, may be observed by burnishing the surface with a flat dissecting needle, or, on live material, by rubbing with a finger tip. If wax is present the surface will take a polish. Thin leaf glaucescence occurs in *S. nigra* throughout the area of sympatry with *S. caroliniana* (Fig. 13), but it is absent in *S. nigra* outside of this region. North of the range of *S. caroliniana* it was observed in *S. nigra* only in Chester Co., Pennsylvania, a short distance north of the present range of *S. caroliniana.* The correlation of the mean of this characteristic and stipe length (Fig. 6) is statistically significant, but because there is wide overlap with stipe length in *S. nigra* with non-glaucous leaves, it is interpreted as normal variation in *S. nigra.*

Since leaf glaucescence is so easily lost in the drying of herbarium specimens, it is possible that some plants included in my sample as thinly glaucous *S. nigra* are *S. caroliniana* in which most of the wax has been lost. A field study of plants with both mature leaves and fruits is necessary to test the significance of the stipe length data presented here.

14. Branchlet indumentum (Fig. 9). The pubescence of branchlets in *S. nigra* varies from glabrous to very sparsely villous to moderately so, or, rarely, densely tomentose. In the Southern and Coastal Plain Region glabrous or glabrate branchlets and pubescent branchlets occur in about equal frequency. In the Northern and Interior Region the frequency of glabrous branchlets increases to about 57% and in the Ozark Region to 66%. Pubescent and glabrous branchlets have been observed on the same plant, and this characteristic is not taxonomically useful.

15. Ament position. See *S. caroliniana,* no. 1, for comment.

Taxonomic Significance of Infraspecific Variation in Salix nigra

In the Southeastern States it is not possible to apply formal taxonomic nomenclature to the infraspecific variation in *S. nigra.* There is a certain amount of geographical variation in the species in the leaf length-width ratio, petiole length, petiole length-leaf width ratio, amphistomatous leaves, and branchlet indumen-

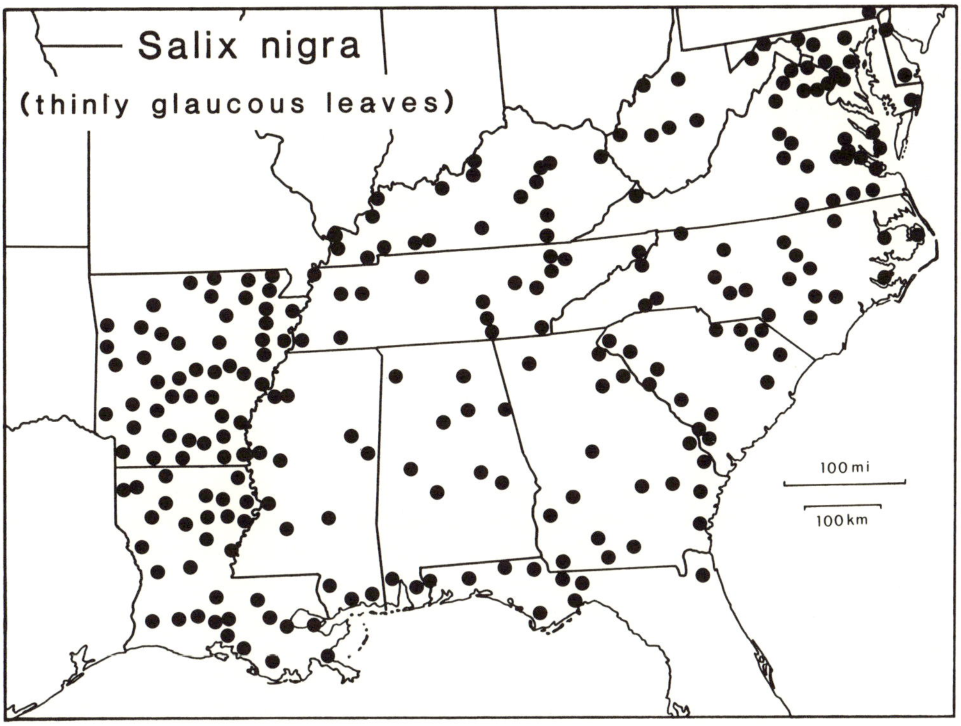

FIG. 13. Distribution of specimens of *Salix nigra* with thinly glaucous leaves. No specimens have been mapped from outside of the study area, except for one record from Pennsylvania. Solid dots, herbarium specimens.

tum; but these characters either vary clinally or are too variable to be taxonomically useful. There are, however, two taxa to the west of the study area that have been recognized by some authors and that deserve consideration: *S. nigra* var. *lindheimeri* C. Schneider and *S. gooddingii.*

Schneider (1918, 1919b) described *S. nigra* var. *lindheimeri* from Texas and Mexico and suggested that it may also occur in Oklahoma and Louisiana. It differs from var. *nigra* in narrower leaves with a more attenuate leaf base; comparatively longer petioles; young branchlets and petioles that are glabrous or glabrate; stipules with glands on the adaxial surface; larger capsules (6–7 mm long); longer stipes (up to 3 mm long); and loose fruiting aments. Ball (1961) also recognized this variety in Arkansas (Izard and Crittenden counties) and in Kansas. I have not been able, however, to recognize this variety in the Arkansas specimens that I have studied.

Herbarium specimens with long, loose fruiting aments are known from Louisiana, Arkansas, Oklahoma, and occasionally from eastern populations. In these plants the loose aments result not from long stipes, as described for var. *lindheimeri* and *S. caroliniana,* but from an elongation of the ament axis. I have included these plants in *S. nigra* s.l., but their possible relationship to var. *lindheimeri* requires further study. The type of *S. nigra* var. *lindheimeri* (*Lindheimer 1189,* Comal Co., Texas, isotype: CAN!) strongly resembles plants typical of *S. nigra* in having non-glaucous leaves and dark reddish brown branchlets, but its fruits are borne on stipes up to 2 mm long and the fruiting aments are very loose, thus suggesting *S. gooddingii.*

Salix gooddingii C. Ball occurs from Kansas south to Texas and Mexico and west to California. Some authors treat it as a distinct species (Schneider 1919b; Ball 1961), whereas others consider it to be either *S. nigra* var. *vallicola* Dudley (Benson & Darrow 1981) or *S. nigra* var. *venulosa* (Andersson) Bebb (Goodrich 1983) or include it in *S. nigra* s.l. (Little 1971). For a discussion of *S. nigra* var. *venulosa* see *S. caroliniana* (no. 1). *Salix gooddingii* differs from *S. nigra* in its long stipes (2–3 mm long), pistils and stipes that are usually glabrous but sometimes pilose, and in its usually tenacious, pale yellowish brown branchlets. The stipes of *S. nigra* are short (0.5–1.5 mm long), the pistils always glabrous, and the branchlets are dark reddish brown or yellowish brown. Dorn (1976) notes that it also differs from *S. nigra* in its flavonoid chemistry, but these data are preliminary and must be used with caution.

Salix gooddingii is sympatric with *S. nigra* var. *lindheimeri,* and it is possible that plants referred to var. *lindheimeri* are hybrids between *S. gooddingii* and *S. nigra.* Schneider (1918) noted that var. *lindheimeri* in Mexico is also intermediate between *S. nigra* and *S. humboldtiana,* from which it is distinguished by its looser aments. The relationship between *S. nigra, S. gooddingii,* and *S. humboldtiana* deserves study.

Hybridization

In the Southeastern States *S. nigra* evidently hybridizes with *S. caroliniana* (see the latter species, no. 1, for discussion) and *S. amygdaloides* (Glatfelter 1894, 1897; Bebb 1895a; Fernald 1950; Braun 1961; Fowles 1965). I have not seen the hybrid *S. amygdaloides* × *nigra* (*S.* ×*glatfelteri* C. Schneider) from the study area, but it is to be expected near the confluence of the Ohio and Mississippi rivers, where the species are sympatric. It is likely that these closely related species do hybridize, and Glatfelter (1894) gives a good account of them in the vicinity of St. Louis, Missouri.

Hybrids between *S. nigra* and *S. lucida* (Fernald 1950) and *S. alba* (Bebb 1895c; Schneider 1921b; Fernald 1950) are reported in the literature, but I am unable to confirm them. On the basis of Bebb's discussion and illustration of *S. alba* × *nigra* (*S.* ×*hankensonii* Dode) I suspect that the plants are *S. amygdaloides* × *nigra.*

3. Salix amygdaloides Andersson, Öfvers. Förh. Svenska Vetensk.-Akad. 15: 114. 1858. *Salix nigra* [var.] ***S. amygdaloides* (Andersson) Andersson, *Monogr. Salicum* 22. 1867.—Type: South Dakota. Fort Pierre, 1833, *Prince Wied-Neuwied s.n.;* type material: "Hb. vindob." W, probably lost to fire during World War II, additional material was not located at BR.—Peachleaf willow.

Trees 4–20 m tall, often with 2 or 3 boles; bark reddish brown, deeply furrowed, the wide ridges thick-scaled and connected by narrower lateral ridges (Sudworth 1934). Branches yellowish to grayish, glabrous; branchlets slender, yellowish brown to reddish brown, becoming yellowish or grayish, glabrous, usually tenacious at base but sometimes somewhat brittle; buds yellowish brown, glabrous, apex sharply pointed, bud scale margin free and overlapping. Leaf blades ovate-lanceolate or narrowly so, the larger ones 5.5–8.7–13 cm long, 1–2.4–3.7 cm wide, length/width 2.8–4–6, apex gradually to abruptly acuminate to caudate, base rounded to acute, margin glandular-serrulate; immature leaf blades yellowish green or reddish, almost glabrous or sparsely pilose with caducous hairs, hairs white and sometimes intermixed with ferruginous hairs; mature leaf blades glabrate above, thickly glaucous and glabrous beneath, sometimes amphistomatous; petioles yellowish, 7–12–21 mm long, glabrous to sparsely tomentose adaxially, small glands at distal end, sometimes eglandular; stipules minute or small glandular-margined lobes, caducous, prominent and reniform on sprout shoots. Aments coetaneous, borne on leafy flowering branchlets, leaves often deciduous after flowering. Staminate aments 3.8–6 cm long, on flowering branchlets 1–2 cm long; stamens 3–7, filaments pilose at base, dry anthers 0.5–0.6 mm long; nectary 1, adaxial, or several slender or blunt lobes surrounding stamens; bract tawny, about 1.5 mm long, obovate to oblong, apex often slightly irregular, sparsely villous. Pistillate aments loosely flowered, 2.5–9 cm long, borne on flowering branchlets 0.4–3.5 cm long; pistils greenish brown, pyriform, short-beaked, glabrous, capsules brownish, about 3 mm long, styles (including stigmas) 0.4–0.6 mm long, stipes 1.6–2–3.2 mm long; nectary 1, adaxial, short and broad, sometimes partly surrounding stipe, 0.2–0.4–0.5 mm long, ⅛ to ⅓ as long as stipe; ovules 9 per carpel; bracts as in staminate flowers but deciduous after flowering. Chromosome number 2n = 38. Fig. 14.

Phenology. No phenological data are available for the Southeastern States, but in Wisconsin *S. amygdaloides* flowers from early April to mid-May (Argus 1964).

Distribution. (Fig. 14-inset, 15). *Salix amygdaloides* is a widespread Great Plains species that barely enters the Southeastern States. It is known here from a single collection from Wickliffe, Kentucky, at the confluence of the Mississippi and Ohio rivers [Ballard Co., *Palmer 16502* (MO, NA, US)]. The report of the species in east-central Kentucky (Little 1971) has not been confirmed. Its occurrence in adjacent southern Missouri and Illinois suggests that it is to be expected in Tennessee and Arkansas, as well as in Kentucky. The report of the species from Crittenden Co., Arkansas (Wilcox 1973; Smith 1978), was based on a specimen of *S. eriocephala,* and the mapped record for Franklin Co., Arkansas (McGregor & Barkley 1977), is an error for Franklin Co., Kansas (R. L. McGregor, pers. comm.).

Salix amygdaloides occurs in wetlands along rivers, streams, and lakes, where it grows on sandy, silty, or gravelly substrates, in marshes, and wet sand dune slacks.

This species is very rare, if not extirpated, in Kentucky. It is listed as of undetermined status in Kentucky by Branson et al. (1981).

Recognition. Salix amygdaloides is a small tree characterized by yellowish branchlets, broadly lanceolate leaves with an abruptly acuminate or caudate apex, usually inconspicuous stipules, and glabrous pistils borne on long stipes.

Taxonomy. This species is related to *S. caroliniana, S. nigra,* and *S. floridana,* native to our area, and to the western *S. gooddingii* (see Table 1).

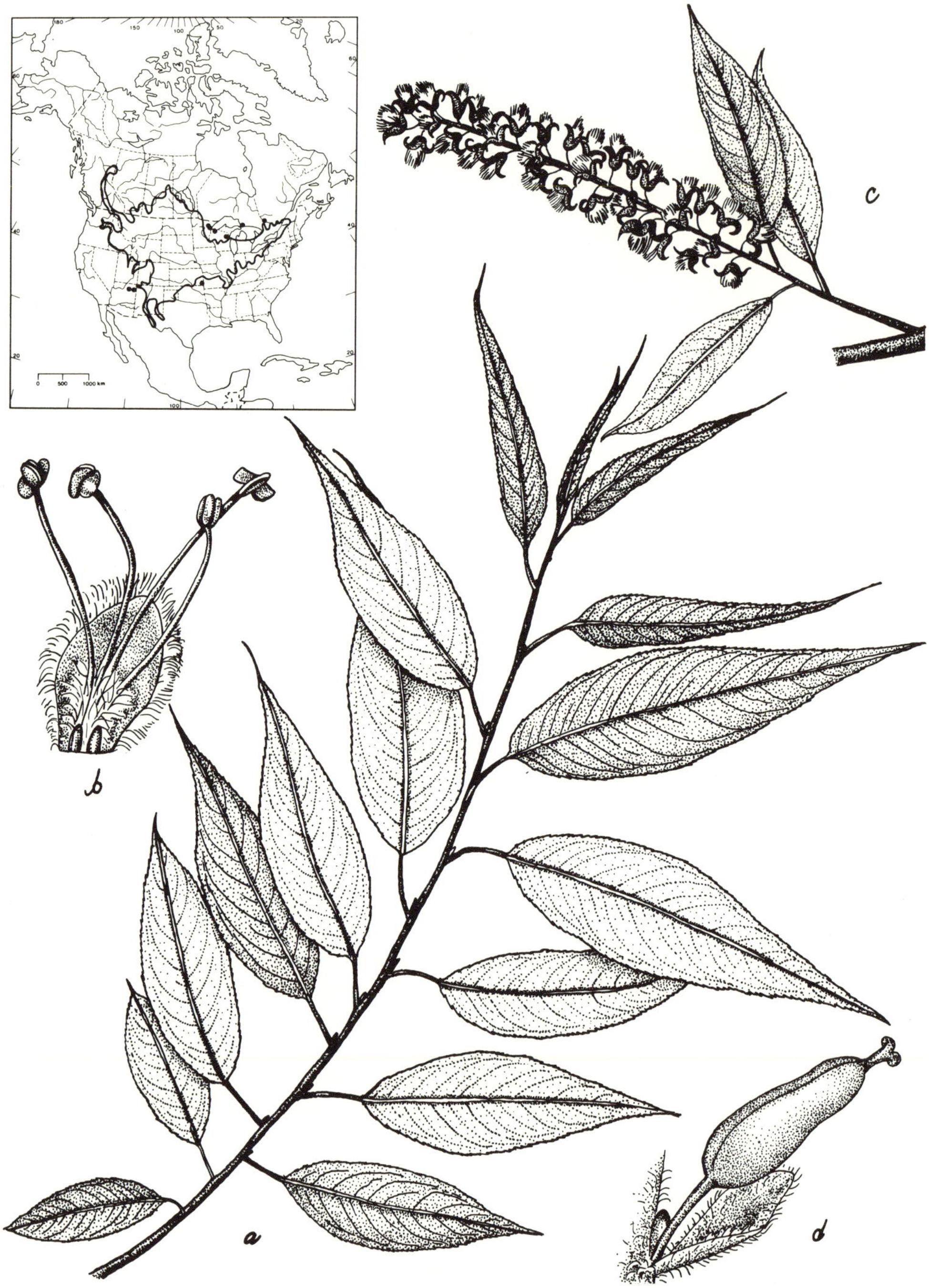

FIG. 14. *Salix amygdaloides*. a. Vegetative branchlet (×0.65). b. Staminate flower (×10). c. Pistillate ament in fruit (×0.65). d. Young pistillate flower (×10). Inset: map of total distribution, based on Argus (1964, 1983), Braun (1961), Brayshaw (1976), Dorn (1970), Little (1971), McGregor and Barkley (1977), Mohlenbrock (1980), Soper and Heimburger (1982), Steyermark (1963), Wherry et al. (1979), and herbarium specimens.

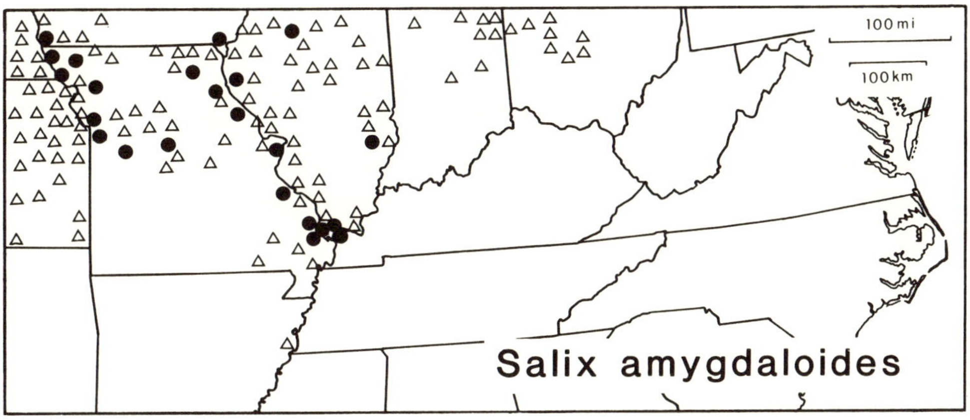

FIG. 15. Distribution of *Salix amygdaloides* in the Southeastern and peripheral states. Solid dots, herbarium specimens; open triangles, literature reports.

Hybridization. Salix amygdaloides is reported to hybridize with *S. caroliniana* (no. 1) and *S. nigra* (no. 2) (see these species for discussion). Fernald's (1950) report of hybridization between *S. amygdaloides* and *S. eriocephala* (as *S. rigida*) has not been confirmed and is doubtful.

4. **Salix floridana** Chapman, *Fl. South. U.S.* 430. 1860. *Salix chapmanii* Small, *Man. Southeast. fl.* 414, 1504. 1933, superfluous name.—TYPE: FLORIDA. Jackson Co., Marianna, *Chapman s.n.* (holotype: GH!; isotype: NY!).—Florida willow.

Salix astatulana Murrill & E. J. Palmer, J. Arnold Arbor. 22: 580. 1941.—TYPE: FLORIDA. Lake Co., Astatula, 7 Mar 1941, *Murrill & Palmer s.n.* (lectotype, here designated: A!; isolectotypes: FLAS! MO! NCU! NY!).

Slender, spindly trees up to 6 m tall; bark grayish, smooth with long sinuous cracks. Branches pubescent to glabrate, very brittle throughout; branchlets reddish brown or greenish, velutinous to pubescent with white and ferruginous hairs; buds greenish brown, puberulous or glabrous, apex sharply pointed, bud scale margin free and overlapping. Leaf blades elliptical to oblong, the larger ones 10–17 cm long, 3.8–5.5 cm wide, length/width 2.5–3 (–4), apex obtuse, acute, or short apiculate, base rounded, cordate on vigorous leaves, margin glandular-serrate or glandular-denticulate; immature leaf blades reddish, translucent, densely villous to sparsely so, midrib and veins velutinous; mature leaf blades glossy and glabrate above, sparsely tomentose with short, rigid hairs and glaucous beneath, veins and midrib prominent, yellowish, tomentose with short, rigid hairs; petioles 13–20 mm long, yellowish, puberulent, glandular-lobed or -dotted at distal end; stipules on first leaves as minute glandular lobes, on later leaves semi-ovate, glandular-denticulate, and up to 10 mm long. Aments coetaneous, borne on short, leafy flowering branchlets. Staminate aments 4–5.5 cm long, on flowering branchlets 0.5–1 cm long; stamens 3–7, filaments pilose on proximal half, dry anthers 0.4–0.5 mm long; nectaries 2, adaxial and abaxial; bracts tawny, 2.4–3.6 mm long, oblong, apex

FIG. 16. *Salix floridana.* a. Branch bearing vegetative branchlets and a pistillate ament in fruit, the leaves on the flowering branchlet have fallen (×0.50). b. Mature leaf (×0.50). c. Young pistillate flower (×5).

rounded, abaxial side densely villous or glabrate. Pistillate aments (2.5–) 5–7.5 cm long, loosely flowered, on flowering branchlets 0.8–3 cm long; pistils greenish, obclavate, short-beaked, glabrous, capsules reddish brown, 6–7 mm long, styles (including stigmas) very short, 0.3–0.4 mm long, stipes 3.2–5.6 mm long; nectary 1, adaxial, square, ⅕ to ⅐ as long as stipe; ovules 2 per carpel; bracts as in staminate flowers, about 2 mm long, persistent in fruit. Chromosome number unknown. Fig. 16.

Phenology. Flowering from mid-February to early April; based on three records.

Distribution. (Fig. 17). *Salix floridana* is a subtropical endemic of northern and central Florida and southern and central Georgia. It is mainly associated with springs in calcareous areas (Kurz & Godfrey 1962), and grows in swamps and on the marshy shores of small, usually spring-fed, streams and rivers running through woodlands.

Between about 1860 and 1984 this rare plant was found at about 15 localities in 9 counties, but it has been collected in only four of these counties since 1948. Collections have not been made in the peripheral populations in Georgia and Florida (Jackson, Levy, Orange, and Lake counties) since 1948, and the species' range seems to be contracting (Fig. 17). It is likely to have been extirpated from some of these stations, particularly from those in the spring areas near Marianna, where it was first discovered, and along the Chattahoochee River, where its habitat may have been eliminated by river control and energy projects. In 1943, at a time when it was known from only seven counties, Ball (1943) regarded it to be on the verge of extinction. Ward (1979) thought this was an overstatement in light of the five additional localities found since 1943, but he agreed that it is rare and that some of its stations may no longer be extant. He is now of the opinion that it may be rarer than he thought (D. Ward, pers. comm.). The species is definitely rare, and its habitat may be highly susceptible to stream and river control measures. It is, however, an inconspicuous species that grows in a habitat that is difficult to explore. It is most easily spotted in the field in the spring when its aments are present. Additional localities should be sought for it along spring-fed streams throughout northern Florida and southern Georgia.

REPRESENTATIVE SPECIMENS. U.S.A. FLORIDA: Alachua Co., 5 Dec 1942, *Murrill s.n.* (GH, US); Columbia Co., *Godfrey 55350* (DUKE, FLAS, FSU, GH, NA, NCSC), *Ward 3430* (FLAS, FSU, GH, USF); Jackson Co., ca 1860, *Chapman s.n.* (GH, NY); Jefferson Co., *Argus 11480* (CAN), *Godfrey 73634* (FLAS, FSU); Lake Co., *Judd 2778* (FLAS), 5 Feb 1941, *Murrill s.n.* (FLAS, GH, US); Levy Co., 23 Feb 1941, *Murrill s.n.* (A); Marion Co., *Clovis 1373* (WVA), *Hansen 5430* (LTU, USF), *Ward 3449* (FLAS, FSU, GH, NCU, US, USF); Orange Co., 20 Mar 1930, *West s.n.* (FLAS); Putnam Co., *Argus 11455* (CAN), *Nelson 888* (FLAS, FSU, NA).—GEORGIA: Early Co., *Harper 3634* (GH, NY, PH, US), *Thorne 8581* (GEO, NA, US); Pulaski Co., *Harper 1381* (GH, MO, NY, US).

Recognition. Salix floridana has 3–7 stamens; glabrous, stipitate pistils; conical, sharply pointed buds; and bud scales with free overlapping margins. In these characteristics it is similar to *S. caroliniana* and *S. nigra* from which it differs in its large, broadly elliptical to oblong leaves that are glandular-denticulate on the margin and sparsely and persistently tomentose beneath with short, rigid hairs, especially on the prominent veins and midrib. In addition, its branches are brittle along their entire length, not just at the base; the bracts of the pistillate flowers

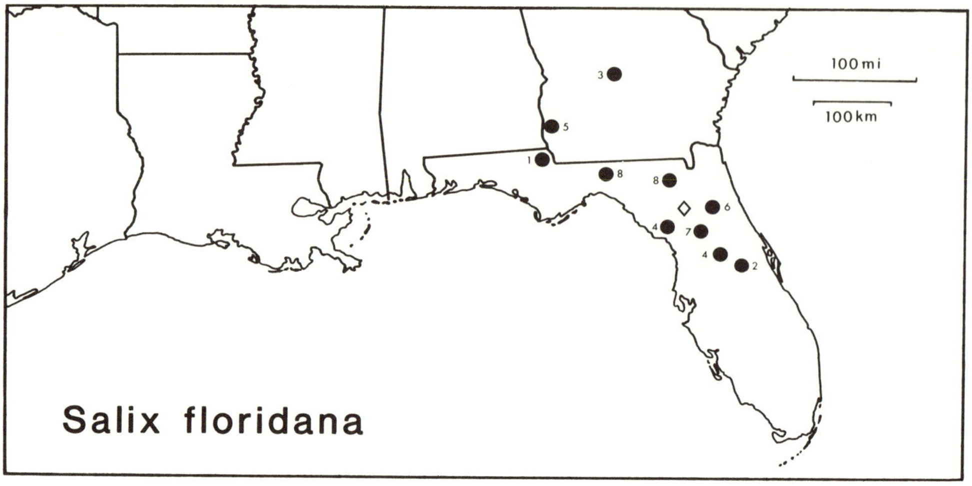

FIG. 17. Total range of *Salix floridana*. The date of most recent collection made in each county is indicated as follows: 1, prior to 1860; 2, 1930; 3, 1938; 4, 1941; 5, 1948; 6, 1977; 7, 1979; 8, 1984. Solid dots, herbarium specimens; open diamond, cultivated.

are persistent after flowering; its capsules are larger; and its ovaries contain only two ovules per carpel compared to the six to nine in *S. caroliniana* and *S. nigra*.

Nomenclature. The typification of *Salix floridana* Chapman and some related problems were discussed by Ball in 1943. In this paper he adequately reviewed the history of the name and its synonyms, but he did not correctly understand the type material or the problem caused by an apparent error in the labelling of one of the type specimens.

Type material of *S. floridana* was searched for in 14 of the 19 herbaria known to contain Chapman specimens (Lanjouw & Stafleu 1954). Specimens were found in only two of these herbaria: the Gray Herbarium and that at the New York Botanical Garden. These two sheets probably represent the total type material, but since one of them lacks Chapman's name, it is possible that other anonymous specimens went unrecognized and may still be found.

The two known specimens of type material of *S. floridana* are morphologically and developmentally identical and appear to have come from the same individual. They are similar in having what Chapman described as the "copious wool of the seeds," referring to the dense woolly frass in which the capsules are embedded. This "wool" was produced by a seed-eating weevil (*Elleschus ephippiatus* Say) that lays its eggs on living willow catkins (A. Howden, pers. comm.) and which evidently had parasitised the specimen before Chapman collected it. This weevil is reported to be common on *Salix*. In the field I have observed similar infestations on *S. caroliniana*, but I have rarely seen such an infestation on herbarium specimens. Its presence on both specimens in question supports the view that they were collected from the same plant even though the locality name written on the labels is different.

The label on the specimen in the Gray Herbarium bears the locality names "Marianna," and "Florida," and the notes "In wet places," "6–10 ft. high," and "mas. fl. not seen." It lacks the collector's name, but it is in Chapman's handwriting. The locality of Marianna is in western Florida on the Chipola River, about 40 km west of the Apalachicola River, near the Georgia and Alabama state lines. The label on the specimen at the New York Botanical Garden bears the locality "Middle Florida" and the name "Dr. Chapman." This label was not written by Chapman himself, and the general locality designation "Middle Florida" is at variance with Chapman's use of the expression "West Florida" for the locality of *S. floridana* in his original description. It is probable that the person who wrote the label for the NY specimen did not know that Chapman (1860) used the Apalachicola River as the dividing line between western and middle Florida. The use of the phrase "West Florida" by Chapman and the appearance of "Middle Florida" on the specimen led to an unfortunate misunderstanding of the type material.

J. K. Small included *S. floridana* in his 1903 *Flora of the Southeastern United States* and gave its range as "middle and western Florida" as though it had been collected in two places. In 1913, in *Florida Trees,* he extended the range of *S. floridana* to include southern Georgia to account for Harper's collections (Harper 1904). Later (1933) he decided, for unexplained reasons, that *S. floridana* (including only the western Florida specimen) was the same as *S. caroliniana* (as *S. longipes*) and coined a new name, *S. chapmanii,* for the specimens from Middle Florida and Georgia. Incidentally this shows that he considered the specimen from Marianna in the Gray Herbarium as the type of Chapman's name. His decision to treat these two identical specimens as different species is difficult to understand; it is most likely, however, that he had not seen the GH specimen, and that he was simply following Schneider's decision (1918, 1919b) to treat *S. floridana* as a synonym of *S. caroliniana* (as *S. longipes*). In accepting the view that the GH specimen, which he regarded to be the holotype of *S. floridana,* was *S. caroliniana,* and having specimens on hand of quite a different species, he proceeded to publish the superfluous name *S. chapmanii.* I have shown that although the locality names on the GH and NY specimens are different they were probably collected from the same individual. Therefore the specimen at GH is the holotype of *S. floridana,* and the specimen at NY is an isotype.

Taxonomy. Salix floridana is specifically distinct from *S. caroliniana,* Schneider's (1918, 1919b) treating them as the same species and Kral's (1983) uncertainty about the identity of *S. floridana* notwithstanding. Some of the problems that Kral has with these two species may be the result of confusing them in the field. His comment that *S. floridana* reaches 75 ft tall was based on a misidentification of *S. caroliniana* (R. K. Godfrey, pers. comm.).

Salix floridana has two ovules per carpel. This ovule number is uncommon in *Salix* (Chmelar 1977; Argus, unpubl.), but it is the basic number in *Populus* (still retained by *P. alba* Linnaeus; Hjelmqvist 1948) and is probably also the basic number in *Salix*. In *Salix* it is known to be typical only of *S. cardiophylla* Trautvetter & G. Meyer (sect. *Urbanianae*); *S. babylonica, S. matsudana* Koidzumi and *S. pierotii* Miquel (sect. *Subalbae*); *S. heterochroma* Seemen (sect. *Chingianae* Hao); and *S. tetrasperma* Roxburgh and *S. floridana* (sect. *Humboldtiana*). It also is typical of *Chosenia,* a monotypic genus that links *Salix* and *Populus.*

The *Salix* species with two ovules per carpel are all of Asian origin. Of them *S. floridana* seems to be most closely related to *S. tetrasperma,* a widespread tropical species that ranges from Pakistan, tropical and subtropical India, Indochina, Malacca, Java, and Sumatra to southern China (Skvortsov 1969). *Salix floridana* resembles *S. tetrasperma* in having an ovule number of two per carpel, bud scales with free, overlapping margins, persistent floral bracts, and glabrous, stipitate capsules. *Salix tetrasperma,* however, differs from *S. floridana* in having staminate flowers with 12–13 stamens and longer aments. It has been suggested that the affinities of *S. floridana* may be with *S. rosthornii* Seemen (sect. *Wilsonianae* Hao), a species of subtropical central China (Zhao Shidong, pers. comm.). Species in this section have free overlapping bud scale margins, but no data are available on ovule number. This alternative hypothesis requires study.

Dorn (1976) reported that the bud scales of *S. floridana* were sometimes fused in the lower two-thirds and that the staminate flowers sometimes had only two stamens. These features, as well as the persistent floral bracts and the broader leaves, suggested to him that the species may be a link between subg. *Salix* and subg. *Vetrix.* I have not been able to confirm his observations of stamen numbers, but his observation on the partial fusion of bud scale margins was confirmed. Twenty buds from four specimens were collected in the field. Of these the bud scales were open to the base in 15 and fused near the base in five. Kimura (1969) and Chmelar (1978) reported that hybrids between species with imbricate bud scales and those with caplike bud scales sometimes produced intermediate type bud scales that are partially fused below. An alternate hypothesis to Dorn's suggestion is that *S. floridana* originated through hybridization between a member of sect. *Humboldtiana* and a species with a caplike bud scale and flowers with two stamens.

If *S. floridana* is to be segregated from the other members of sect. *Humboldtiana,* as was done by Dorn (1976) who placed it in the monotypic sect. *Floridana,* it would be better to place it, together with *S. tetrasperma* and perhaps some other species, into sect. *Tetrasperma* (Andersson) C. Schneider in Sargent (1917, *Plantae Wilsonianae* 3: 93. Cambridge). I prefer, for the present, to retain these species in sect. *Humboldtiana* because of their strong morphological affinities with the other species in this section.

The origin of *S. floridana* is speculative, but if its affinities are with *S. tetrasperma* or *S. rosthornii* it may be a relict of, or may have been derived from, the Arcto-Tertiary flora that migrated southward in eastern North America following the Oligocene. Thorne (1949) and James (1961) suggest that it may have found refuge in northern Florida and southern Georgia, where some populations may have survived the Pleistocene on "islands" including the Marianna Red Hills (Jackson Co.) and the Lake Wales Ridge (Lake Co.). Little is known about the ecological requirements of *S. floridana* or how far north it could survive, but its brittle branches provide a good mechanism for dispersal along rivers and streams. Populations associated with the Chattahoochee and Suwannee rivers could have migrated southward from northern refugia, while populations in the "Lake District" south of Interlachen and along the Ichetucknee River may have been derived from the Lake Wales Ridge refugium.

II. Salix subgenus **Salix** section **Salicaster** Dumortier, *Fl. belg.* 14. 1827.—Type: *Salix pentandra* Linnaeus.

5. Salix lucida Muhlenberg, Ges. Naturf. Freunde Berlin II. 4: 239. 1803. *Salix pentandra* var. *lucida* (Muhlenberg) Kuntze, *Revis. gen. pl.* 2: 643.

1891.—TYPE: PENNSYLVANIA. Lancaster, *Muhlenberg 572* (holotype: PH!).—Shining willow.

Shrubs or small trees up to 6 m tall. Branches reddish brown or yellowish, glabrous and highly glossy; branchlets glabrous, sometimes pubescent; buds yellowish brown, glabrous, apex blunt. Leaf blades lanceolate to broadly so, the larger ones 4–14 cm long, 1.5–3.5 (-4.5) cm wide, length/width 2.8–5, apex caudate on later leaves, acute to acuminate on earlier ones, base acute to rounded, margin glandular-serrate; immature leaf blades reddish, glabrous or with caducous white and ferruginous hairs; mature leaf blades glabrous and dark green above, glabrous and pale green beneath, mostly hypostomatous; petioles yellowish brown, 5–16 mm long, glabrous, glandular-lobed at distal end; stipules reniform to semi-circular, 1–6 mm long, margin glandular. Aments coetaneous, borne on leafy flowering branchlets. Staminate aments 1.7–4 cm long, flowering branchlets 1–2.5 cm long; stamens 3–6, filaments pilose near base, dry anthers 0.6–0.8 mm long; nectaries 2, more or less cuplike; bracts tawny, 2–3 mm long, oblong, apex acute to rounded, entire or dentate, pilose on both sides or glabrate adaxially. Pistillate aments densely flowered, 1.8–2.5–5 cm long, borne on flowering branchlets 1.3–2.5 cm long; pistils greenish or brown, pyriform, short-beaked, glabrous, capsules light brown, 5–7 mm long, often deciduous after dehiscence, styles 0.5–0.8 mm long, stigmas 0.2–0.4 mm long, stipes 0.5–2 mm long; nectary 1, adaxial, short and broad, 0.2–0.5 mm long, ⅕ to ⅓ as long as stipe; ovules 8–11 per carpel; bracts as in staminate flowers but deciduous after flowering. Chromosome number $2n = 76$. Fig. 18.

Phenology. No data on flowering time are available from the Southeastern States, but in Wisconsin *S. lucida* flowers from early May to early June (Argus 1964).

Distribution. (Figs. 18-inset, 19). This species is a northern willow that barely enters the Southeastern States. Populations in northern Delaware, Maryland, and West Virginia are peripheral disjunct occurrences of the species. The central West Virginia record mapped by Little (1977) is apparently based on a misidentification.

In Roanoke Co., Virginia, is an interesting population that was first collected in 1942 by C. E. Wood, Jr. (1944), and relocated during this study. The species forms a thicket along about 1.1 km of Bottom Creek in the vicinity of Airpoint P.O., where it grows in mucky soil between the creek and the road. The population is a pistillate "clone" that flowers profusely but does not bear seed. An informant living across from the thicket on Bottom Creek told me that the previous occupant of the house introduced many plants and may have introduced this species as well. It is more likely that this population is a naturalized introduction rather than "a sporadic extension from Pennsylvania down the Appalachian Plateau into S. W. Virginia," as suggested by Massey and Ball (1944: 19).

Salix lucida grows along wet margins of streams and lakes and in wet meadows and boggy places.

Salix lucida is very rare in the Southeastern States. This study confirms its rarity and possible extirpation in Delaware (Tucker et al. 1979), where it was last collected in 1890, and in Maryland (Broome et al. 1979), where it was collected in Cecil Co. in 1913 and 1927, and in Garrett Co. where collections were made in 1966. It is also rare in West Virginia where it was last collected in Preston Co. in

FIG. 18. *Salix lucida*. a. Vegetative branchlet (×0.75). b. Staminate flower (×7.5). c. Pistillate ament (×1). d. Young pistillate flowers (×7.5). Inset: map of total distribution, based on herbarium specimens.

FIG. 19. Distribution of *Salix lucida* in the Southeastern and peripheral states. Solid dots, herbarium specimens; open triangles, literature; solid diamond, naturalized.

1940, and in Virginia where it occurs only in Roanoke Co., albeit as a probable introduction.

REPRESENTATIVE SPECIMENS. U.S.A. DELAWARE: New Castle Co., 23 Jul 1890, *Commons s.n.* (PH), 2 Jun 1868, *Commons s.n.* (DOV).—MARYLAND: Cecil Co., *Long 1329* (CAN, PH); Garrett Co., 1 Oct 1966, *Brown s.n.* (NCU).—VIRGINIA: Roanoke Co., *Argus 10948* (CAN), *Wood 5817* (CAN, GH, NCU, PH).—WEST VIRGINIA: Preston Co., *Davis 3652* (WVA).

Recognition. Salix lucida is characterized by shiny leaves that are green on both sides and have an abruptly caudate apex; stipules are prominent and glandular, and the petiole has prominent glands at the base of the blade. Young leaves are clothed with caducous, ferruginous hairs and the branchlets are brownish and glabrous.

It is similar to *S. pentandra,* an introduced species that rarely escapes cultivation. The latter species differs from *S. lucida* by its acute to acuminate leaves and the usual absence of stipules, even on young shoots.

Hybridization. Salix lucida is reported to hybridize with *S. alba* (Bebb 1895c; Braun 1961) and *S. nigra* (Schneider 1921b). The hybrid *S. alba* × *lucida* has not been recognized in this study, but it is possible that plants here attributed to *S. alba* × *pentandra* could be this hybrid (see *S. alba,* no. 7). Crosses with *S. nigra* have not been seen, and it is unlikely that they occur.

6. Salix pentandra Linnaeus, *Sp. pl.* 2: 1016. 1753.—Bay-leaved willow.

Shrubs or trees up to 7 m tall. Branches and branchlets dark reddish brown, glabrous, glossy; buds yellowish brown, glabrous, apex blunt. Leaf blades broadly lanceolate to ovate-oblong, the larger ones 5.5–9.5–13.5 cm long, 2.3–3.2–4.3 cm wide, length/width 2–4, apex acute to abruptly short acuminate, base rounded or acute to cuneate, margin glandular-serrate; immature leaf blades reddish, glabrous; mature leaf blades glabrous and glossy above, glabrous and pale green beneath, hypostomatous; petioles yellowish, 5–15 mm long, glabrous, glossy, with prominent glandular processes at distal end, often extending down petiole; stipules small glandular lobes, rarely up to 3 mm long, caducous. Aments coetaneous, borne on leafy flowering branchlets. Staminate aments 2.5–8.5 cm long; flowering

FIG. 20. *Salix pentandra*. a. Vegetative branchlet (×0.75). b. Staminate flower (×7.5). c. Young pistillate flowers (×0.75).

branchlets 1.2–3 cm long; stamens 4–9, filaments pilose near base, dry anthers 0.5–0.6 mm long; nectaries 2; bracts tawny, 2–4 mm long, oblong to obovate, apex acute to rounded, entire to dentate, pilose with white or ferruginous hairs. Pistillate aments densely flowered, 4–7 cm long; flowering branchlets 4.5–9 cm long; pistils glabrous, pyriform, long-beaked, capsules brownish, 8–9 mm long, styles 0.4–0.6 mm long, stigmas 0.6 mm long, stipes 0.8–1.6 mm long; nectary 1, adaxial, or 2, adaxial and abaxial, rectangular and slender, 0.4–0.8 mm long, ½ to ¾ as long as stipe; ovules 12 per carpel; bracts as in staminate flowers but deciduous after flowering. Chromosome number 2n = 76. Fig. 20.

Distribution. (Fig. 21). *Salix pentandra* is a Eurasian species that is cultivated in North America. It occasionally escapes from cultivation in the Northern States and Canada (Fernald 1950), but in the Southeastern States it is known to be naturalized only in Montgomery Co., Virginia (Massey & Ball 1944). It is reported to be "established" in Garrett and Allegany counties, Maryland (Brown & Brown 1972); the Allegany record is unconfirmed.

REPRESENTATIVE SPECIMENS. U.S.A. DISTRICT OF COLUMBIA: *Ball 43* (GH, NY).—KENTUCKY: Kenton Co., 19 May 1946, *Davies Herbarium s.n.* (NCU); Lyon Co., *Eggleston 4555* (NY).—MARYLAND: Garrett Co., *Ball 3016* (CAN, NA, NY, MARY, PH, US).—NORTH CAROLINA: Orange Co., 2 Jun 1909, *Coker s.n.* (NCU); Watauga Co., 5 June 1918, *Holmes s.n.* (NCU).—VIRGINIA: Fauquier Co., *Ball 2762* (NA); Montgomery Co., 15 Jun 1936, *Davis s.n.* (VPI).

Recognition. This species is similar to *S. lucida,* no. 6, which should be seen for discussion.

Hybridization. See *S. alba* (no. 7) for *S. alba* × *pentandra* (*S.* × *ehrhartiana* G. Meyer).

III. Salix subgenus **Salix** section **Salix.**

7. Salix alba Linnaeus, *Sp. pl.* 2: 1021. 1753.—White willow.

Trees 10–25 m tall, narrow, ascending crown. Branches grayish brown to reddish brown, glabrate; branchlets olive, yellowish to reddish brown, often pendulous, densely long-sericeous, becoming glabrate, somewhat brittle at base; buds yellowish brown, sericeous, oblong, flat adaxially, ribbed laterally, flat-beaked. Leaf blades narrowly lanceolate, the larger ones 7–9–13 cm long, 1.3–1.5–2.5 cm wide, length/width 5–5.5–6.5, apex gradually to abruptly short-acuminate to acute, base attenuate to cuneate, margin minutely and regularly glandular-serrulate, the first leaf blades entire; immature leaves densely long-sericeous, especially beneath; mature leaf blades sparsely sericeous to glabrate above, glaucous and long-sericeous beneath, amphistomatous; petioles yellowish, 3–6–9 mm long, long-sericeous, several glandular dots or processes near base of blade; stipules narrowly lanceolate, caducous. Aments coetaneous, borne on short, leafy flowering branchlets, leaves deciduous after fruiting. Staminate plants not seen from the study area. Staminate aments 3–3.5 cm long, stamens 2, filaments pilose, dehiscent anthers 0.5–0.7 mm long (Skvortsov 1968); nectaries 2; bracts as in pistillate flowers, but persistent. Pistillate aments 3.5–6 cm long, densely flowered, flowering branchlets 1–2.5 cm long; pistils greenish brown, pyriform, short-beaked, base

FIG. 21. Distribution of *Salix pentandra* in the Southeastern States. Solid dots, naturalized; open diamonds, cultivated; solid diamonds, naturalization or cultivation status unknown.

cordate, glabrous, capsules greenish brown, 3.5–5 mm long, styles and stigmas each about 0.2 mm long, stipes about 0.2 mm long; nectary 1, adaxial, square, 0.3–0.4 long, equalling or two times as long as stipe; ovules 3–5 per carpel; bracts tawny, 1.6–2.8 mm long, ovate to oblong, apex acute, sparsely crisp-villous to glabrate adaxially, deciduous after flowering. Chromosome number 2n = 76. Fig. 22.

Distribution. (Fig. 23). *Salix alba* is an introduced Eurasian species that is widely cultivated and naturalized in North America, especially in the Northeast, as far south as the northern part of the Southeastern States.

REPRESENTATIVE SPECIMENS. U.S.A. ARKANSAS: Carroll Co., *Moore 5* (UARK).—DELAWARE: New Castle Co., *Tatnall 2115* (DOV).—DISTRICT OF COLUMBIA: *Ball 35* (CAN, GH, NY, NA).—GEORGIA: Meriwether Co., *Harbison 6317* (NCU); Rabun Co., *Hardin 16478* (GA).—KENTUCKY: Ballard Co., *Palmer 16512* (NA, US); Campbell Co., *Braun 3977* (A, CFR); Grabes Co., *Argus 11540* (CAN); Greenup Co., *Demaree 11257* (A, MO); Jefferson Co., *Davies 202* (DHL); Kenton Co., *Horsey 1001* (A); Madison Co., *McFarland 3855* (NA); Menifee Co., *McFarland 4078* (NA); Perry Co., *McFarland 3935* (NA).—MARYLAND: Allegany Co., *Westerfeld 5052* (PAC); Anne Arundel Co., *Tidestrom 6704* (NA); Baltimore Co., 5 May 1910, *Churchill s.n.* (GH); Calvert Co., *Reed 95892* (CFR); Cecil Co., *Tidestrom 11698* (NA); Frederick Co., 15 Aug 1904, *s.c.* (MARY); Garrett Co., *Downes 7168* (NCU); Howard Co., 27 Aug 1969, *s.c.* (MARY); Montgomery Co., 28 Jul 1904, *Morton s.n.* (MARY); Prince Georges Co., *Dickerson 13652* (NA); Talbot Co., *Earle 4088* (PH); Washington Co., 24 Apr 1938, *Freeman s.n.* (NA); Wicomico Co., *Reed 94459* (CFR).—NORTH CAROLINA: Durham Co., *Meyer 16579* (NA); Haywood Co., *Walton 3597* (DUKE); Jackson Co., *Pittillo 4464* (FUGR); Macon Co., *Radford 6372* (NCU); Madison Co., *Wharton 5685* (KY); Mitchell Co., 6 Aug 1940, *Ashe s.n.* (NCU); Orange Co., 30 Aug 1923, *Alexander s.n.* (NCU); Polk Co., *Peattie 1062* (NCU).—TENNESSEE: Franklin Co., *Schipper 25* (NCU); Knox Co., *Mazzeo 12945* (NA); Sevier Co., *Sharp 4219* (TENN).—VIRGINIA: Arlington Co., 12 Jun 1942, *Morse s.n.* (NA); Botetourt Co., 25 Aug 1934, *Totten s.n.* (NCU); Fairfax Co., 12 Oct 1912, *Bissett s.n.* (NA); Giles Co., *Daly 114* (VA); James City Co., *Grimes 3417* (NY): Montgomery Co., *Kral 12251* (FSU, WVA, VDB, VPI); Page Co., *Ball 3002* (MTL, NA, VPI); Roanoke Co., *Harbison 5060* (NCU, UARK); Rockingham Co., *Ball 2708* (NA, NCSC, TENN, US, VPI); Shenandoah Co., *Allard 9893* (GH, US, VPI); Smythe Co., *Meyer 15897* (CAN); Tazewell Co., *James 6467* (NCU); Warren Co., 4 Nov 1978, *Brinkman s.n.* (LFCC).—WEST VIRGINIA: Berkeley Co., 1965, *Thompson s.n.* (WVA); Cabell Co., 23 Jun 1947,

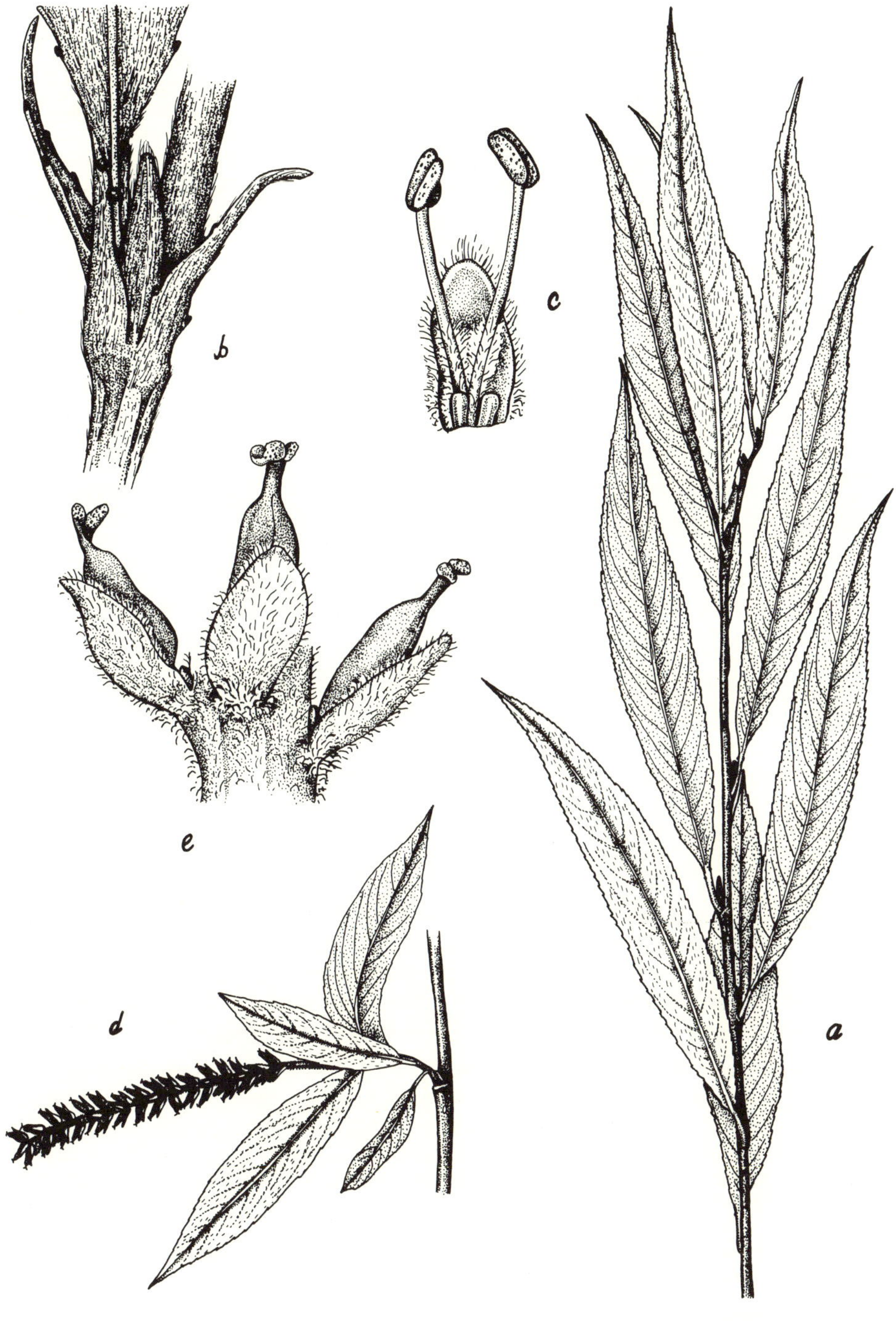

FIG. 22. *Salix alba*. a. Vegetative branchlet (×0.75), b. Leaf base showing stipules, bud, and petiolar glands (×7.5), c. Staminate flower (×10), d. Pistillate ament (×1), e. Young pistillate flowers (×10).

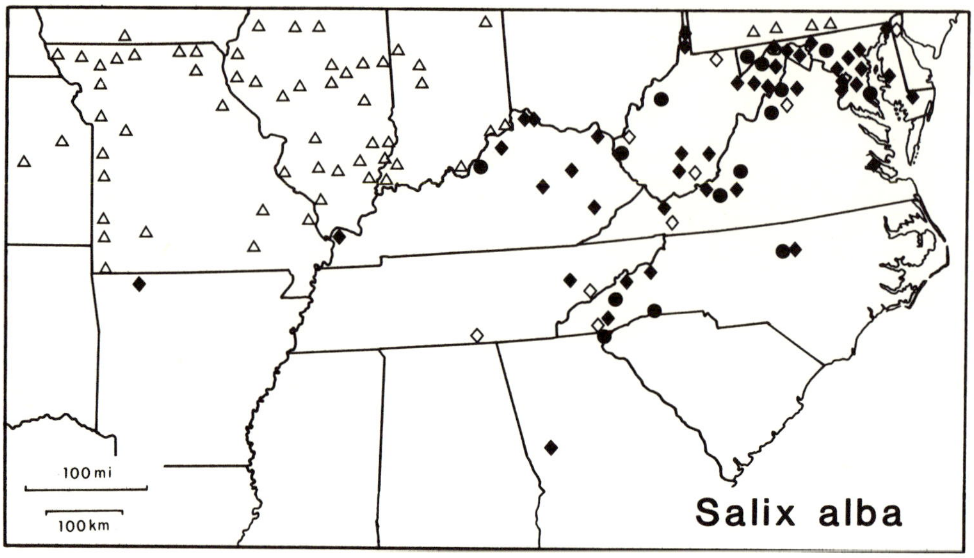

FIG. 23. Distribution of *Salix alba* in the Southeastern and peripheral states. Solid dots, naturalized; open diamonds, cultivated; solid diamonds, naturalization or cultivation status unknown; open triangles, literature reports.

(NA); Greenbrier Co., *Harbison 7114* (GH); Hampshire Co., *Ball 2603* (NA); Hardy Co., 25 May 1942, *Wilson s.n.* (WVA); Marshall Co., 24 Jun 1938, *West s.n.* (WVA); Mineral Co., Aug 1956, *Brown s.n.* (WVA); Monongalia Co., *Millspaugh 55* (NY, WVA); Morgan Co., *Argus 10820* (CAN); Ohio Co., 14 Jun 1878, *Guttenberg s.n.* (CM); Raleigh Co., *Tosh 918* (US, WVA); Summers Co., *Boone 805* (WVA); Tucker Co., *Clarkson 949* (WVA); Wayne Co., *Plymale 472* (NA, WVA); Wirt Co., *Bartholomew 43* (WVA).

Recognition. Salix alba is similar to two other widely cultivated tree willows, *S. fragilis* and *S. babylonica.* The cultivation of hybrids of these species sometimes makes the identification of specimens problematical, but they may generally be recognized by the characters in Table 3. *Salix alba* may be distinguished from other native arborescent willows by its amphistomatous leaves, a single capshaped bud scale, the presence of petiolar glands (at least as dots) at the base of the blade, and two stamens (rarely 4–8 in atavisms) (Skvortsov 1968).

Cultivated Varieties and Cultivars. Several varieties and cultivars are grown in the study area. The most popular cultivated form is *S. alba* var. *vitellina* (Linnaeus) Stokes [*S. vitellina* Linnaeus, *S. alba* subsp. *vitellina* (Linnaeus) Arcangeli], a plant with bright yellow branchlets. This form has been cultivated for centuries and was already widely distributed in Europe before it was described as a species by Linnaeus (Skvortsov 1968; Bean 1980). This variety is more properly a cultivar than a botanical variety, because it occurs only as a group of clones (mainly pistillate) that have been cultivated for ornamental purposes (Bean 1980).

Salix alba var. *vitellina* 'Pendula' [*S. alba* var. *tristis* (Seringe) Gaudin, *S.* ×*chrysocarpa* Dode] is a cultivar with pendulous, yellow branchlets. It is often mistaken in North America for *S. babylonica* from which it may be distinguished

Table 3. Comparison of *Salix alba, S. fragilis,* and *S. babylonica.*

	S. alba	*S. fragilis*	*S. babylonica*
Branchlets	long-sericeous, somewhat brittle at base, ascending, sometimes pendulous	glabrous or glabrate, very brittle at base, stout and ascending	glabrous or glabrate, brittle at base, slender and pendulous
Leaf blade shape	narrowly lanceolate, L/W 5–6.5	narrowly lanceolate, L/W 4–7	very narrowly lanceolate, L/W 6.5–13
Leaf margin	serrulate with pointed teeth	coarsely serrate-crenate, teeth rounded or bent	spinulose-serrulate with elongated teeth
Stipules	narrowly lanceolate, eglandular above	broad, half-cordate, eglandular above	narrowly-oblong, glandular above
Petioles	sericeous 3–6 mm long	glabrous 10–20 mm long	tomentose 7–12 mm long
Pistillate aments (length)	3.5–6 cm	6–8 cm	1.8–2.2 cm
Stipes	0.5 times as long as nectary	2–3 times as long as nectary	absent or 0.5 times as long as nectary
Floral bracts	deciduous after flowering	deciduous after flowering	persistent

by its sericeous leaves (see Table 3 and *S. babylonica,* no. 9). It is not certain whether the pendulous nature of this cultivar is derived from mutation or from hybridization with *S. babylonica* (Bean 1980).

Salix alba var. *caerulea* (Smith) Smith [*S. alba* subsp. *caerulea* (Smith) Rechinger f., *S. alba* 'Caerulea', *S. alba* var. *calva* G. Meyer] is characterized by dark brown branchlets and leaves that are only slightly silky, becoming glabrous.

Salix alba 'Britzensis' (*S. alba* var. *britzensis* Spaeth, *S. alba* var. *chermesina* Hartig) is a cultivar with bright red branchlets, and *S. alba* f. *argentea* Wimmer (*S. alba* var. *sericea* Gaudin) is a form with very densely sericeous leaves. Both are cultivated in our area, but they are not known to be naturalized.

Hybridization. Three hybrids involving *S. alba* have been introduced and naturalized in the Southeastern States. Many of the following comments on these hybrids are based on Bean (1980), Bailey et al. (1976), Chmelar and Meusel (1976), and Meikle (1975).

Salix alba × *babylonica—S.* ×*sepulcralis* Simonkai, Oesterr. Bot. Zeitschr. 40: 424. 1890 [1889 according to Bean, 1980]; *S. salomonii* hort. Fig. 24a.

A group of hybrids that closely resemble *S. babylonica* is in cultivation. These plants differ from *S. alba* in their narrower leaves and pendulous branchlets that are not as bright yellow. From *S. babylonica* they differ in their vigorous growth and less pendulous branchlets. The young leaves and branchlets are more sericeous, and the petioles are shorter and not as densely tomentose as in *S. babylonica.* The name *S.* ×*chrysocarpa* Dode is attributed to the cross *S. alba* var. *vitellina* 'Pendula' × *S. babylonica* (Chmelar & Meusel 1976; Bean 1980). This clone has not been recognized in our area.

Representative Specimens. U.S.A. Alabama: Talladega Co., *Harmon 50* (VDB).—Arkansas: Cleburne Co., *Thomas 10209* (NCU, NLU).—Delaware: New Castle Co., *Kelsey 41* (NA).—District of Columbia: 7 Apr 1966, *McConnell s.n.* (MARY).—Kentucky: Clay Co., *Reed 110309* (CFR); Fayette Co., *Northcutt 23* (KY); Henry Co., *Gentry 628* (KY, NCU, NY); Oldham Co., May 1938, *Smith s.n.* (DHL).—Louisiana: Sabine Par., *Thomas 76654* (NLU).—Maryland: Anne Arundel Co., *Meyer 15108* (CAN, NA); Baltimore Co., *Mazzeo 1200* (NA); Garrett Co., *Downs 7936A* (NCU); Talbot Co., *Earle 3232* (PH).—North Carolina: Buncombe Co., *Pittillo 3296* (NCU); Jackson Co., *Pittillo 3143* (NCU); Stokes Co., *Leonard 501* (NCU).—Tennessee: Gibson Co., *Sharp 12376* (TENN); Lewis Co., *Kral 55506* (VDB); Monroe Co., *Sharp 17042* (TENN).—Virginia: Campbell Co., *Hooton 223* (LYN); Giles Co., 10 Aug 1956, *Thorne s.n.* (DUKE, FSU).—West Virginia: Brooke Co., 5 June 1960, *Bonar s.n.* (WVA); Mineral Co., *Downes 7553* (NCU); Mingo Co., 1955, *White s.n.* (WVA); Raleigh Co., Jul 1966, *Molina s.n.* (WVA).

Salix alba × *fragilis—S.* ×*rubens* Schrank, *Baier. fl.* 1: 226. 1789. Fig. 24b.

Several clones of this hybrid are in cultivation (Chmelar & Meusel 1976). The hybrids are generally intermediate but variable. They differ from *S. alba* in their sparsely short-sericeous immature leaves and branchlets that soon become glabrate, coarsely serrate mature leaves, and more distinctly stipitate pistils (stipes 1.5–2 times as long as the nectaries). From *S. fragilis* they differ in the caducous sericeous hairs on the young branchlets and leaves. Mang (1980) reports also that their buds are elongate, more or less of uniform width and thickness, anteriorly

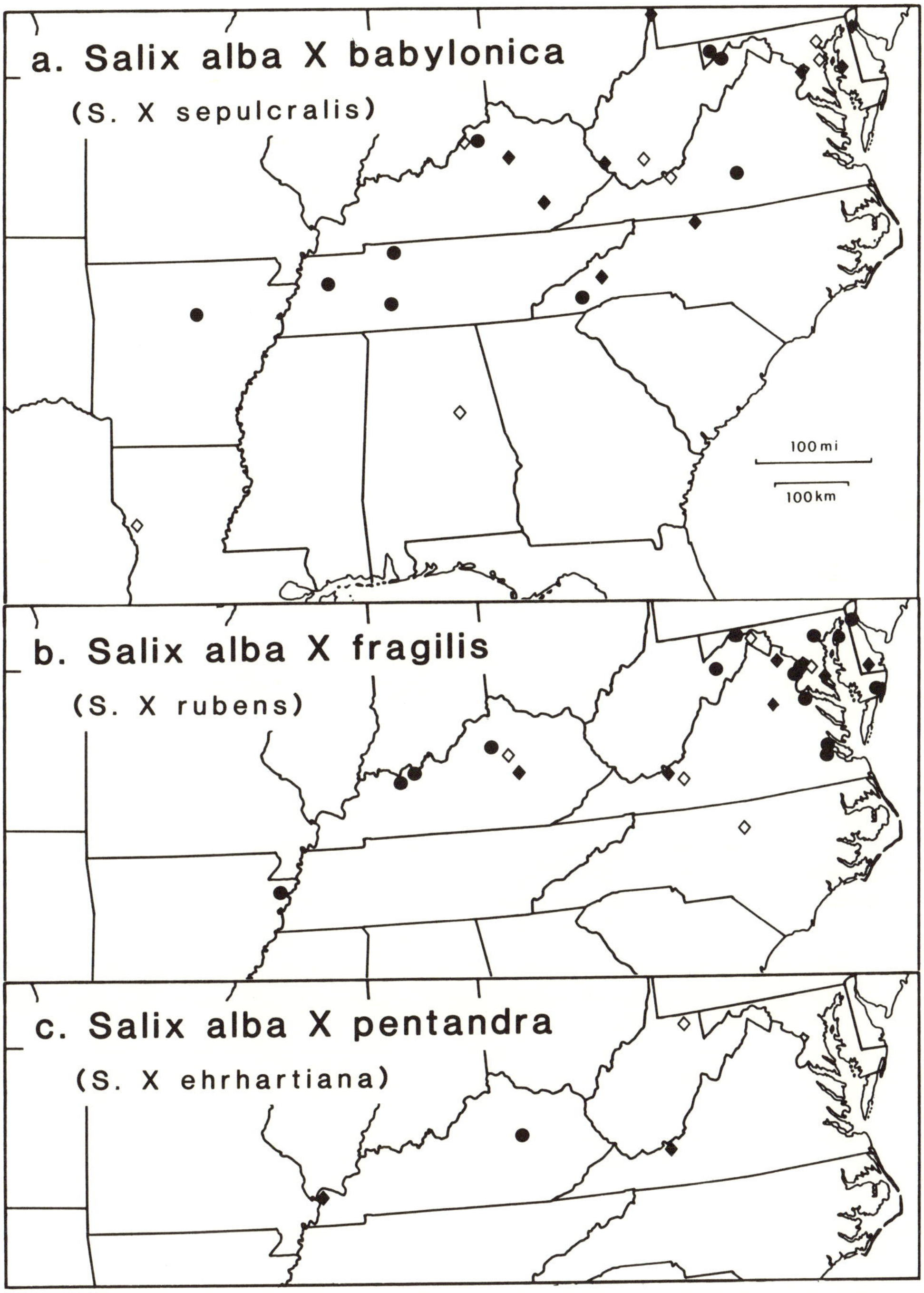

FIG. 24. Distribution of hybrids of *Salix alba* in the Southeastern States. No data are available for peripheral states. a. *S. alba* × *babylonica*. b. *S. alba* × *fragilis* c. *S. alba* × *pentandra*. Solid dots, naturalized; open diamonds, cultivated; solid diamonds, naturalization or cultivation status unknown.

rounded and, at first, pubescent, and that the glands on the leaf margin are displaced to the middle of the teeth. I could not confirm the latter character on herbarium specimens.

REPRESENTATIVE SPECIMENS. U.S.A. ARKANSAS: Mississippi Co., *Demaree 27455* (NA, UNA).—DELAWARE: New Castle Co., *Kelsey 136* (NA); Sussex Co., *Reed 8839* (CRF).—DISTRICT OF COLUMBIA: *Ball 12* (GH, NA, NY).—KENTUCKY: Daviess Co., *Palmer 17825* (A); Fayette Co., *Jarvis 89* (AUA); Franklin Co., *Wharton 10276* (NA); Hancock Co., *Palmer 17197* (A, US); Madison Co., *Reed 19857* (CFR).—MARYLAND: Allegany Co., *Reed 79001* (CFR); Baltimore Co., *Reed 98145* (CFR); Calvert Co., 9 May 1938, *Norton s.n.* (MARY); Kent Co., 11 Jun 1965, *Gladu s.n.* (MARY); Prince Georges Co., 1 May 1967, *Gleason s.n.* (NCU); Worcester Co., *Ball 2735* (DUKE, NA, NCSC, VPI).—VIRGINIA: Arlington Co., 25 Oct 1929, *Norton s.n.* (MARY); Fairfax Co., *Bradley 5641* (NLU); Giles Co., *Fogg 14970,* (NY); James City Co., *Mikula 339* (WILLI); King George Co., *Painter 1246* (NA); Loudon Co., *Allard 20432* (WVA); Montgomery Co., *Argus 10934* (CAN); Orange Co., 30 Aug 1923, *Alexander s.n.* (NCU); Surry Co., *Mikula 5506* (WILLI).—WEST VIRGINIA: Grant Co., *Downes 10808* (NCU, UNCC); Marshall Co., 23 Aug 1967, *Zwolensky s.n.* (WVA).

Salix alba × *pentandra*—*S.* ×*ehrhartiana* G. Meyer, *Chloris han.* 486. 1836. Fig. 24c.

Specimens identified as this hybrid resemble *S. alba.* They differ from it in having ferruginous hairs on the immature leaves and in mature leaves that are broader, glossy above, more abruptly acuminate, and with stipitate petiolar glands. The characteristics that suggest *S. pentandra* could also come from *S. lucida,* and the possibility that they are *S. alba* × *lucida,* as maintained by Bebb (1895c) and Braun (1961), cannot be completely ruled out. The specimen from Ballard Co., Kentucky, named *S. lucida* by C. R. Ball, differs from that species in its amphistomatous leaves and acute stipules, characteristics that could have come from *S. alba.* I have attributed this and other similar southeastern specimens to *S. alba* × *pentandra* because of their close similarity to European specimens of this hybrid, and because a cross involving two European species is more likely to be in cultivation than a cross involving a native species and a European one.

SPECIMENS EXAMINED. U.S.A. KENTUCKY: Ballard Co., *Palmer 15115* (A, US); Clark Co., *Beckett 332* (KY, NA), *388* (KY).—VIRGINIA: County unknown, Goodwin's Ferry, 13 Jul 1938, *Carter s.n.* (VSC).—WEST VIRGINIA: Monongalia Co., 15 Jul 1966, *Molina s.n.* (WVA).

8. Salix fragilis Linnaeus, *Sp. pl.* 2: 1017. 1753.—Crack willow, brittle willow.

Trees 10–25 m tall, crown broad. Branches olive-brown, glabrous; branchlets yellowish brown, glabrate or sparsely pilose, very brittle at base; buds yellowish brown, glabrate, oblong-ovate, convex on both sides, ribbed laterally, apex blunt. Leaf blades lanceolate, the larger ones 10–11.5–14 cm long, 1.7–2–3 cm wide, length/width 4–5.5–7, apex long-acuminate, base cuneate to obtuse, margin coarsely and irregularly glandular-serrate, the first blades entire; immature leaf blades usually glabrous or very sparsely sericeous with long hairs; mature leaf blades glabrous, glossy above, glaucous and glabrous beneath, amphistomatous with scattered stomata or hypostomatous; petioles yellowish, 10–15–20 mm long, glabrous, glandular dots or processes at distal end; stipules half-ovate, glandular

FIG. 25. *Salix fragilis*. a. Vegetative branchlet (×0.75). b. Staminate flower (×10). c. Young pistillate flowers (×10).

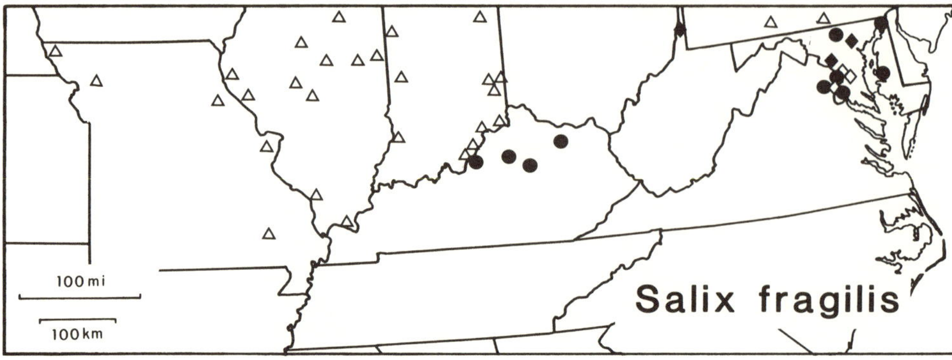

FIG. 26. Distribution of *Salix fragilis* in the Southeastern States. Solid dots, naturalized; open diamonds, cultivated; solid diamonds, naturalization or cultivation status unknown; open triangles, literature reports.

adaxially. Aments coetaneous, borne on long, leafy flowering branchlets, 1–1.5 cm long, rachis pilose. Staminate aments 5–6 cm long; stamens 2, filaments pilose at base, dry anthers 0.4–0.6–0.8 mm long; nectaries 2, abaxial and adaxial, bracts tawny, 1–2 mm long, oblong, apex acute, with long, straight hairs on both sides, crispate at base. Pistillate aments moderately densely flowered, 6–8 cm long; pistils greenish brown, pyriform, short-beaked, glabrous, capsules brownish, about 5 mm long, styles 0.6–0.8 mm long, stigmas 0.2–0.4 mm long, stipes about 0.8 mm long; nectary 1, adaxial, rarely 2, 0.3–0.6 mm long, about ½ to ⅓ as long as stipe; ovules 4 per carpel; bracts as in staminate flowers, but deciduous after flowering. Chromosome number 2n = 76. Fig. 25.

Distribution. (Fig. 26). *Salix fragilis* is occasionally introduced in North America and sometimes naturalized in the northern part of our area. It is native to northern and northwestern Asia Minor (Turkey, Caucasus Mts, USSR, Iran, and Iraq) and extensively naturalized in Europe from where it was introduced to North America (Skvortsov 1973).

REPRESENTATIVE SPECIMENS. U.S.A. DISTRICT OF COLUMBIA: *Ball 52* (GH, NA).—KENTUCKY: Fayette Co., *Wharton 10335* (KY, NA); Fleming Co., *Reed 14546* (CFR); Franklin Co., *Wharton 10276* (KY); Jefferson Co., *McFarland 6* (MO, NA).—MARYLAND: Baltimore Co., *Reed 91805* (CFR); Carroll Co., *Reed 2504* (CFR); Cecil Co., *Reed 26209* (CFR); Charles Co., 6 Oct 1929, *Freeman s.n.* (NA); Kent Co., 30 Jun 1965, *Gladu s.n.* (MARY); Montgomery Co., *Terrell 14* (MARY); Prince Georges Co., *Meyer 13682* (NA); Talbot Co., *Reed 10015* (CFR).—VIRGINIA: Arlington Co., *Fosberg 30978* (GH, NY); Fairfax Co., 1 Aug 1925, *Freeman s.n.* (NA); Prince William Co., *Tidestrom 7635* (NA).—WEST VIRGINIA: Ohio Co., 2 May 1877, *Guttenberg s.n.* (CM).

Recognition. Salix fragilis is related to *S. alba* and *S. babylonica* and may be distinguished from them by the characteristics listed in Table 3.

Hybridization. Hybrids with *S. alba* (*S.* ×*rubens*) and *S. babylonica* (*S.* × *blanda*) are discussed under those species (nos. 7, 9). *Salix fragilis* is often confused with *S. alba* × *fragilis* (Neumann & Polatchek 1972; Meikle 1975). The comment is made by Skvortsov (1973) that all of the American specimens called

S. fragilis that he has seen are only hybrids; nevertheless the species does seem to occur in our area, albeit infrequently.

IV. Salix subgenus **Salix** section **Subalbae** Koidzumi, Bot. Mag. Tokyo 27: 88. 1913.—Type: *Salix pierotii* Miquel.

9. Salix babylonica Linnaeus, *Sp. pl.* 2: 1017. 1753.—Weeping willow.

Trees to 20 m tall, with long pendulous branches almost reaching the ground. Branches yellowish brown, slender, glabrate; branchlets yellowish to yellowish brown, slender, glabrate but persistently tomentose at nodes, brittle at base; buds yellowish or brownish, sparsely sericeous to glabrate, ovate, ribbed laterally, flat-beaked. Leaf blades narrowly lanceolate or linear-lanceolate, the larger ones 7–11–14 cm long, 0.9–1.3–1.8 cm wide, length/width 6.5–9–13, apex gradually long-acuminate, base attenuate to cuneate, margin spinulose-serrate; immature leaf blades densely to sparsely short-sericeous to glabrous except for midrib, tomentose near base; mature leaf blades glabrate, rarely sparsely sericeous above, glaucous and glabrate beneath, amphistomatous; petioles yellowish, densely tomentose adaxially, eglandular or with only two small dots near base of blade, 7–12 mm long; stipules narrowly oblong to half-ovate, apex acuminate, glandular adaxially. Aments coetaneous, borne on short, leafy flowering branchlets with 2 to 4 small leafy bracts. Staminate aments 2.5–3.5 cm long, on flowering branchlets about 0.3 mm long; stamens 2, filaments pilose near base, dry anthers 0.6 mm long; nectary 1, adaxial, or several slender glands surrounding filaments; bracts tawny, 1.2–2 mm long, triangular to ovate, apex obtuse to apiculate, pubescent with short straight hairs, glabrate adaxially. Pistillate aments densely flowered, 1.8–2.2 cm long, flowering branchlets 0.4–1 cm long; pistils greenish brown, ovate-conic, swollen below style, short-beaked, base cordate, glabrous, capsules tawny-brown, 2.8–3.8 mm long, styles 0.15–0.2 mm long, stigmas brushlike, 0.3 mm long, stipes 0–0.3 mm long; nectary 1, adaxial, 0.5–0.7 mm long, always longer than stipe; ovules 2 per carpel; bracts as in staminate flowers. Chromosome number $2n = 76$. Fig. 27.

Distribution. (Fig. 28). *Salix babylonica* is an Asian species with a center of distribution in central and northern China. It was introduced into Europe from the Near East in the late 17th or early 18th centuries, probably along the silk route through Tadjikistan, USSR, and from there to North America (Skvortsov 1968; Bean 1980). Most European specimens are said to be from a pistillate clone with pendulous branchlets, but there are at least three pendulous clones, one of which, in Central Asia (Tadjikistan), is staminate. A great variety of cultivated forms exists in the Far East, but where it is indigenous, non-pendulous forms predominate (Skvortsov 1968).

This species is cultivated throughout the Southeastern States and has been widely naturalized. Both staminate and pistillate plants occur here, as well as one non-pendulous individual (*Ball 2601,* District of Columbia, NA). *Salix babylonica* cv. crispa (var. *annularis* Forbes), a cultivar with curled leaves, is sometimes cultivated in Maryland, Virginia, Tennessee, and North Carolina (specimens at NA).

FIG. 27. *Salix babylonica*. a. Pendulous vegetative branchlet (×0.75). b. Staminate flower (×10). c. Young pistillate flowers (×10).

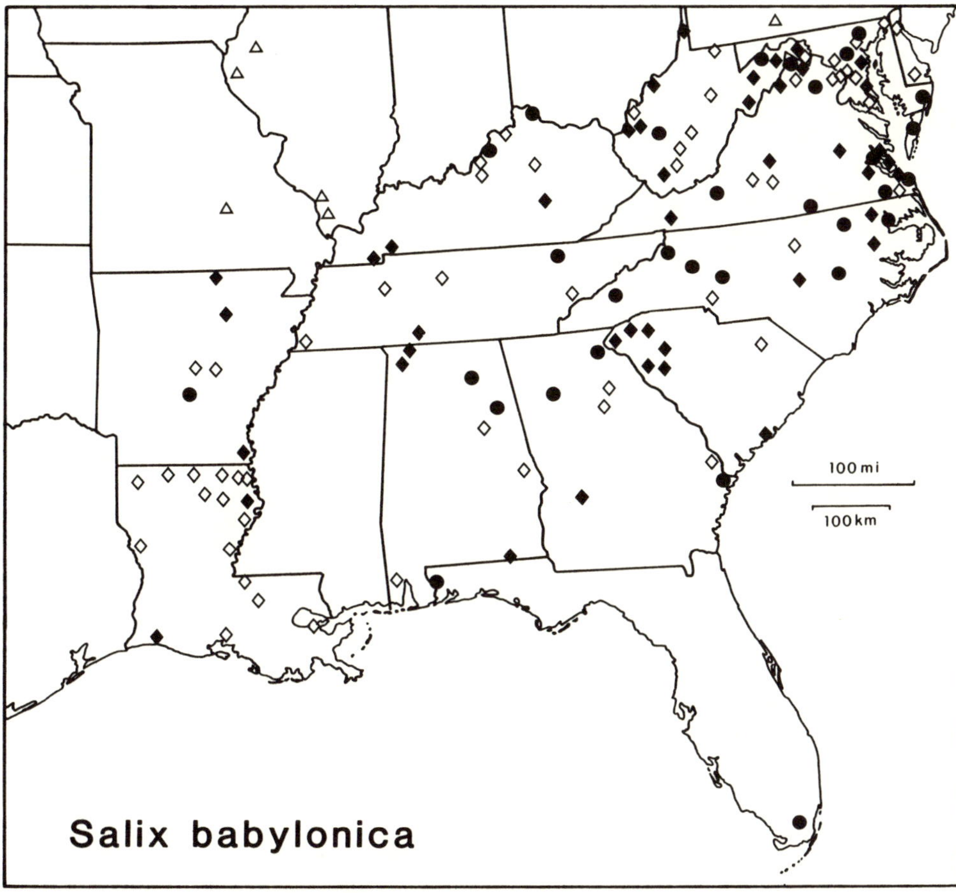

FIG. 28. Distribution of *Salix babylonica* in the Southeastern and peripheral states. Solid dots, naturalized; open diamonds, cultivated; solid diamonds, naturalization or cultivation status unknown; open triangles, literature reports.

REPRESENTATIVE SPECIMENS. U.S.A. ALABAMA: Calhoun Co., *Clark 2496* (NCU); Colbert Co., 19 Feb 1950, *T.F.H. s.n.* (MUR); Geneva Co., 11 Jul 1959, *Farmer s.n.* (NCU); Lauderdale Co., 16 Apr 1969, *Emerson s.n.*(NCU); Lee Co., *Lankford 41* (AUA); Marshall Co., *Brown 70* (UNA); Mobile Co., *Meyer 10427* (NA); Talladega Co., *Atkinson 50* (AUA).—ARKANSAS: Chicot Co., *Mohead 97* (APCR); Fulton Co., *Lipscomb 2938* (UARK); Grant Co., *Lawson 218* (NLU); Independence Co., *Thomas 9111* (NLU); Lonoke Co., *Tucker 4468* (NCU); Pulaski Co., *Meyer 18456* (NA).—DELAWARE: New Castle Co., 15 May 1897, *Canby s.n.* (DOV); Sussex Co., *Tidestrom 3248* (US).—DISTRICT OF COLUMBIA: *Ball 36* (GH, MO, NA, NY).—FLORIDA: Dade Co., *Jack 8245* (A, US); Escambia Co., *Thomas 6860* (NLU).—GEORGIA: Chatham Co., 26 Jul 1958, *Mellinger s.n.* (NCU); Clarke Co., *Meyer 16415* (CAN, NA); Cobb Co., 27 Jun 1959, *Jones s.n.* (FSU); Effingham Co., *Mazzeo 1974* (NA); Habersham Co., *Duncan 1838* (MUHW); Morgan Co., 15 Jul 1959, *Faircloth s.n.* (VSC); Sumpter Co., *Parker 96* (VSC).—KENTUCKY: Bullitt Co., *Gunn 1252* (KY); Calloway Co., *Wyatt 171* (MUR); Campbell Co., 12 Apr 1971, *Staley s.n.* (NLU); Carroll Co., *Argus 11546* (CAN); Fayette Co., *Gentry 1094* (KY); Jefferson Co., 17 Apr 1942, *McGuire s.n.* (DHL); Oldham Co., *Matthews 608* (DHL); Rockcastle Co., May 1968, *Reppert s.n.* (EKY); Trigg Co., 28 Jul 1966, *Wofford s.n.* (NLU, TENN).—LOUISIANA: Bossier Par., *Richardson 11* (NLU); Cameron Par., *Massey 49* (LSUS); Claiborne Par., *Thomas 59660* (NLU); Concordia Par., *Thomas 22636* (NLU); East Baton

Rouge Par., *Brown 1910* (NLU); East Carroll Par., 28 Oct 1973, *Crewes s.n.* (NLU); Iberia Par., *Thomas 19708* (NLU); Madison Par., *Crow 387* (NLU); Morehouse Par., *Thomas 35877* (NCU, NLU, TENN); Orleans Par., 1 Apr 1932, *Aydell s.n.* (NO); Ouachita Par., *Smith 1592* (NLU); Richland Par., *White 222* (NLU); Sabine Par., *Meyer 11744* (NA); Tensas Par., *Thomas 69674* (NLU); Union Par., *Brantley 38* (NLU); West Carroll Par., 10 Nov 1968, *Johnson s.n.* (NLU); West Filiciana Par., *Thomas 74810* (NLU).—MARYLAND: Anne Arundel Co., *Meyer 15113* (CAN, NA); Baltimore Co., *Keller 14* (CFR); Calvert Co., Sep 1935, *Norton s.n.* (MARY); Cecil Co., *Tidestrom 11695* (NA); Howard Co., 4 Jul 1965, *Engh s.n.* (MARY); Montgomery Co., 22 Apr 1967, *Higgins s.n.* (MARY); Prince Georges Co., *Wingate 33* (MARY); Saint Mary's Co., *Benedict 4387* (MARY); Worcester Co., *Argus 11375* (CAN).—NORTH CAROLINA: Alexander Co., *Keever 215* (DUKE); Chowan Co., *Ahles 39759* (NCU); Durham Co., *Baker 32* (LYN); Halifax Co., *Leonard 1303* (AUA, CLEMS, CM, DHL, FLAS, FSU, FUGR, GH, LYN, MARY, MISS, MUR, NCU, NLU, NY, PAC, TENN, UNA, UNCC, VSC, WILLI); Harnett Co., *Bostick 290* (USF); Haywood Co., *Walton 35147* (DUKE); Hertford Co., 1 Jul 1967, *Carroll s.n.* (NCU); Martin Co., *Radford 32247* (NCU); Mecklenburg Co., 21 Apr 1970, *DeLany s.n.* (UNCC); Rowan Co., 26 Jul 1979, *Holshauser s.n.* (UNCC); Wetauga Co., *Carpenter 4125* (CM, FLAS, FSU, GH, LSU, MO, NCU, TENN, USF, VPI); Wayne Co., *Brunton 141* (NCSC).—SOUTH CAROLINA: Abbeville Co., *Counts 15* (NCU); Charleston Co., *Rodgers 75154* (FUGR, USF); Darlington Co., *Smith 1627* (NCU); Greenville Co., *Rodgers 77002* (CLEMS, FUGR); Greenwood Co., *Ridlehoover 958* (FUGR); Laurens Co., *Grace 10* (FUGR); Oconee Co., *Harbison 14024* (NCU, PAC); Pickens Co., *Rodgers 69231* (FUGR).—TENNESSEE: Benton Co., *Mitchell 15* (MUR); Blount Co., 11 Sep 1965, *Thomas s.n.* (TENN); Campbell Co., 25 Jul 1934, *Lamb s.n.* (TENN); Davidson Co., *Abel 98* (NCU); Lawrence Co., *Jones 37783* (CAN); Shelby Co., 24 May 1952, *Schramm s.n.* (DUKE).—VIRGINIA: Accomack Co., *Reed 36428* (CFR); Amherst Co., *Heck 15* (LYN); Arlington Co., 12 Jun 1942, *Morse s.n.* (NA); Bedford Co., *Gray 27* (LYN); Campbell Co., *Herbruck 4* (LYN); Clarke Co., 27 Sep 1978, *Morrison s.n.* (LFCC); Fairfax Co., 21 Sep 1970, *Grimshaw s.n.* (WILLI); Fauquier Co., *Creel 17* (LFCC); Frederick Co., 19 Nov 1977, *Hatt s.n.* (LFCC); Gloucester Co., *Greaves 1589* (WILLI); Henrico Co., 29 May 1928, *Freeman s.n.* (NA); James City Co., *Meyer 15431* (CAN, NA); Mecklenburg Co., *Seaman 7035* (NCU); Montgomery Co., *Argus 10935* (CAN); Nansemond Co., *Waters 30* (LYN); Norfolk Co., *Clevenger 68* (UNA); Princess Anne Co., *Caldwell 65* (VPI); Shenandoah Co., *Artz 1200* (VPI); Smythe Co., *Meyer 15899* (CAN, NA); Surry Co., *Meyer 15211* (CAN, NA); Warren Co., 4 Nov 1978, *Brinkman s.n.* (LFCC); York Co., *Salle 23* (NCU, VPI, WILLI).—WEST VIRGINIA: Berkeley Co., 28 Apr 1967, *Thompson s.n.* (WVA); Cabell Co., *Stryker 49* (MUHW); Fayette Co., 12 Aug 1973, *Grafton s.n.* (WVA); Grant Co., 17 Sep 1976, *Tomblin s.n.* (WVA); Hampshire Co., 11 May 1940, *Frye s.n.* (WVA); Jefferson Co., 31 Jun 1964, *Thompson s.n.* (WVA); Kanawha Co., *Crum 14* (MUHW); Mason Co., *Schramm 10* (MUHW); Mineral Co., 4 Aug 1950, *Brown s.n.* (WVA); Monongalia Co., *Davis 3967* (WVA); Nicholas Co., 8 Apr 1966, *Milroy s.n.* (WVA); Ohio Co., *Bartholomew 827* (WVA); Pendleton Co., *Argus 10827* (CAN); Putnam Co., *Collins 18* (MUHW); Raleigh Co., *Nor 35* (MUHW); Upshur Co., 16 Apr 1966, *Brewer s.n.* (WVA); Wood Co., *Battin 41* (MUHW); Wyoming Co., *Lafferty 534* (MUHW).

Recognition. The weeping form of *S. babylonica* is very distinctive, but in the northern part of our range it is often confused with the pendulous cultivar of *S. alba,* var. *vitellina* 'Pendula'. From this cultivar it may be distinguished by its more olivaceous or yellowish brown branchlets, its narrower leaves, and its unique brushlike stigmas. The petioles are eglandular or bear only two minute dots near the base of the blade, and the aments are compact and borne on very short branchlets (usually less than 1 cm long). Other characters that will distinguish it from *S. alba* and *S. fragilis* are given in Table 3.

Hybridization. Salix babylonica × *fragilis*—*S.* ×*blanda* Andersson, *Monogr. Salicum 50.* 1867. This hybrid occurs in the Southeastern States only in cultivation. It resembles *S. alba* × *babylonica* but differs from it in its glabrous petioles and more coarsely serrate leaves (Meikle 1975). See also *S. alba* (no. 7) and *S. fragilis* (no. 8) for discussion of hybrids.

V. **Salix** subgenus **Salix** section **Longifoliae** Pax in Engler & Prantl, *Nat. Pflanzenfam.* 3(1): 36. 1888.—TYPE: *Salix exigua* Nuttall.

10. Salix exigua Nuttall, *N. Amer. Sylv.* 1: 75. 1842. *Salix longifolia* var. *exigua* (Nuttall) Bebb, Willows Calif. (preprint from *Salix,* S. Watson, Bot. Calif. 2: 85.) 1879. *Salix fluviatilis* var. *exigua* (Nuttall) Sargent, *Silva N. Amer.* 9: 124. 1896.—TYPE: Oregon Territory, "Columbia banks," *Nuttall s.n.* (holotype: BM!).—Sandbar willow.

Salix longifolia Muhlenberg, Ges. Naturf. Freunde Berlin N. S. 4: 238. 1803, non Lamarck, 1778. *Salix interior* Rowlee, Bull. Torrey Bot. Club 27: 253. 1900. *Salix longifolia* var. *interior* (Rowlee) M. E. Jones, *Willow Fam.* 25. 1908. *Salix exigua* subsp. *interior* (Rowlee) Cronquist, *Vasc. Pl. Pac. Northwest* 2: 51. 1964.—TYPE: PENNSYLVANIA. Lancaster, *Muhlenberg 573* (holotype: PH!).

Salix longifolia var. *sericans* Nees ex Wied-Neuwied, *Reise Int. N. Amer.* 2: 448. 1841. *Salix fluviatilis* var. *sericans* (Nees) J. Boivin, Naturaliste Canad. 93: 436. 1966.—TYPE: unknown. The specimens described were "collected on the Missouri, probably in eastern Montana about July 8" (Schneider 1919a: 346).

Salix interior var. *wheeleri* Rowlee, Bull. Torrey Bot. Club 27: 253. 1900. *Salix wheeleri* (Rowlee) Rydberg in Britton, *Man. fl. n. states,* ed. 2, 1061. 1905. *Salix longifolia* var. *wheeleri* (Rowlee) C. Schneider, Bot. Gaz. 67: 342. 1919. *Salix longifolia* f. *wheeleri* (Rowlee) C. Schneider, J. Arnold Arbor. 3: 65. 1921. *Salix interior* f. *wheeleri* (Rowlee) Rouleau, Naturaliste Canad. 71: 268. 1944. *Salix exigua* f. *wheeleri* (Rowlee) C. Reed, Phytologia 50: 462. 1982.—TYPE: PENNSYLVANIA. Presque Isle, 2 July 1868, *Porter s. n.* (lectotype, designated by Schneider in schedula: F!).

Colonial shrubs, up to 4–6 m tall, forming thickets by root shoots (soboliferous). Branches reddish brown, glabrous or glabrate, sometimes villous; branchlets slender, yellowish brown to reddish brown, tomentose or villous, becoming glabrate; buds reddish brown, sericeous becoming glabrate, oblong, opening by splitting at apex and often remaining at base of shoot, apex blunt. Leaf blades linear or very narrowly so, sides parallel, the larger ones 6–16 cm long, 0.4–1.1 cm wide, length/width (6.5–9–) 11–19 (–31), apex acute to subacuminate, base narrowly cuneate, margin distantly glandular-denticulate, teeth up to 0.6 mm long; immature leaf blades sericeous on both sides, often reddish; mature leaf blades sparsely villous becoming glabrate on both sides, non-glaucous beneath, amphistomatous, veins prominent and running parallel to margins; petioles yellowish brown, 1–5 (–7–9) mm long, glabrous or sparsely villous; stipules usually small glandular lobes, sometimes absent, occasionally linear and up to 2–5 mm long; seedling leaves somewhat pinnatifid. The first aments coetaneous, borne on leafy flowering branchlets 2–2.5 cm long, later aments serotinous and developing throughout the season from buds of the year, these borne on long, leafy flowering branchlets 4–18 cm long; aments often branched. Staminate aments 2–5 cm long; stamens 2, filaments villous on proximal half, dry anthers 0.4–0.7 mm long; nectaries 2, adaxial and abaxial; bracts tawny to greenish tawny, 1.5–3.5 mm long, oblong to obovate, apex usually acute, rounded or erose, sometimes apicu-

FIG. 29. *Salix exigua*. a. Vegetative branchlet (×0.75). b. Seedling showing lyrate lower leaves (×0.50). c. Staminate flower (×10). d. Young pistillate flowers (×10). e. Pistillate flower after fertilization, its bract, style, and stigmas are deciduous (×10). Inset: map of total distribution, based on Argus (1973), Little (1976a), Porsild and Cody (1980), Rousseau (1974), Soper and Heimburger (1982), and herbarium specimens.

FIG. 30. *Salix exigua*. a. Pistillate ament borne on a leafy branchlet, sylleptic shoots have emerged from the axils of the uppermost two leaves (×7.5). b. Staminate ament borne on a leafy branchlet, the ament is branched and secondary aments appear near its base (×7.5).

late and glandular-toothed, pilose on both sides, becoming glabrate abaxially. Pistillate aments 4–10 cm long, loosely flowered; pistils reddish brown, sparsely long-sericeous, capsules brown or greenish, 6–9 mm long, obclavate to pyriform, long-beaked, densely sericeous to glabrate, styles rudimentary to 0.1–0.2 mm long, stigmas four recurved lobes 0.5–0.7 mm long, deciduous after flowering, stipes villous or glabrous, 0.5–0.8 mm long; nectary 1, adaxial usually slender and obclavate or bifurcate, sometimes cuplike, completely surrounding stipe and bearing several slender lobes, 0.5–0.8 mm long, usually as long or slightly shorter than stipe, ovules 9–15 per carpel; bracts as in staminate flowers but deciduous after flowering. Chromosome number 2n = 38. Figs. 29, 30.

Phenology. Flowering during two periods, the first from early March to mid-May, with a mean in late April, and a then again from early June and late August, with a mean in late June; based on 182 records.

Distribution. (Figs. 29-inset, 31). This is a wide-ranging North American species. It enters the Southeastern States from the north and west via the Ozark Plateau, the Mississippi and Ohio river valleys, and along drainages eastward to

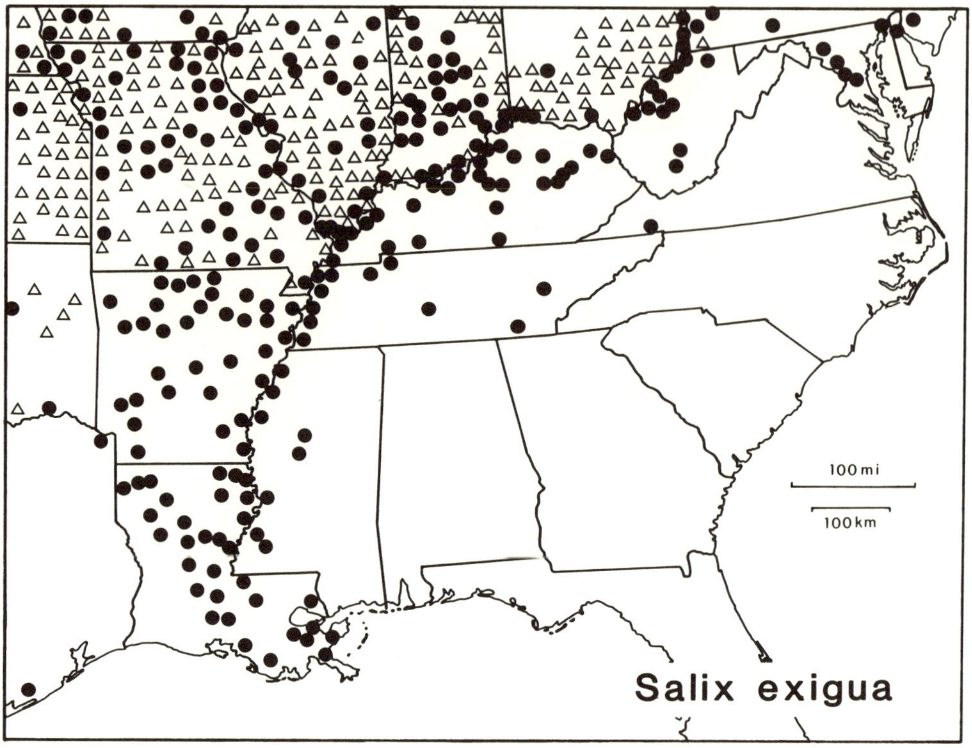

FIG. 31. Distribution of *Salix exigua* in the Southeastern and peripheral states. Solid dots, herbarium specimens; open triangles, literature reports.

the Atlantic. It occurs southward in Kentucky to the rim surrounding the Blue Grass Basin and then at scattered localities in West Virginia, Kentucky, Tennessee, North Carolina, and Mississippi.

It commonly forms thickets on river floodplains and on the wet, open margins of streams and ponds, drainage ditches, and roadsides. It is a pioneer on river bars where recent substrate is available for colonization. This habitat, which is subject to drastic, recurrent flooding and site instability, often prevents the species from reaching its full potential. The species is tolerant of flooding, but it does not compete well with species less tolerant of flooding and, therefore, it rarely extends to higher levels (Lindsey et al. 1961).

Salix exigua is rare in Delaware, Maryland, the District of Columbia, and Virginia. Collections seen from Delaware all came from New Castle Co. and were collected between 1894 and 1898. The 1978 report of the species from Sussex Co. (Tucker et al. 1979) was based on a sight record (A. Tucker, pers. comm.) and has not been confirmed.

Five collections have been made in Maryland from Cecil, Frederick, Montgomery, and Prince Georges counties; one of these was collected in 1967, one in 1928, and the others in 1894 and 1902. I have been unable to confirm recent reports from Allegany and Garrett counties (Broome et al. 1979). Only five collections have been made in the District of Columbia, the most recent in 1948.

In Virginia the species is known to occur only in Washington County. The report of the species by Massey and Ball (1944) from Arlington, Fairfax, and Princess Anne counties were based either on predictions that have not come true or on misidentifications. None of Eiser's (1961) records of the species for Virginia have been confirmed and were probably based on misidentifications. Kearney's (1901) report of the species (as *S. fluviatilis* Nuttall) from sand dunes at Cape Henry, Norfolk Co., was based on misidentified *S. caroliniana.*

REPRESENTATIVE SPECIMENS. U.S.A. ARKANSAS: Baxter Co., *Lipscomb 179* (UARK); Boone Co., *Demaree 30208* (NA); Chilcot Co., *Demaree 14067* (MO, NA, NY, TENN); Craighead Co., *Demaree 28741* (NA, NCSC, NCU, TENN); Crittenden Co., *Demaree 12991* (CAN, DUKE, GH, MO, NA, NY); Desha Co., *Demaree 14234* (MO, NA, NY); Drew Co., *Demaree 16660* (NA); Franklin Co., *Barber 945* (UARK); Fulton Co., *Demaree 26326* (NA, NCU); Garland Co., *Scully 1320* (US); Hempstead Co., *Palmer 24978* (A, MO, UARK); Hot Springs Co., *Demaree 14266* (A, MO, NA, NY); Howard Co., *Lawson 1710* (NLU); Independence Co., *Demaree 17054* (GH, MO, NA, NY); Izard Co., *Thomas 9324* (NLU); Jackson Co., *Demaree 20377* (MO, NA); Jefferson Co., *Demaree 14451* (A, MO, NA, NO, NY); Johnson Co., 12 Sep 1974, *Calloway s.n.* (APCR); Lafayette Co., 5 Aug 1972, *Knesel s.n.* (NLU); Lawrence Co., *Demaree 28894* (NA); Marion Co., *Demaree 66955* (MO, WILLI); Mississippi Co., *Moore 524* (UARK); Newton Co., *Nee 655* (NCU); Phillips Co., *Demaree 15225* (A, DAO, MO, MUHW, NA, NY); Pike Co., *Demaree 65060* (WILLI); Poinsett Co., *Marsh 2024* (UARK); Pope Co., 17 May 1966, *Graham s.n.* (APCR); Prairie Co., *Demaree 15542* (MO, NA, NY); Pulaski Co., *Demaree 16620* (MO, NA); Saint Francis Co., *Demaree 21555* (MO, NA, NY); Stone Co., *Thomas 7877* (NLU, TENN); Van Buren Co., *Demaree 25462* (GH); Washington Co., 20 Sept 1949, *Guess s.n.* (UARK).—DELAWARE: New Castle Co., 22 Jun 1896, *Commons s.n.* (DOV, GH, MO, NY, PH); 26 Jul 1894, *Commons s.n.* (A, MO); Jun 1894, *Tatnall s.n.* (DOV, PH); *Tatnall 1* (GH); 13 Jul 1894, *Tatnall s.n.* (DOV).—DISTRICT OF COLUMBIA: *Kelso 5901* (GH).—KENTUCKY: Ballard Co., *McFarland 40* (NA); Boone Co., *Braun 4311* (A, CFR); Bourbon Co., *Browne 4834* (KY); Breckenridge Co., *McFarland 17* (MO, NA); Bullitt Co., *Gunn 1198* (KY); Campbell Co., *Thieret 53869* (CAN); Carlisle Co., *McFarland 50* (NA); Carroll Co., *Argus 11547* (CAN); Crittenden Co., *McFarland 30* (NA); Cumberland Co., *Thomas 57835* (NLU); Elliot Co., *Reed 15378* (CFR); Estill Co., *Argus 10843* (CAN); Franklin Co., *McFarland 71* (NA); Fulton Co., *Conrad 1230* (KY, MO); Gallatin Co., *Braun 4546* (A, US); Hancock Co., *Palmer 17798* (A, US); Henry Co., *Gentry 457* (KY, NCU, NY); Jefferson Co., *Argus 11544* (CAN); Kenton Co., *Applegarth 102* (DUKE, KNK, VDB); Lawrence Co., *McFarland 3375* (NA); Livingston Co., *Windler 1265* (NCU); Madison Co., *Lassetter 1986* (EKY); McCracken Co., *McFarland 36* (NA); McLean Co., *Conrad 602* (KY, NCU, WKU); Meade Co., *Haskinson 149A* (DHL, NCU); Mennifee Co., *Higgins 1258* (DHL); Nelson Co., *McFarland 70* (A, DUKE, MO, NA, NY, PH, TENN, US, WVA); Oldham Co., *Woodward 3056* (DHL, NCU); Powell Co., *McFarland 3321* (NA); Taylor Co., *Reed 81661* (CFR); Todd Co., *McFarland 62* (NA); Trigg Co., *Ellis 1341* (APSC, NCU, TENN); Washington Co., *Frey 108* (MUR); Wayne Co., *Plymale 760* (WVA).—LOUISIANA: Avoyelles Par., *Thieret 24966* (FSU); Bienville Par., *Williams 1170* (LTU); Bossier Par., 2 Feb 1979, *Elston s.n.* (LTU); Caddo Par., *Thomas 32789* (NCU, NLU, TENN); Catahoula Par., *Thomas 41245* (NLU); Concordia Par., *Thieret 23842* (DUKE, FSU, US); East Baton Rouge Par., *Brown 5188* (A, LSU); East Carroll Par., *Thomas 38229* (NLU, TENN, VDB); Evangeline Par., *Thomas 60948* (NLU); Grant Par., *Thomas 28842* (LSU, LYN, NCU, NLU, TENN, UARK, USF, VDB); Jefferson Par., *Riedlinger 127* (LTU); Lafayette Par., *Thieret 25112* (DUKE, FSU); LaSalle Par., *Thomas 50516* (NLU); Madison Par., *Thomas 67504* (NLU); Morehouse Par., *Thomas 52409* (NLU, TENN); Natchitoches Par., *Ewan 21283* (MO, NCU); Orleans Par., *Ball 46* (GH, NA, NY, US); Plaquemines Par., *Rich 373* (NLU, NY); Rapides Par., *Ball 425* (CAN, GH, NA, NO, US); Red River Par., *Thomas 67826* (NLU); Richland Par., *Thomas 59071* (NLU); Saint Bernard Par., *Thomas 40615* (NCU, NLU); Saint Charles Par., *Smith 551* (NLU, TENN); Saint Landry Par., *Rogers 1998* (MISS, NCU, VDB); Saint Martin Par., Jun 1973, *Smith s.n.* (NLU); Saint Mary Par., *Montz 4288* (LSU, NO); Saint Tammany Par., *Thomas 64800* (NLU); Tensas Par., *Thieret 23970* (DUKE, FSU, US); Terrebonne Par., 14 Jul 1913, *Wurzlow s.n.* (LSU); Webster Par., *Byam 150* (LTU); West Carroll Par., *Demaree 13823* (A, MOAR, NA, NY, PH); West Feliciana Par., *Haynes 1550* (LTU); Winn Par., *Thomas 73170* (NLU).—MARYLAND: Cecil Co., 10 Jun 1928, *Tatnall s.n.* (DOV, PH); Frederick Co., 13 Jul 1894, *Tatnall s.n.*

(DOV); Montgomery Co., *Ball 61* (NA); Prince Georges Co., *Meyer 13671* (NA).—MISSISSIPPI: Bolivar Co., 10 Jun 1954, *Davis s.n.* (AUA); Carroll Co., *McDaniel 13327* (FLAS, VDB); Coahoma Co., *Argus 11527* (CAN); Franklin Co., *Glasscock 601* (MO); Holmes Co., *Pullen 691114* (MISS); Jefferson Co., *Jones 6335* (MISS, NCU); Tunica Co., *Pullen 66385* (MISS); Warren Co., *Crow 399* (NLU).—TENNESSEE: Dyer Co., *Demaree 30783* (FSU, GH); Hamilton Co., *Argus 10859* (CAN); Henry Co., *Davenport 72* (UNA); Lake Co., *Argus 11539* (CAN); Lauderdale Co., *Argus 11534* (CAN); Maury Co., *Kral 39752* (VDB); Obion Co., *Eyles 8527* (GH, NA); Roane Co., *DeSelm 29974* (TENN); Shelby Co., *Demaree 19690* (MO, NA); Stewart Co., *Chester 2837* (APSC); Tipton Co., *Warrington 173* (FUGR, NCU).—VIRGINIA: Washington Co., 19 Dec 1973, *Glover s.n.* (LFCC).—WEST VIRGINIA: Brooke Co., *Bartholomew 12* (WVA); Calhoun Co., 31 Aug 1964, *Bartholomew s.n.* (WVA); Fayette Co., 29 Aug 1976, *Grafton s.n.* (WVA); Jackson Co., *Bartholomew 98* (CM, WVA); Marion Co., *Reed 79975* (CFR); Marshall Co., *Bartholomew 106* (CFR, CM, MUHW, NA, NCSC, TENN, VPI, WVA); Mason Co., *Argus 10838* (CAN); Ohio Co., *Bartholomew 952* (GH, US); Pleasants Co., 18 Jun 1930, *Strausbaugh s.n.* (WVA); Raleigh Co., 13 Jul 1968, *Phillips s.n.* (NLU); Roane Co., 28 Aug 1964, *Bartholomew s.n.* (WVA); Tyler Co., 23 Jun 1932, *Core s.n.* (WVA); Wetzel Co., *Bartholomew 58* (NLU); Wirt Co., *Bartholomew 1728* (CM, WVA); Wood Co., *Bartholomew 244* (WVA).

Recognition. Salix exigua is characterized by linear leaves that are amphistomatous, green on both sides, and toothed with widely spaced and elongate denticulations. The aments are often branched (Argus 1964, and Fig. 31). It is sometimes confused with *S. nigra* from which it can be distinguished by its leaves with widely spaced denticulations and by its caplike bud scales. From *S. caroliniana* and *S. eriocephala* it can be distinguished by leaves that are green on both sides as well as by its leaf denticulation. Clustered willow galls are produced by the gall midge *Rhabdophaga racemi* Felt (Felt 1940).

Nomenclature. Salix exigua has a very extensive synonymy, and synonyms are given here only as they relate to the species as it occurs in the Southeastern States.

The epithet "*wheeleri*" has been applied to plants with densely long-villous leaves and branchlets, and prominent, lanceolate stipules up to 1.2 mm long. These characteristics are exhibited by foliage produced late in the year, usually as second growth from buds of the year. This second flush of growth is sometimes stimulated by insect defoliation or other physical damage or simply by favorable growing conditions in late summer. The characteristics that distinguish this taxon are developmental or environmental artifacts and are of no taxonomic importance.

Taxonomy. This species is a member of sect. *Longifoliae,* which also includes *S. fluviatilis, S. melanopsis* Nuttall, *S. sessilifolia* Nuttall, and *S. taxifolia* Kunth (Schneider 1919a; Dorn 1976). Some of these species are taxonomically distinct, but others are highly variable and intergrade in a perplexing manner.

The history of the names applied to the taxa in the section was dealt with in detail by Schneider (1919a) and will not be repeated here. In his monograph Schneider recognized eight species: *S. exigua, S. interior* Rowlee (as *S. longifolia* Muhlenberg), *S. melanopsis,* and *S. fluviatilis.* In *S. exigua* he included five varieties and in *S. interior* three varieties. He was aware that these taxa were difficult to separate and noted that "the main characters, glabrousness or pubescence of the ovaries and leaves, are more liable to variation" (p. 311). In 1964 Cronquist treated *S. interior* and *S. melanopsis* as subspecies of *S. exigua,* and in 1966 Boivin also combined *S. exigua* and *S. interior,* but this time under *S. fluviatilis,* as var. *sericans,* including *S. melanopsis* as a synonym of var. *fluviatilis.*

I agree with Cronquist and Boivin that *S. exigua* and *S. interior* are conspecific; but, on the basis of my work with western Canadian species (Argus 1983), I would not include *S. melanopsis* in this circumscription. I am here treating *S. interior* and *S. exigua* as conspecific but without recognizing infraspecific taxa.

In general, the eastern phase (*S. interior*) differs from the western phase (*S. exigua*) in having leaves less densely sericeous, more distinctly toothed, and more veiny. The capsules of the eastern phase tend to be longer (5–9 mm) than those of the western phase [3–5 (–6) mm], and the aments are more loosely flowered due to longer stipes in the eastern plants (0.4–1.5 mm). All of these characteristics, however, are highly variable, especially where the phases are sympatric. In this highly variable group of species varieties and forms have been recognized in some species that differ from each other in the same characters that supposedly separate one species from another. Until this section can be carefully studied it seems preferable to treat *S. exigua* in a broad sense and not to recognize infraspecific taxa.

Hybridization. Natural hybrids have not been reported between *S. exigua* and any other species outside of sect. *Longifoliae.* Synthetic hybrids, however, have recently been created by A. Mosseler (pers. comm.) between *S. exigua,* from southwestern Ontario, and *S. discolor, S. eriocephala,* and *S. petiolaris.* The morphology of the F_1 seedlings is clearly intermediate, but some of the plant are weak and depauperate (vouchers deposited at CAN). These unexpected results indicate that the possibility of natural hybridization exists and should be looked for in the field.

VI. Salix subgenus **Vetrix** (Dumortier) Dumortier, Bull. Soc. Roy. Bot. Belgique 1: 1412. 1862.—TYPE: *Salix caprea* Linnaeus.

VIA. Salix subgenus **Vetrix** section **Vetrix** subsection **Substriatae** Görz, Feddes Repert. Beih. 52: 140. 1928.—TYPE: *Salix starkeana* Willdenow.

11. Salix bebbiana Sargent, Gard. & Forest 8: 463. 1895. *Salix rostrata* Richardson in Franklin Journ., App. 753. 1823, non Thuillier, 1799. *Salix vagans* var. *occidentalis* Andersson, *Salices bor.-amer.* 16. 1858. *Salix vagans* [var.] 1. *S. rostrata* (Richardson) Andersson, *Monogr. Salicum* 87. 1867. *Salix livida* var. *occidentalis* (Andersson) A. Gray, *Manual,* ed. 5, 464. 1867. *Salix livida* [var.] *rostrata* (Richardson) Dippel, *Handb. Laubholzk.* 2: 255. 1891. *Salix depressa* subsp. *rostrata* (Richardson) Hiitonen, Memoranda Soc. Fauna Fl. Fenn. 25: 82. 1950. *Salix starkeana* subsp. *bebbiana* (Sargent) Youngberg, Rhodora 72: 549. 1970, nom. illeg.—TYPE: SASKATCHEWAN. *Richardson 13* (lectotype, fruiting specimens here designated: K!).—Long-beaked willow, Gray willow.

Salix perrostrata Rydberg, Bull. New York Bot. Gard. 2: 163. 1901. *Salix rostrata* var. *perrostrata* (Rydberg) Fernald, Rhodora 16: 177. 1914. *Salix bebbiana* var. *perrostrata* (Rydberg) C. Schneider, J. Arnold Arbor. 2: 71. 1920.—TYPE: SOUTH DAKOTA. Hermosa, Black Hills, 23 June 1892, *Rydberg 1018* (holotype: NY!).

Shrubs or small trees 0.5–6 m tall. Branches dark reddish brown, persistently gray-pilose to glabrate; branchlets reddish brown, densely pubescent; buds brown-

FIG. 32. *Salix bebbiana*. a. Branch bearing vegetative branchlets and a pistillate ament in early fruit (×0.75). b. Mature leaf (×1). c. Staminate flower (×7.5). d. Young pistillate flower (×7.5). Inset: map of range in North America, based on Argus (1973), Braun (1961), Brayshaw (1976), Eilers (1971), Little (1976a), McGregor and Barkley (1977), Mohlenbrock (1980), Porsild and Cody (1980), Wherry et al. (1979), and herbarium specimens.

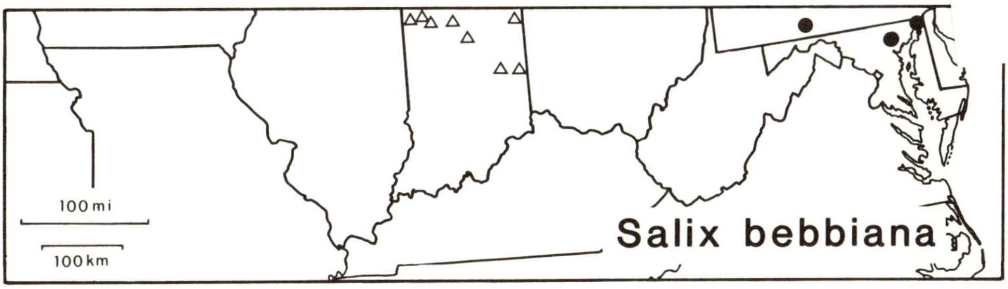

FIG. 33. Distribution of *Salix bebbiana* in the Southeastern and peripheral states. Solid dots, herbarium specimens; open triangles, literature reports.

ish, sericeous, ribbed laterally, flat-beaked. Leaf blades elliptic, elliptic-ovate, oblanceolate, or rarely broadly elliptic, the larger ones (2–) 4–8 cm long, 1.3–3.3 cm wide, length/width 2–3.8, apex abruptly acute, sometimes tapering or rarely obtuse, base acute to obtuse, margin crenate or irregularly serrate to entire; immature leaf blades pilose or sericeous-tomentose; mature leaf blades pilose, sparsely sericeous or glabrate, impressed veiny and dull green above, glaucous, sericeous-tomentose to sparsely so and from rugose-veiny to smooth beneath, hypostomatous; petioles (2–) 5–7 (–10) mm long, pubescent; stipules small, usually less than 2 mm long, caducous except on vigorous shoots. Aments coetaneous or subprecocious, borne on short, leafy flowering branchlets. Staminate aments 1.2–2.5 cm long, on flowering branchlets 2–6 cm long or subsessile; stamens 2, filaments glabrous, distinct or coalescent at base, dry anthers 0.6–0.8 mm long; nectary 1, adaxial; bracts tawny, about 2 mm long, narrowly oblong, sparsely pilose to long hairy. Pistillate aments loosely flowered, often lax, 2.5–6 cm long, on leafy flowering branchlets 0.3–1.5 cm long; pistils greenish brown, pyriform, long-beaked, gray short-sericeous, capsules brown to tawny, 5–9 mm long, styles nearly rudimentary, 0.2–0.4 mm long, stigmas short, mostly 0.3–0.4 (–0.5) mm long, stipes 2–5 mm long; nectary 1, adaxial, $\frac{1}{10}$ as long as stipe; ovules 3–8 per carpel; bracts as in staminate flowers. Chromosome number 2n = 38. Fig. 32.

Phenology. No data are available for the Southeastern States, but in Wisconsin *S. bebbiana* flowers from early April to the end of June, with a mean in early May (Argus 1964).

Distribution. (Figs. 32-inset, 33). *Salix bebbiana* is widespread in northern and western North America; it just enters the Southeastern States in northern Maryland and possibly also in Delaware. Specimen-documented records are known only from Maryland [Cecil Co., *Long 8035* (PH), and Baltimore City, Jun 1880, *Smith s.n.* (CFR)]. The reports of the species from Talbot Co., Maryland (Little 1976a; Broome et al. 1979), are based on misidentified specimens of *S. humilis*. The report of the species from Sussex Co., Delaware (Tucker et al. 1979), is based on an unconfirmed sight record (A. O. Tucker, pers. comm.).

It is a species of upland deciduous woods, wet thickets, and prairies.

Recognition. Salix bebbiana is a spreading shrub with divaricate branches and elliptic leaves that are often rugose-veiny beneath and impressed veiny above. The pistillate aments have long stipes and appear loosely flowered. The pistils are long beaked and subtended by pale, tawny bracts.

Hybridization. The species is reported by Fernald (1950) to hybridize with *S. discolor, S. humilis, S. petiolaris* (as *S. gracilis* Andersson), and *S. eriocephala* (as *S. rigida*). Successful artificial crosses have been made with *S. bebbiana* and *S. petiolaris* (Argus 1974) and *S. humilis* (Argus, unpubl.), but not with *S. discolor* (Argus 1974). No hybrids involving this species have been seen from the study area.

VIB. Salix subgenus **Vetrix** section **Vetrix** subsection **Laeves** Camus & Camus, *Classif. saules Europe* 1: 45. 1904.—Type: *Salix caprea* Linnaeus.

12. Salix discolor Muhlenberg, Ges. Naturf. Freunde Berlin II. 4: 234. 1803.—Type: Pennsylvania. Lancaster, *Muhlenberg 563* (lectotype, specimen with older leaves here designated: PH!).—Pussy willow.

Salix prinoides Pursh, *Fl. Amer. sept.* 2: 613. 1814. *Salix discolor* [var.] ***S. prinoides* (Pursh) Andersson, *Monogr. Salicum* 86. 1867.—Type: unknown.

Salix crassa J. Barratt, *Salic. amer.,* sp. 7. 1840.—Type: Connecticut. Middlctown, *Barratt s.n.* (holotype: NY!).

Salix discolor [f.] -*rigidior* Andersson, *Monogr. Salicum* 84. 1867. *Salix discolor* [f.] 2° *angustifolia* Andersson in DeCandolle, Prodr. 16(2): 225. 1868. *Salix discolor* var. *rigidior* (Andersson) C. Schneider, J. Arnold Arbor. 2: 4. 1920.—Type: unknown.

Salix discolor [f.] -*virescens* Andersson, *Monogr. Salicum* 85. 1867.—Type: unknown.

Salix squamata Rydberg in Britton, *Man. fl. n. states* 317. 1901.—Type: New Jersey. Sussex Co., Sparta Junction, 17 Sep 1887, *Porter & Britton s.n.* (holotype: NY!).

Salix discolor var. *overi* C. Ball, Rhodora 26: 137. 1924.—Type: South Dakota. Roberts Co., 12 Aug 1922, *Over 14483* (holotype: NA, but not found there in 1982; isotype: SDU; paratypes: collected by Ball and Over in 1923, CAN!).

Shrubs 2–4 (–6) m tall. Branches dark brown to yellowish, glabrate or sometimes persistently villous; branchlets moderately villous, soon glabrate; buds yellowish to reddish brown, pubescent with short white and ferruginous hairs, becoming glabrate, apex blunt; decorticated wood smooth or with a few short, indistinct ridges. Leaf blades narrowly lanceolate to oblanceolate, elliptic or obovate, the larger ones 3–10 cm long, 2–3 cm wide, length/width 2.3–4.5, apex acute, base narrowly to broadly cuneate, margin flat, crenate to coarsely serrate; immature leaves reddish, thin, usually sparsely pubescent with ferruginous hairs; mature leaf blades glabrate and dull or sometimes glossy above, glabrate and glaucous beneath, sometimes with sparse, persistent, ferruginous hairs, hypostomatous; petioles yellowish, 6–17 mm long, tomentose above; stipules often minute but large on vigorous shoots, apex acute, margin coarsely toothed. Aments precocious, sessile with several small, leafy bracts at base, or rarely on a short, leafy flowering branchlet. Staminate aments 2–3 cm long; stamens 2, filaments glabrous or pilose at base, dry anthers 0.5–1 mm long; nectary 1, adaxial; bracts dark brown to black, 1.5–2.5 mm long, oblong, apex obtuse, villous with long wavy hairs. Pistillate aments 4–8 cm

FIG. 34. *Salix discolor*. a. Vegetative branchlet (×0.75). b. Staminate flower (×7.5). c. Branch bearing pistillate aments in late flower and emerging vegetative shoots (×1). d. Young pistillate flower (×7.5). Inset: map of total distribution, based on Argus (1957, 1983), Brayshaw (1976), Dorn (1970), Eilers (1971), Little (1976a), McGregor and Barkley (1977), Mohlenbrock (1980), Steyermark (1963), and herbarium specimens.

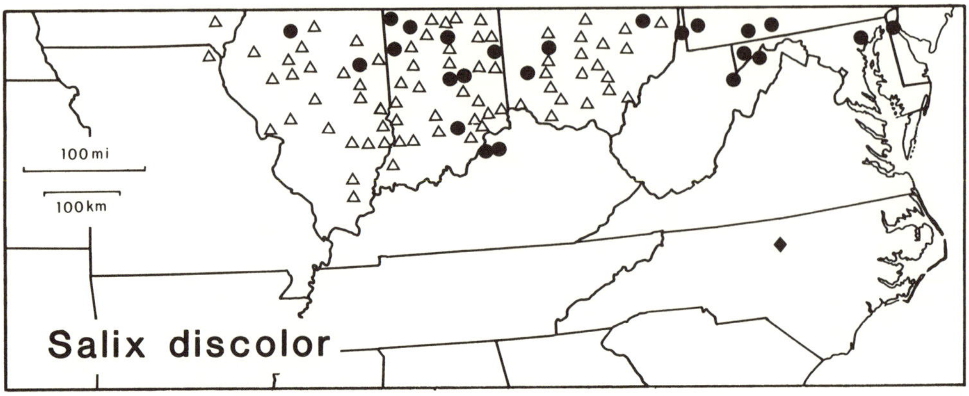

FIG. 35. Distribution of *Salix discolor* in the Southeastern and peripheral states. Solid dots, herbarium specimens; open triangles, literature reports; solid diamond, naturalized.

long, densely flowered but loose in fruit; pistils greenish gray, pyriform, long-beaked, sericeous with short, refractive hairs, capsules greenish brown, sericeous, 6–11 mm long, styles 0.3–1 mm long, stigmas 4-lobed usually as long as style or longer, stipes 2–2.7 mm long; nectary 1, adaxial, rectangular and about ⅐ as long as stipe; ovules 3– (6–) 8 per carpel; bracts as in staminate flowers. Chromosome numbers 2n = 76, 95, 114. Fig. 34.

Phenology. No data are available for the Southeastern States, but to the north *S. discolor* flowers from late March to mid-May.

Distribution. (Figs. 34-inset, 35). *Salix discolor* is a widespread northern species that just reaches the northern edge of the Southeastern States. Its occurrence in Kentucky, Maryland, and West Virginia represents the expected southern limits of its range. Collections from Tucker Co., West Virginia, were made by H. Allard and H. Davis in the botanically interesting Canaan Valley (Allard & Leonard 1952). Its cold, northern environment provides a suitable habitat for this and other northern species, which are also here at the southern limits of their ranges in eastern North America (Clarkson 1966).

Reports of the species from Virginia, North Carolina, and Georgia are problematical and some are erroneous. The Bedford Co., Virginia, record is based on a mixed collection of *S. discolor, S. humilis,* and *S. sericea* (*Curtiss 7517,* MO) made in 1871. In the absence of other evidence of the species's occurrence in the state it is assumed that the specimen of *S. discolor* came from another place. In North Carolina the collection from Orange Co. (*Eller 56,* WCHU) was made at Mason Farm and is probably an introduction (J. Moore, pers. comm.). I have not been able to document the report of *S. discolor* from along mountain streams in North Carolina, as reported by Ashe (1918); the Macon County record (Coker & Totten 1937) also cannot be traced and may have been reidentified. A report of the species from the Marshall Forest, Rome, Georgia, (Lipps & DeSelm 1969) has not been confirmed and is probably based on a misidentification. Similarly, the localities in Tennessee and Kentucky mapped by Little (1976a) are apparently based on misidentifications of *S. humilis* and *S. eriocephala.* His record from

Table 4. Comparison of *Salix discolor* and *S. humilis*.

	S. discolor	*S. humilis*
Leaf margins	flat	revolute
Branches	glabrous, sometimes persistently villous	tomentose with gray, white or brownish hairs, becoming glabrate
Stipules	very small or broad	narrow
Pistillate aments (length)	4–8 mm	2–3.5 mm
Staminate aments (length)	2–3 cm	1–2 cm
Styles (length)	0.3–1 mm	0.2–0.4 mm
Stipes (length)	2–2.7 mm	1.2–2 mm
Mature leaves	glabrate on both sides or with persistent, ferruginous hairs beneath	glabrate above, persistently tomentose beneath

Grant Co., West Virginia, was based on *S. humilis,* and the Fayette Co. record was based on *S. caroliniana.*

This species forms thickets in marshes and boggy areas and along creeks and rivers.

It is rare in Delaware, Maryland, West Virginia, and Kentucky; it does not, however, appear on any state list of rare plants.

Representative Specimens. U.S.A. Delaware: New Castle Co., 19 Sep 1921, *Otis s.n.* (PH).—Kentucky: Henry Co., *Gentry 606* (KY, NCU, NY); Oldham Co., *Matthews 710* (DHL).—Maryland: Baltimore Co., *Redman 531* (NCU); Garrett Co., Aug 1964, *Brown s.n.* (WVA).—North Carolina: Orange Co., *Eller 56* (WCUH).—West Virginia: Mineral Co., Aug 1956, *Brown s.n.* (WVA); Tucker Co., *Allard 9536* (US), *Davis 6127* (NA, WVA).

Recognition. Salix discolor flowers precociously and has large, usually sessile aments. Its leaves are crenate to coarsely serrate, soon glabrate above, but sometimes with persistent ferruginous hairs beneath. It is similar to the native *S. humilis* and the naturalized *S. cinerea* and *S. caprea.*

It is larger in all respects than *S. humilis,* and its mature leaves and branchlets are usually glabrate. Specimens with sparsely pubescent branchlets or leaves have leaves with flat margins compared to the strongly revolute margins in *S. humilis.* Flowering branches that lack leaves may be distinguished by their glabrescence and larger aments (see Table 4 and discussion of hybrids below).

Salix discolor may be distinguished from *S. cinerea* and *S. caprea* (Table 5) by the presence of ferruginous hairs on the young leaves and sometimes also on the underside of mature leaves. *Salix cinerea* subsp. *oleifolia* also has ferruginous hairs on the leaves. *S. discolor* may be separated from it by its smooth or indistinctly ridged wood (see *S. cinera,* no. 15, and Table 6). The immature leaves of *S. discolor* are usually sparsely sericeous and lack the lanate hairs found in *S. cinerea* and *S. caprea.*

Conspicuous pine-cone galls are produced by the gall midge *Rhabdophaga strobiloides* Walsh (Felt 1940).

Nomenclature. I have not been able to typify the Andersson names *rigidior, angustifolia,* and *virescens,* or the Pursh name *S. prinoides.* Schneider (1920) included them all in *S. discolor* and, on the basis of their descriptions, that is where they seem to belong.

Hybridization. Hybrids are reported to occur between *S. discolor* and *S. humilis, S. bebbiana,* and *S. eriocephala* (as *S. rigida*) (Schneider 1921b; Fernald 1950). Synthetic hybrids have been made between *S. discolor* and *S. humilis* in Ontario, Canada (Argus, unpubl.), but no synthetic hybrids could be made with *S. bebbiana* (Argus 1974). Hybrids with *S. eriocephala* have not been observed.

Salix discolor × *humilis—S.* ×*conifera* Wangenheim, *Beytr. teutsch. holz. forstw. nordamer. holzart.* 123. 1787.—TYPE: NEW YORK. *von Wangenheim,* no specimen preserved. LECTOTYPE: illustration (Wangenheim, Fig. 72) supplemented by diagnosis and description (Argus 1984a).

Probable synonyms:

Salix discolor [f.] – *latifolia* Andersson, *Monogr. Salicum* 84. 1867. *Salix discolor* [f.] 1° *latifolia* Andersson in DeCandolle, Prodr. 16(2): 225. 1868. *Salix discolor* var. *latifolia* (Andersson) C. Schneider, J. Arnold Arbor. 2: 5. 1920. TYPE: unknown.

Salix discolor [f.] 3° *hirsuta* Andersson in DeCandolle, Prodr. 16(2): 225. 1868.—TYPE: "Hb Vindob," not located by Schneider (1920).

Salix discolor and *S. humilis* hybridize easily, and natural hybrids seem to occur wherever the habitats of the species are in contact. The hybrids, however, are more commonly found growing with *S. discolor* than with *S. humilis.* The following characterization of the hybrids is based on a preliminary study of 22 synthetic hybrids made in Ontario (Argus, unpubl.).

Salix ×*conifera* is known in the Southeastern States only from Tucker Co., West Virginia [*Allard 10087* (NA, NY, US) and *Davis 15216, 15217* (WVA)]. It differs from *S. discolor* in having densely villous branchlets and branches that are persistently villous. The immature leaves are densely tomentose on both sides with a mixture of white and ferruginous hairs; the mature leaves are densely to sparsely tomentose beneath and sparsely tomentose above. From *S. humilis* it differs in being a taller shrub, up to 4 m tall, with coarsely toothed leaves. The petioles (5–9–16 mm long), styles (0.3–0.4–0.5 mm long), and pistillate aments (2.7–4–5.5 cm long) are generally intermediate between the parental species.

The identity and typification of *S.* ×*conifera* is discussed elsewhere (Argus 1984a).

13. Salix humilis Marshall, *Arbust. amer.* 140. 1785.—TYPE: unknown.

Branches reddish brown, tomentose with gray, white, or brownish hairs, becoming glabrate; branchlets velutinous-tomentose or villous becoming glabrate; buds reddish brown, puberulent or densely sericeous with grayish and ferruginous hairs, oblong, flat-beaked, humped abaxially, edge ridged, splitting open on adaxial face, apex blunt. Leaf blade apex tapering, acute to obtuse and abruptly acute at tip, base narrowly cuneate, rarely obtuse, margin entire, undulate, distantly glandular or glandular-crenate, revolute; immature leaves densely tomentose with white hairs sometimes intermixed with ferruginous hairs, sometimes sparsely tomentose or soon glabrate, blade margin distinctly revolute, midrib velutinous, hypostomatous; petiole tawny; stipule apex acute. Aments precocious, sessile with 2 or 3 green bracts at base, aments often numerous on each branch, densely flowered. Stamens 2, filaments glabrous or sparsely pilose, dry anthers usually purplish, rarely yellowish, 0.4–0.6 mm long; nectary 1, adaxial; bracts brown or black at tip, usually paler toward base, oval to oblong, apex rounded, villous with long wavy or straight hairs. Pistils greenish tawny, often suffused with red, pyriform, long-beaked, sericeous with refractive hairs, sometimes intermixed with ferruginous hairs, capsules tawny, styles 0.2–0.4 mm long, stigmas 2 or 4 slender lobes, 0.3–0.5 mm long, equalling or slightly exceeding styles, stipes 1.2–2 mm long, pilose; nectary 1, adaxial, square, narrowly rectangular or bifurcate, ⅕ to ⅓ as long as stipe; ovules 6 (rarely 2 or 4–5) per carpel; bracts as in staminate flowers.

Taxonomy. Salix humilis s.l. is closely related to *S. discolor* and the western American *S. scouleriana* J. Barratt ex Hooker. In 1920 Schneider placed these species in different sections: *S. humilis* in sect. *Griseae* Borrer with *S. sericea* and *S. petiolaris,* and *S. discolor* and *S. scouleriana* in sect. *Discolorae* J. Barratt ex Hooker along with three Mexican species. Dorn's (1976) placement of these three species in sect. *Vetrix,* subsect. *Laeves* better reflects their relationship.

Infraspecific Classification of Salix humilis

Salix humilis and its diminutive variety *microphylla* have been treated taxonomically in two ways. They have been regarded to be distinct species by several authors including Ball (1910, 1938, 1952), Schneider (1920), Gleason (1952), and Braun (1961), who followed the early treatments of Barratt (1840), Torrey (1847), and Andersson (1858a, 1858b, 1867, 1868). In 1971 Skvortsov reaffirmed this view, but he overstated the case when he wrote that they differ from each other morphologically, ecologically, and geographically. In 1981 Kral also expressed the view that these two taxa may be distinct species. Although there are morphological differences between them (see below), their ranges are sympatric (Figs. 37, 38), and they often occupy the same habitat.

Griggs (1905: 301) treated them as varieties and proposed the name *S. humilis* var. *tristis.* He held that they could not be regarded as species because "there is not a single constant character by which the two differ and what differences there are, are such as would be likely to be caused by differences in environment." In 1946 Fernald adopted this view, but he used the nomenclaturally correct name of varietal rank, var. *microphylla.* Although Schneider (1920: 23) kept the two as separate species, he was close to agreement with Griggs when he wrote that "probably it [*S. tristis*] is nothing but an ecological subspecies of *S. humilis.*" In some recent floras (Steyermark 1963; Mohlenbrock 1980) this varietal treatment is also followed. In 1964 I also treated them as conspecific, but did not recognize the variety. At that time I commented that the two variants may be ecotypes or ecophenes, but that this hypothesis needed to be tested.

In the present study I have examined the ecology, distribution, and morphology of these taxa in order to determine their appropriate rank.

Ecology. Little is known about the comparative ecology of the two variants of *S. humilis.* A survey of habitats of the two taxa, based on data from herbarium labels, shows that they occupy the same general range of habitats but that they usually occur separately. They have been observed in the field growing side by side on open, grassy balds (Buffalo Mtn, Virginia); on other balds (Yellow Mtn, North Carolina) var. *microphylla* formed low colonies in exposed places, and var. *humilis* occurred only at the edges of nearby thickets. They also are reported to grow together on serpentine outcrops (C. Reed, pers. comm.). The two were not observed growing together in woodland habitats.

Distribution. The ranges of var. *humilis* (Figs. 37-inset, 38) and var. *microphylla* (Figs. 39-inset, 40) are completely sympatric.

Morphology. The usefulness of the presence or absence of stipules in separating var. *humilis* and var. *microphylla* has long been recognized; this character is mentioned in the original description of the latter taxon, as a species. A study of *S. humilis* s.l. in the field and in the herbarium confirmed that stipules are a useful diagnostic character. In an attempt to determine whether it was correlated with other characters, specimens of *S. humilis* s.l. from the Southeastern States were sorted into two classes: stipulate (255 specimens) and exstipulate (142 specimens).

The presence or absence of stipules was observable on most specimens; but stipule scars had to be used as evidence of the occurrence of stipules on young flowering specimens and on specimens in early leaf. Stipules are not usually associated with the first leaves on the shoots of *S. humilis* var. *humilis* and often only begin to appear at the fifth to fifteenth leaf position. On plants that lack later leaves, it is therefore necessary to examine older braches for stipule scars.

A number of vegetative and floral characteristics were studied in specimens representing the two stipule classes. Plants with well-developed leaves were measured for leaf length, leaf width, and petiole length. Two ratios were calculated: leaf length/leaf width and leaf length/petiole length, and branchlet thickness was observed. Reproductive characteristics, including ament length, capsule length, stipe length, style length, and floral bract length, were measured on fruiting pistillate specimens or staminate specimens in anthesis. For all specimens the available data on plant height and evidence of vegetative reproduction were recorded. The data are presented in Fig. 36 and Table 4.

The differences in the means of leaf length, leaf width, and petiole length in the two classes are highly significant (Fig. 36). The differences in the means of the ratios are less significant, but statistically they represent different populations. The leaves of exstipulate plants tend to be smaller, narrower, and longer in relation to the petiole than those of stipulate plants.

Data on plant height and mode of vegetative reproduction were available for few herbarium specimens; the following data are based mainly on field studies. Stipulate plants tend to be mainly mid-sized shrubs ranging up to 3 m tall. Exstipulate plants are lower growing, ranging from 10–55 cm up to 70 cm tall. Both classes reproduce by layering, but exstipulate plants more commonly form clonal populations, sometimes 1.5–5 m in diameter. On mountain balds, where plants of the two stipule classes are growing together, the stipulate plants were reduced in stature to about 40–60 cm tall, but even in exposed situations stipulate plants may grow to 1.5 m tall.

Branchlet thickness appears to be somewhat correlated with the stipule classes (exstipulate plants have very slender branchlets); but there is considerable variation in this characteristic in stipulate plants, and no attempt was made to quantify it.

Floral and fruiting characteristics are not strongly correlated with the stipule

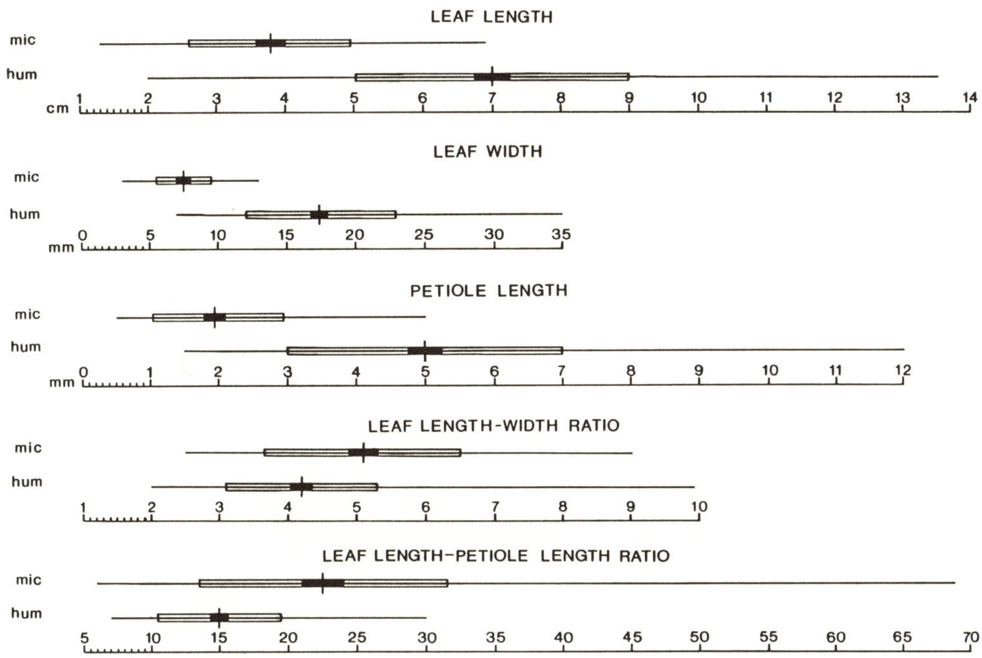

FIG. 36. Comparison of leaf length, leaf width, leaf length-width ratio, petiole length, and leaf length-petiole length ratio in *Salix humilis* var. *humilis* and *S. humilis* var. *microphylla* in the Southeastern States.

classes, but relatively few herbarium specimens in comparable developmental stages were available for study; the following data, therefore, are preliminary. Fruiting ament length in stipulate plants (31 specimens) ranges from (0.8–) 2–3.5 (–5.5) cm long and in exstipulate plants (17 specimens) from 1–1.8 cm long. The length of staminate aments in anthesis ranges from 1–2 cm long in stipulate plants (11 specimens) and from 0.5–1.1 cm long in exstipulate plants (9 specimens). Floral bract length in stipulate plants varies from 1.4–1.6 mm long, and in exstipulate plants from 0.8–1.4 mm long.

On the basis of herbarium population samples the correlation between the presence or absence of stipules and leaf size, plant height, ament length, and floral bract length is strong enough to warrant the recognition of two taxa. In the field, however, there are variations that must be accounted for in their taxonomy.

Populations of stipulate plants in northern Florida (Jefferson and Taylor Cos.), southern Georgia (Grady Co.) and central Mississippi (Kemper Co.) consisted of plants that were up to 2 m tall, but which had leaves averaging 4 cm long (the mid-range of var. *microphylla*). Similar populations also occur in Alabama, Louisiana, Missouri, Minnesota, and Manitoba, and probably elsewhere. A population from Taylor Co., Florida [*Argus 11467* (CAN), 15 specimens, and *Perkins 901* (CAN, FLAS), 11 specimens], also showed variation in the presence or absence of stipules. Most of these plants lacked stipules, but some were stipulate and others had stipule scars on old branches, although the current year's growth was completely exstipulate. These plants ranged up to 2 m tall and formed clones by layering. Their leaves averaged 3.8 cm long and 9 mm wide and had petioles 2.5 mm long, all well within the range of plants in the exstipulate class. Strictly on the basis of stipules some of these plants would be identified as one taxon and others as another, but in the field they clearly represented a single interbreeding population.

On the basis of these herbarium data and field observations it seems best to recognize two taxa at the varietal level: *S. humilis* var. *humilis* and *S. humilis* var. *microphylla.* The morphological differences between these taxa are mainly quantitative but comprise a number of characteristics, including leaf length, leaf width, petiole length, ament length, and floral bract length. The only qualitative characteristic that separates the two is the presence or absence of stipules, and that characteristic may be more variable than has been appreciated. The question raised by Griggs (1905) about possible environmental effects on the quantitative characteristics remains unanswered. The cultivation of these taxa in a common garden and their intensive study under a variety of environmental conditions are required to resolve this question.

Key to the Varieties of Salix humilis

Leaves stipulate; leaf blades 5–7–9 cm long, petioles 3–5–7 cm long; staminate aments 1–2 cm long; bracts 1.4–1.6 mm long; pistillate aments usually more than 2 cm long.
13a. *S. humilis* var. *humilis.*

Leaves exstipulate; leaf blades 2.5–4–5 cm long; petioles 0.5–2–3 mm long; staminate aments 0.5–1.1 cm long; bracts 0.8–1.4 mm long; pistillate aments less than 2 cm long.
13b. *S. humilis* var. *microphylla.*

13a. Salix humilis Marshall var. **humilis.**—Upland willow.

Salix conifera Muhlenberg, Ges. Naturf. Freunde Berlin II. 4: 240. 1803, non Wangenheim, 1787.—Type: Pennsylvania. Lancaster, *Muhlenberg 567* (holotype: PH!).

Salix muhlenbergiana var. *angustifolia* J. Barratt, *Salices amer.,* Sp. 1. 1840.—Type: Connecticut. Middletown, Aug 1834, *Barratt s.n.* (holotype: NY!).

Salix humilis [f.]-*grandifolia* J. Barratt ex Andersson, *Monogr. Salicum* 111. 1876. (Based on *S. muhlenbergiana* var. *grandifolia* J. Barratt, *Salices amer.,* Sp. 1. 1840, nomen nudum.) *Salix humilis* [var.] α *grandifolia* (J. Barratt ex Andersson) Andersson in DeCandolle, Prodr. 16(2): 236. 1868.—Type: Connecticut. Middletown, *Barratt s.n.* (holotype NY!).

? *Salix humilis* [f.]-*longifolia* J. Barratt ex Andersson, *Monogr. Salicum* 111. 1867. (Based on *S. muhlenbergiana* var. *longifolia* J. Barratt, *Salic. amer.,* Sp. 1. 1840, nomen nudum.) *Salix humilis* [var.] β *longifolia* (J. Barratt ex Andersson) Andersson in DeCandolle, Prodr. 16(2): 236. 1868.—Type: unknown.

Salix humilis [f.]-*rigidiuscula* Andersson, *Monogr. salicum* 111. 1867. *Salix humilis* var. *rigidiuscula* (Andersson) Robinson & Fernald ex C. Ball in C. L. Lundell, *Fl. Texas* 3(6): 390. 1961.—Type: unknown.

Salix humilis var. *keweenawensis* Farwell, Ann. Rep. Mich. Acad. Sci. 6: 206. 1904.—Type: Michigan. Keweenaw Co., "Bluffs between N. A. mine and No. 4 shaft house," 18 June 1883, *Farwell 14* (holotype: BLH!).

Salix humilis var. *hyporhysa* Fernald, Rhodora 48: 45. 1946.—Type: New Jersey. Ocean Co., Manahawkin, 23 July 1923, *Long 28011* (holotype: GH!).

FIG. 37. *Salix humilis* var. *humilis*. a. Vegetative branchlet (×0.75). b. Branch with a young subterminal shoot, the staminate aments borne at the nodes below it have fallen (×0.75). c. Staminate flower (×7.5). d. Branch bearing pistillate aments in late flower and vegetative shoots starting to emerge (×1). e. Young pistillate flower (×7.5). Inset: map of total distribution, based on McGregor and Barkley (1977), Soper and Heimburger (1982), and herbarium specimens.

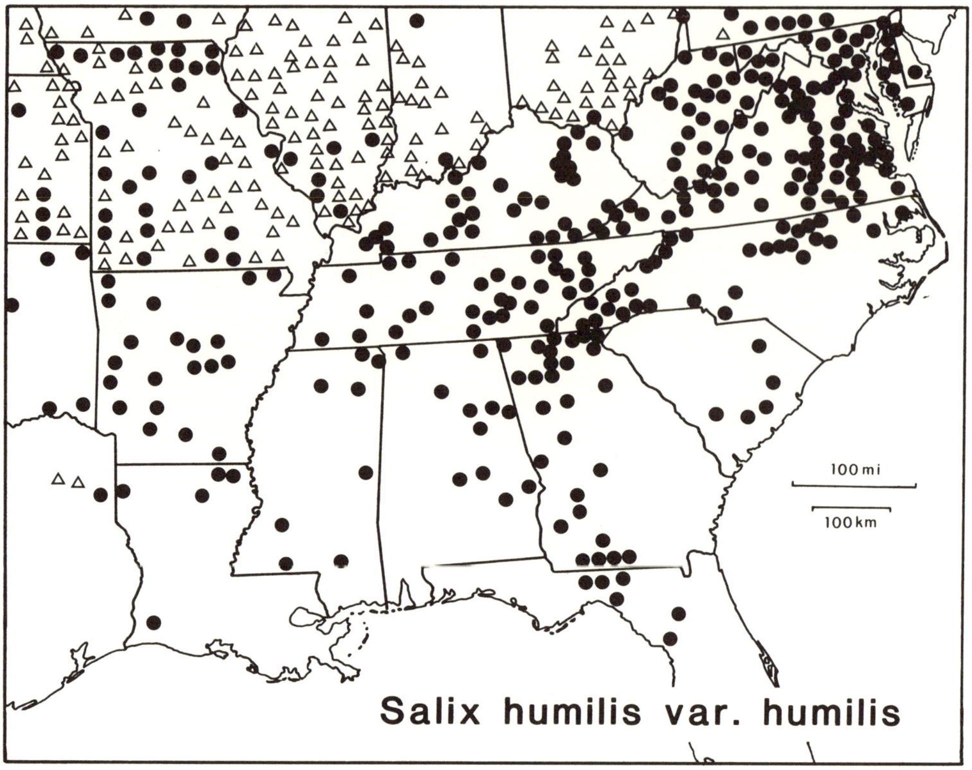

FIG. 38. Distribution of *Salix humilis* var. *humilis* in the Southeastern and peripheral states. Solid dots, herbarium specimens; open triangles, literature reports.

Shrubs 0.3–3 m tall, erect, sometimes decumbent at base and layering. Branchlets greenish brown or reddish brown. Leaf blades oblong to oblanceolate or narrowly so, rarely elliptical, the larger ones (2–) 5–7–9 (–13.5) cm long, (0.7–) 1.3–1.7–2.3 (–3.5) cm wide, length/width 2.5–4–7.5; mature leaf blades rugose and glabrate above, midrib velutinous or glabrate, glaucous and persistently tomentose beneath, sometimes sparsely so or even glabrate; petioles (1.5–) 3–5–7 (–12) mm long, velutinous or sparsely villous; stipules narrowly oblong to half ovate, glandular-serrulate, 1.5–11 mm long, usually appearing at fifth to fifteenth leaf, earlier leaves exstipulate. Staminate aments 1–1.3–2 cm long; bracts 1.4–1.6 mm long. Pistillate aments 0.8–2.25–3.2 cm long; capsules 7–12 mm long. Chromosome number 2n = 76. Fig. 37.

Phenology. Flowering in the South (Georgia to Florida and Louisiana) from late January to late April, and in the North from early March to mid-May, with a mean in mid-April; based on 97 records.

Distribution. (Figs. 37-inset, 38). *Salix humilis* var. *humilis* ranges in eastern North America from Labrador south to Florida and from the Atlantic coast to the Great Plains. In the Southeastern States it occurs most commonly on the Piedmont Plateau westward to the Ozarks. It is less common on the Coastal Plain and in the Mississippi Embayment.

It occurs in dry, open, oak-pine woods, prairies, grassy and rocky heath balds, limestone and serpentine barrens, moist barrens, wet flatwoods and meadows, old pastures, and roadsides. Reported on slopes of old coal mine tailings where it is mycorrhizal (Schramm 1966).

This variety has been described as rare in Alabama by Clark (1971), but it seems to have a relatively wide range in the state.

Representative Specimens. U.S.A. Alabama: Autauga Co., Mar 1942, *Harper s.n.* (NA); Bullock Co., *Kral 56323* (VDB); Calhoun Co., *Harper 3811* (A, GH, NY, PH, US); Cleburne Co., *Kral 62406* (VDB); Cullman Co., 20 Jun 1941, *Morgan s.n.* (A, GH, NA); Elmore Co., *Clark 17149* (NCU); Jackson Co., *Kral 37569* (VDB); Lauderdale Co., 26 Apr 1893, *Mohr s.n.* (UNA); Russell Co., *Kral 62190* (VDB); Saint Clair Co., *Kral 44625* (VDB); Talladega Co., *Kral 53105* (VDB).—Arkansas: Ashley Co., *Moore 172* (UARK); Benton Co., *Demaree 6345* (GH, NO, NY); Calhoun Co., *Demaree 37617* (GH); Clark Co., *Demaree 49046* (AUA, NCU, TENN, VDB); Clay Co., *Bush 2584* (A); Conway Co., *Demaree 62003* (MO); Faulkner Co., *Demaree 57* (MO, UARK); Garland Co., *Palmer 27125* (A, UARK); Howard Co., *Moore 5351* (CAN, UARK); Logan Co., *Demaree 5629* (NCU); Lonoke Co., *Demaree 22476* (MO, NA, NY); Nevada Co., *Moore 44* (UARK); Newton Co., *Redfearn 27929* (NCU, WILLI); Polk Co., *Palmer 12627* (A, MO, US); Prairie Co., *Demaree 22300* (MO, NA, NY); Pulaski Co., 22 Mar 1975, *Shepherd s.n.* (APCR); Randolph Co., *Marshall 16* (US); Scott Co., *Demaree 20768* (MO, NA, NY); Washington Co., *Palmer 23264* (A, MO); White Co., *Demaree 27089* (NA).—Delaware: New Castle Co., *Tatnall 2116* (DOV); Sussex Co., *Tatnall 2879* (DOV, GH).—District of Columbia: *Ball 1606* (CAN, GH, MO, NA).—Florida: Alachua Co., 27 Jul 1932, *West s.n.* (FLAS); Jefferson Co., *Argus 11476* (CAN); Leon Co., *Godfrey 56263* (CFR, CM, DAO, DHL, DUKE, FLAS, GH, MO, NCU, NY, US, WVA, WVW); Levy Co., 1843, *Rugel s.n.* (NA, US); Madison Co., *Godfrey 74979* (FSU, VDB, WIS); Taylor Co., *Godfrey 72331* (FSU, GH, MO, NCU, US, USF, VDB).—Georgia: Bartow Co., *Greear 6431* (GA); Brooks Co., *Faircloth 7092* (MO, NLU, VSC); Calhoun Co., *Harper 3811* (MO); Cherokee Co., *Duncan 8327* (GA); Clarke Co., *Miller 4425* (GA, LSU, PH); Colquitt Co., *Faircloth 7151* (MO, NCU, VSC); Dade Co., *Palmer 35473* (A, MO, NY); Dawson Co., *Duncan 7209* (GA, NCSC); DeKalb Co., *Venard 2134* (GA, GEO); Douglas Co., *Duncan 8434* (GA); Fannin Co., *Duncan 7298* (GA); Floyd Co., ca 1860, *Chapman s.n.* (AUA); Gilmer Co., *Jones 22506* (NA); Grady Co., *Faircloth 238* (GA, NCU, VSC); Habersham Co., *Duncan 6622* (GA); Harris Co., Mar 1966, *Jones s.n.* (AUA); Houston Co., *Duncan 3158* (FLAS, GA, NA); Lee Co., *Thorne 2744* (GA, GEO, GH, MTJB, NA, NY, US); Lowndes Co., *Faircloth 5864* (GA, MO, NCU, VSC); Murray Co., *Duncan 7267* (GA); Pickins Co., *Duncan 6323* (GA); Pike Co., *Duncan 6278* (GA); Rabun Co., *Duncan 2618* (GH, NCSC, UNA); Sumter Co., *Harper 1393* (GH, MO, NY, US); Talbot Co., *Henry 3382* (PH); Thomas Co., *Godfrey 56242* (FSU, GH, NA, USF); Towns Co., *Faircloth 1031* (VSC); Union Co., *Jones 1235* (FSU, MISS); White Co., *Duncan 21722* (LSU, LTU, VDB).—Kentucky: Barren Co., 23 Mar 1887, *Aschman s.n.* (CM); Bath Co., *Wharton 5972* (NA); Bell Co., 5 Jul 1965, *Carr s.n.* (KY); Caldwell Co., *Palmer 17681* (A, MO, US); Carter Co., *Braun 2475* (A, CFR); Casey Co., *Reed 7924* (CFR); Edmonson Co., *Lix 559* (US); Fleming Co., *Braun 3409* (A, CFR); Grayson Co., *Reed 16123* (CFR); Hardin Co., *Reed 96241* (CFR); Harlan Co., *McInteer 1294* (KY); Hart Co., *McFarland 64* (MO, NA); Jefferson Co., *Wharton 5803* (KY, NA); Knox Co., *Reed 7775* (CFR); Letcher Co., *Reed 78928* (CFR); Lewis Co., *Braun 1156* (US); Lincoln Co., 10 Aug 1842, *Short s.n.* (MO); Logan Co., *Shacklette 4S* (NA); Lyon Co., *Eggleston 4829* (NY); Marshall Co., 21 Apr 1973, *Athey s.n.* (VDB); McCreary Co., *Morton 42986* (KY, TENN); Meade Co., *Braun 3872* (A, US); Menifee Co., *McFarland 4485* (NA); Montgomery Co., *Wharton 3078* (NA, NY); Powell Co., *McInteer 436* (NA); Rockcastle Co., *Wharton 2664* (CAN, NA); Trigg Co., *Clebsch 2795* (APSC); Warren Co., *Nicely 2955* (WKU); Washington Co., 2 Apr 1966, *Meyer s.n.* (NA); Whitley Co., *Braun 3801* (A, US); Wolfe Co., *Higgins 1285* (DHL).—Louisiana: Caddo Par., *Thieret 31280* (DUKE, FSU); Calcasieu Par., *Thieret 22975* (DUKE, FSU, NCU); Morehouse Par., *Thomas 51210* (LSU, NLU); Ouachita Par., *Thomas 56101* (LSU, NLU, NO); West Carroll Par., *Gough 591* (NLU).—Maryland: Ann Arundel Co., *Reed 42686* (CFR); Baltimore Co., *Braun 23140* (CFR); Caroline Co., *Tatnall 2469* (DOV, PH); Carroll Co., *Reed 80321* (CFR); Cecil Co., *Long 68355* (PH); Charles Co., *Chrysler 75* (MARY); Dorchester Co., *Reed 28* (CFR); Frederick Co., *Hermann 13752* (NA); Garrett Co., *Braun 21817* (CFR); Harford Co., *Braun 89106* (CFR); Howard Co., *Salomon 408* (MARY); Kent Co., *Reed 41902* (CFR); Montgomery Co.,

Ball 38 (MO, NA); Prince Georges Co., *Hill 8957* (MARY, NY); Queen Annes Co., *Braun 24877* (CFR); Saint Marys Co., *Reed 40626* (CFR); Somerset Co., *Reed 31808* (CFR); Talbot Co., *Earle 2245* (NA, PH); Washington Co., *Harbison 2* (A).—MISSISSIPPI: Alcorn Co., *Jones 37782* (CAN); Copiah Co., *Jones 14453* (MISS); Forrest Co., *Rogers 6832* (GH, MO, NCU, NY); Kemper Co., *Jones 18775* (AUA, FLAS, FSU, MISS, NCU, NLU, TENN); Lafayette Co., 30 Apr 1970, *Holman s.n.* (MISS); Lee Co., *Mann 339* (MISS); Pike Co., *Jones 13182* (MISS, NCU); Tishomingo Co., *Cooley 4602* (USF).—NORTH CAROLINA: Alamance Co., *Parker 89* (NY); Alleghany Co., *Correll 14623* (DUKE); Ashe Co., *Jones 2474* (DUKE, NA); Avery Co., *Rohrer 1712* (NCU); Bertie Co., *Ahles 46177* (VDB); Buncombe Co., *Palmer 35423* (A, MO, NY); Clay Co., *Boufford 14217* (NCU); Durham Co., *Clark 572B* (FUGR); Franklin Co., *Ahles 16462* (NCU); Gaston Co., 20 Jul 1979, *Matthews s.n.* (UNCC); Granville Co., *Batson 754* (DUKE); Guilford Co., Jun 1953, *Melvin s.n.* (MUR, US); Haywood Co., *Wilbur 7116* (DUKE); Henderson Co., 11 Mar 1956, *Pittillo s.n.* (NCU); Jackson Co., *Thomas 53237* (NLU); Macon Co., *Argus 10863* (CAN); Mitchell Co., 4 May 1893, *Ashe s.n.* (NCU, NY); Nash Co., *Godfrey 7021* (DUKE, GH, MO, NY, PH, US); Northampton Co., *Ahles 38393* (NCU); Orange Co., *Boufford 12464* (CAN, CM, FUGR, MO); Pasquotank Co., *Ahles 4008A* (NCU); Person Co., *Bell 14486* (NCU); Polk Co., 10 Apr 1923, *Peattie s.n.* (NCU); Stanley Co., 25 Jul 1917, *Holmes s.n.* (NCU); Swain Co., *Bruhn 76* (TENN); Transylvania Co., *Boufford 21997* (CM, VPI); Union Co., 13 Aug 1956, *Mark s.n.* (DUKE); Vance Co., 30 Apr 1966, *Osborne s.n.* (NCSC); Wake Co., *Fox 2152* (NA, NCSC); Warren Co., *Leonard 1329* (NCU); Watauga Co., *Wood 4159* (CM).—SOUTH CAROLINA: Bamberg Co., *Ahles 26119* (VPI); Berkeley Co., *Porcher 1617A* (CITA); Darlington Co., *Smith 1625* (NCU); Dorchester Co., *Ahles 21851* (LYN, NCU); Williamsburg Co., 2 Apr 1939, *Stewart s.n.* (NCU, NY).—TENNESSEE: Anderson Co., *DeSelm 1873* (TENN); Bledsoe Co., *Shanks 8934* (TENN); Blount Co., *Sharp 3718* (MUHW, TENN); Campbell Co., *Iltis 27810* (TENN); Coffee Co., *Sharp 11222* (GH, TENN); Cumberland Co., *Sharp 7431* (TENN); Davidson Co., 26 Apr 1925, *Shaver s.n.* (VDB); DeKalb Co., *Shanks 3775* (TENN); Fayette Co., *Palmer 17334* (A, GH, MO, US); Franklin Co., *Sasnett 38* (NCU); Grainger Co., 25 Apr 1934, *Jennison s.n.* (TENN); Grundy Co., *Sharp 1242* (NCU, TENN); Henderson Co., *Sharp 9339* (NCSC, TENN); Jefferson Co., *Rugel 2* (A, CAN); Knox Co., 10 Jun 1893, *Kearney s.n.* (OS); Lewis Co., *Kral 23507* (GH, PH, VDB); Maury Co., *Kral 39740* (VDB); McNairy Co., 20 Apr 1941, *Anderson s.n.* (MO); Meigs Co., 7 Mar 1974, *DeSelm s.n.* (TENN); Montgomery Co., *Chester 55* (APSC); Polk Co., *Ruth 154* (US); Putnam Co., *Kral 55923* (VDB); Roane Co., *DeSelm 29973* (TENN); Scott Co., *Morton 42659* (NY, TENN); Sequatchie Co., *Kral 47460* (VDB); Sevier Co., 30 Sep 1965, *Thomas s.n.* (TENN); Stewart Co., *Kral 48347* (VDB); Sumner Co., *Clebsch 14275* (TENN); Van Buren Co., *Shanks 3330* (NCSC, TENN); Wayne Co., *Sharp 9636* (NCSC, TENN); Weakley Co., *Sharp 7797* (GH).—VIRGINIA: Accomax Co., *Reed 36606* (CFR); Alleghany Co., *Sharp 30094* (TENN); Amelia Co., *Mikula 4360* (WILLI); Amherst Co., *Freer 1804* (GH, LYN, NLU); Arlington Co., *Fosberg 16895* (NA, NO, PH); Augusta Co., *James 7874* (NCU); Bedford Co., *Freer 1725* (GH, NLU, LYN); Bland Co., *Nicely 1004* (VPI); Brunswick Co., *Kral 13670* (FSU, VPI); Buckingham Co., *James 6954* (NCU); Caroline Co., *James 13904* (NCU); Charles City Co., *James 4559* (NCU); Charlotte Co., *James 2742* (NCU); Chesterfield Co., *Kiltz 14* (VPI); Craig Co., *Uttal 6363* (LYN, NCU, VPI); Culpeper Co., *James 2817* (NCU); Cumberland Co., *James 7145* (NCU); Dickenson Co., *James 9867* (NCU); Dinwiddie Co., *Argus 10963* (CAN); Fairfax Co., *Ball 2716* (DUKE, NA, NCSC, VPI); Fauquier Co., *Allard 11175* (GH, PH, VPI, WVA); Floyd Co., *Argus 10936* (CAN); Giles Co., *Iltis 2001* (TENN, VA, WIS); Gloucester Co., *Greaves 1731* (WILLI); Goochland Co., *James 6640* (NCU); Greene Co., *Reed 97683* (CFR); Greensville Co., *James 4331* (NCU); Hanover Co., *Smith 508* (WILLI); Henrico Co., *Fernald 6980* (GH, MO, PH); Highland Co., *James 7682* (NCU); James City Co., *Palmer 39783* (GH, MO); King George Co., *James 13743* (NCU); King William Co., *James 5429* (NCU); Lee Co., *Reed 82623* (CFR); Louisa Co., *Reed 89516* (CFR); Lunenburg Co., *James 3604A* (NCU); Madison Co., *Fosberg 16931* (NA, NO, PAC, PH); Mecklenburg Co., *Seaman 3122* (NCU); Middlesex Co., *James 5559* (NCU); Montgomery Co., *Massey 2407* (GH, NA, VPI); New Kent Co., *Gillespie 453* (NCU, WILLI); Norfolk Co., *Fernald 3888* (GH, PH); Nottoway Co., *James 6797* (NCU); Orange Co., *James 13650* (NCU); Page Co., May 1978, *Frey s.n.* (LFCC); Pittsylvania Co., *James 3250* (NCU); Powhatan Co., *James 6288* (NCU); Prince Edward Co., 20 Apr 1878, *Smith s.n.* (US); Prince George Co., *Massey 13156* (VPI); Prince William Co., *Allard 9811* (CM, GH, MO, NA, NY, VPI); Rappahannock Co., *Fosberg 39602* (US); Richmond Co., *Kiltz 13* (VPI); Roanoke Co., *Wood 857* (PH); Rockbridge Co., *Stevens 140,* (LYN); Russell Co., 9 Jul 1959, *Shields s.n.* (VPI); Scott Co., *Peake 698*

(FLAS); Shenandoah Co., *Fosberg 44238* (US); Smythe Co., *Small 6980* (MO); Southampton Co., *Reed 89220* (CFR); Sussex Co., *Fernald 13311* (GH, MO, PH); Warrick Co., *Baldwin 95* (GH), Wise Co., *Reed 78929* (CFR); Wythe Co., *James 8225* (NCU); York Co., *James 4755* (NCU).—West Virginia: Barbour Co., *Davis 5053* (WVA); Braxton Co., 10 Jun 1953, *Boggs s.n.* (WVA); Brooke Co., *Bonar 1* (WVA); Cabell Co., *Gilbert 554* (CM, DUKE, GH, MO, MUHW, PH, TENN, US, WVA); Calhoun Co., 12 May 1933, *Harris s.n.* (MUHW); Fayette Co., 2 Sep 1973, *Grafton s.n.* (WVA); Grant Co., *Rossbach 1071* (US, WVA, WVW); Greenbrier Co., *Henry 2176* (PH); Hampshire Co., *Palmer 42587* (A); Hardy Co., 15 Oct 1962, *Wratchford s.n.* (WVA); Lewis Co., *Bartholomew 53* (LSU); Mercer Co., *Uttal 9659* (LSU, NLU, VPI); Mineral Co., 1 Sep 1956, *Brown s.n.* (NA, WVA); Monongalia Co., 29 Sep 1937, *Davis s.n.* (WVA); Monroe Co., 1 Aug 1940, *Fox s.n.* (WVA); Nicholas Co., *Bartholomew 31* (WVA); Pendleton Co., *Rossbach 3963* (WVA, WVW); Pocahontas Co., *Clarkson 185* (NCU, WVA); Preston Co., *Davis 3964* (WVA); Raleigh Co., *Zeto 135* (MUHW); Randolph Co., *Rossbach 1157* (NLU, WVW); Ritchie Co., *Bartholomew 66* (CM, WVA); Tucker Co., *Allard 12050* (MT, PH, US, WVA); Upshur Co., *Rossbach 2505* (WVW); Webster Co., 10 Aug 1940, *Fox s.n.* (WVA); Wetzel Co., Nov 1940, *Dillaman s.n.* (WVA); Wirt Co., *Bartholomew 4143* (WVW).

Recognition. Salix humilis var. *humilis* is a small to mid-sized shrub that grows as scattered individuals or in small populations produced by layering. It is precocious in flowering and has short, sessile aments borne on grayish tomentose branches. Its leaves are stipulate, tomentose beneath with curly hairs, and have a revolute margin that varies from entire to glandular-crenate.

Salix humilis var. *humilis* can usually be distinguished from S. *discolor* by the dense tomentum on the underside of its leaves. Plants with glabrate leaves may be distinguished by their revolute leaf margins; the margins of immature leaves of *S. humilis* are particularly revolute, whereas the leaves of *S. discolor* are either flat or sometimes folded abaxially. The recurved aments of *S. humilis* may be diagnostic (Fernald 1950), but they are not always present or unequivocal. See also *S. discolor,* no. 12.

Immature flowering specimens of *S. humilis* var. *humilis* are very similar to and often confused with *S. sericea.* If vegetative shoots are emerging, the tomentose pubescence on the leaves of *S. humilis* readily distinguishes it from the sericeous pubescence of *S. sericea;* but if not, the winter bud scales may be diagnostic. In *S. humilis* the bud scales are usually densely gray-pubescent whereas in *S. sericea* they are usually glabrate. Immature pistillate flowers of *S. humilis* can be distinguished by their long and distinctly beaked pistils; those of *S. sericea* have shorter, truncate pistils. The stigmatic lobes of *S. humilis* spread apart later in the development of the pistil than do those of *S. sericea;* by that time the style is no longer obscured by hairs as it is in *S. sericea.* In the field, *S. sericea* can be separated from *S. humilis* by its brittle branches, compared to the flexible and tenacious branches of *S. humilis.*

Pine cone galls are produced by the gall midge *Rhabdophaga strobiloides* Walsh (Felt 1940).

Nomenclature. Apparently no type material of *S. humilis* is in existence. According to Ewan (1967) Marshall did not preserve herbarium specimens. A search of the Darlington Herbarium (DWC), where some specimens from Marshall's garden are located, and at the Philadelphia Academy of Natural Sciences did not turn up any specimens of *Salix* collected by him. Marshall (1785: 140) described the species as one with entire, villous leaves that "seldom rises above three or four feet, with greenish, somewhat downy stalks. The leaves are larger than the other kinds [*S. nigra* and *S. sericea*], entire, oblong, somewhat oval, and glaucous or whitish beneath." This description does not fit any other species of *Salix* in the region in which Marshall lived, and it has been used here to fix the name.

Salix conifera sensu Muhlenberg (1803) is the same as *S. humilis* (see *S. discolor,* no. 12, for a discussion of the identity of *S.* ×*conifera* Wangenheim).

In 1840 Barratt named three varieties of *S. muhlenbergiana* Willdenow: var. *angustifolia,* var. *grandifolia,* and var. *longifolia.* The first two have been typified and are the same as *S. humilis.* The latter name is a nomen nudum, and its application is uncertain.

The Anderson (1867) "form" *S. humilis rigidiuscula* has not been typified by specimens, but, on the basis of its description, it seem to be *S. humilis.* Similarly, the name *S. tristis* [var.] α *glabrata* Andersson has not been typified by specimens, but it too may be *S. humilis* (Schneider 1920).

The name var. *rigidiuscula* was used by Ball (1961) for the narrow-leaved variant of *S. humilis.* He attributed the varietal combination to Robinson and Fernald, but in 1946 Fernald wrote that they had used Andersson's name in the wrong sense and described *S. humilis* var. *hyporhysa* for the taxon that he and Robinson had had in mind. The latter variety is, according to Fernald, the commonest variety of *S. humilis* throughout the Southeastern States. He characterized it by leaves that are glabrate to sparsely puberulent and rugose-reticulate beneath; but leaf indumentum and leaf surface irregularity vary in the species thoughout its range and do not seem to be taxonomically important. Fernald also describes the variety as having staminate aments 1–3 cm long and fruiting aments 2–8 cm long. In my studies I have not seen any specimens of *S. humilis* with aments this large, and I suspect that Fernald was describing the hybrid *S. discolor* × *humilis* (*S.* ×*conifera* Wangenheim), which could account for both the larger aments and the glabrate leaves.

Hybridization. Salix humilis var. *humilis* is reported to hybridize with *S. bebbiana, S. discolor, S. eriocephala* (as *S. rigida*), *S. humilis* var. *microphylla* (also as *S. tristis*), *S. petiolaris* (as *S. gracilis*), and *S. sericea* (Schneider 1921b; Fernald 1950).

Controlled hybridizations have been made in Ontario between *S. humilis* and *S. discolor, S. petiolaris,* and *S. bebbiana* (Argus, unpubl.). *S. humilis* and *S. discolor* are easily crossed, and natural hybrids are common in nature wherever the habitats of these species are contiguous (see *S.* ×*conifera* and *S. discolor,* no. 12). The cross *S. humilis* × *petiolaris* was successfully made, and some viable seedlings were produced. Crosses of *S. humilis* and *S. bebbiana* yielded few seeds and the seedlings soon died; the natural occurrence of this hybrid is in doubt.

Hybrids of *S. humilis* and *S. sericea* were not recognized in specimens from the Southeastern States even though their ranges are sympatric; it is doubtful that this hybrid occurs. Hybrids between *S. humilis* var. *humilis* and var. *microphylla* were not recognized in this study, nor were intermediates observed in the field where the varieties were growing together. The presence of a single qualitative character, however, namely stipule presence or absence, presents a problem in recognizing intermediates. A similar problem is posed in the recognition of hybrids between *S. caroliniana* and *S. nigra.*

13b. Salix humilis var. **microphylla** (Andersson) Fernald, Rhodora 48: 46. 1946. *Salix tristis* var. *microphylla* Andersson, *Salices bor.-amer.* 21. 1858.—Type: unknown.—Dwarf upland willow, Sage willow, Dwarf gray willow.

Salix occidentalis Walter, *Fl. carol.* 243. 1788.—Type: "Carolina." No precise locality given, *Herb. Walter p. 93* (holotype: BM, photo: CAN!).

Salix alpina Walter, *Fl. carol.* 243. 1788, non Scopoli, 1760.—Type: "Carolina." No precise locality given, *Herb. Walter p. 93* (holotype: BM, photo: CAN!).

Salix tristis Aiton, *Hort. kew.* 3: 393. 1789. *S. humilis* var. *tristis* (Aiton) Griggs, Proc. Ohio Acad. Sci. 4: 301. 1905.—Type: Pennsylvania. Introduced by W. Young, 1765, and cultivated at the Royal Botanic Gardens, Kew (holotype: BM!).

Salix longirostris Michaux, *Fl. bor.-amer.* 2: 226. 1803.—Type: "Hab. a Canada ad Carolinam," *Michaux s.n.* (lectotype, here designated: P–Herb. Michaux!).

Salix muhlenbergiana Willdenow, *Sp. pl.* 4: 692. 1806.—Type: Pennsylvania. Lancaster, *Muhlenberg s.n.* (holotype: B–W no. 18182, photo: CAN!).

Salix tristis var. *monadelphia* J. Barratt, *Salic. amer.*, Sp. 2. 1840.—Type: Connecticut. Middletown, *Barratt s.n.* (holotype: NY!).

Salix tristis f. *curtifolia* Fernald, Rhodora 36: 195. 1934. *Salix humilis* var. *microphylla* f. *curtifolia* Fernald, Rhodora 48: 46. 1946.—Type: Massachusetts. Sandwich, 23 Oct 1922, *Fernald s.n.* (holotype: GH!).

Salix tristis f. *festiva* Fernald, Rhodora 36: 195. 1934. *Salix humilis* var. *microphylla* f. *festiva* Fernald, Rhodora 48: 46. 1946.—Type: Massachusetts. Bourne, 21 Aug 1919, *Fernald & Long 17002* (holotype: GH!).

Salix humilis var. *microphylla* f. *tortifolia* Fernald, Rhodora 48: 46. 1946.—Type: Massachusetts. Plymouth, *Oakes s.n.* (holotype: GH!).

Shrubs 0.3–0.7 (–1) m tall, decumbent and forming clones by layering. Branchlets light brown or yellowish, tomentose, very slender. Leaf blades narrowly oblong to narrowly oblanceolate, the larger ones (1.3–) 2.5–4–5 (–7) cm long, 0.3–0.7–1.3 cm wide, length/width 2.5–5–9; mature leaf blades pilose to velutinous and glossy above, gray-tomentose or sparsely so beneath, midrib yellowish; petioles 0.5–2–3 (–6) mm long; stipules absent or present only as minute glandular lobes or hairs, sometimes present on sprout shoots. Staminate aments 0.5–0.7–1.1 cm long; bracts 0.8–1.4 mm long. Pistillate aments 1–1.4–1.8 cm long; capsules 5–9 mm long. Chromosome number unknown. Fig. 39.

Phenology. Flowering from early March to late May with a mean in mid-April; based on 94 records.

Distribution. (Fig. 39-inset, 40). Variety *microphylla* has its center of occurrence in the Appalachian System and ranges from the Appalachian Plateau in southern Ohio, Indiana, and Pennsylvania to the Piedmont Plateau in North Carolina and Virginia. On the Atlantic Coastal Plain it occurs as disjunct populations in eastern Virginia and South Carolina, but these reports need field confirmation. In the Mississippi Embayment it occurs in central Mississippi, eastern Louisiana, and eastern Arkansas. I did not see the collection of var. *microphylla* from Covington, Tammany Parish, Louisiana, made by Thomas Drummond (K) and reported as *S. tristis* by Brown (1945). Some of the plants on which records in

FIG. 39. *Salix humilis* var. *microphylla*. a. Vegetative branch arising from a decumbent branch (×0.75). b. Staminate flower (×10). c. Branch with young pistillate aments and vegetative branchlets starting to emerge (×1). d. Young pistillate flowers (×10). Inset: map of total distribution, based on Braun (1961), Deam (1940), Mohlenbrock (1980), Seymour (1969), Wherry et al. (1979), E. G. Voss (pers. comm.), and herbarium specimens.

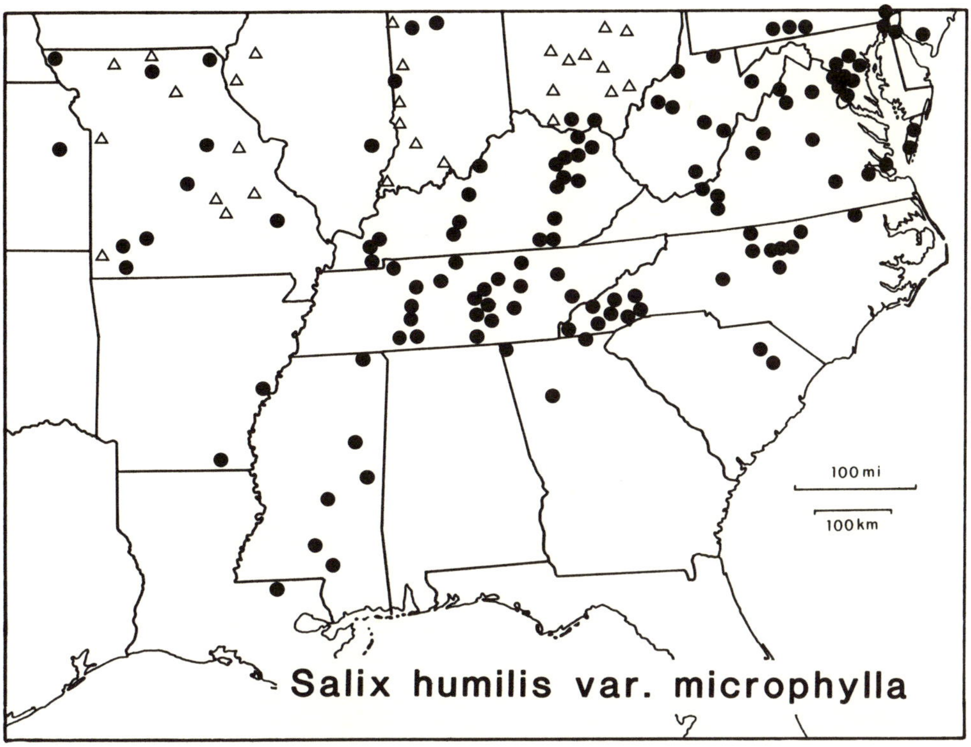

FIG. 40. Distribution of *Salix humilis* var. *microphylla* in the Southeastern and peripheral states. Solid dots, herbarium specimens; open triangles, literature reports.

the Mississippi Embayment are based may be small-leaved, exstipulate plants of var. *humilis* and require further study. Wide disjuncts are reported in Maine, Kansas, and Nebraska (Fig. 39-inset). The Maine report is discounted (Hyland & Steinmetz 1944); the other reports require confirmation.

This taxon occurs on moist limestone and serpentine barrens, open heath balds, open pine woods, moist prairies, swampy places in open deciduous woods, and along streambanks.

Variety *microphylla* is reported to be rare, and possibly threatened or endangered, in Arkansas and Georgia (Thorne 1954, as *S. tristis;* Kartesz & Kartesz 1977). This study confirms its rarity in these states and suggests that it may be rare also in Delaware and South Carolina.

REPRESENTATIVE SPECIMENS. U.S.A. ARKANSAS: Ashley Co., *Demaree 18745* (GH, MO, NA, NY); Phillips Co., *Demaree 18742* (MO, NA, NY, PH, TENN, WIS).—DELAWARE: New Castle Co., 8 Jun 1875, *Commons s.n.* (NA).—DISTRICT OF COLUMBIA: 21 Jul 1900, *Steele s.n.* (GH).—GEORGIA: Cobb Co., 1860, *Cleveland s.n.* (FLAS); Dade Co., *Duncan 5436* (GA); Towns Co., *Pyron 1556* (GA, US).—KENTUCKY: Bath Co., *McFarland 4060* (NA); Calloway Co., *Smith 4193* (GH); Carter Co., *Reed 19718* (CFR); Edmonson Co., 1 Apr 1898, *Price s.n.* (NY); Estill Co., *Wharton 3113* (NA); Hardin Co., *Braun 3458* (A, CFR); Jefferson Co., *Ahles 54392* (NCU); Laurel Co., *Braun 1380* (A, US); Lewis Co., *Braun 2182* (A); Lyon Co., *Clanton 24* (NA); Marshall Co., *Athey 1708* (WKU); McCreary Co., *McFarland 3185* (NA); Montgomery Co., *Wharton 1835* (NA); Powell Co., 27 Jun

1974, *Meijer s.n.* (KY); Rowan Co., *Meijer 1153* (KY, NCU); Warren Co., Jul 1898, *Price s.n.* (NY); Whitley Co., *McFarland 3193* (NA); Wolfe Co., *McFarland 3292* (NA).—LOUISIANA: Saint Helena Par., *Allen 940* (LSU).—MARYLAND: Ann Arundel Co., 10 Apr 1944, *Waite s.n.* (NA); Baltimore Co., *Reed 75375* (CFR); Cecil Co., *Long 34055* (PH); Charles Co., *McVaugh 4861* (NA); Harford Co., *Reed 89109* (CFR); Howard Co., *Reed 24422* (CFR); Montgomery Co., 26 May 1940, *Freeman s.n.* (NA); Prince Georges Co., 18 Sep 1941, *Waite s.n.* (NA, PH).—MISSISSIPPI: Alcorn Co., *Jones 37782* (CAN); Jefferson Davis Co., *Jones 14287* (MISS); Kemper Co., Apr 1859, *Hilgard s.n.* (MO, S); Lamar Co., *Rogers 2513* (VDB); Oktibbeha Co., *McDaniel 2840* (NY); Scott Co., *Jones 19389* (MISS, VDB).—NORTH CAROLINA: Alamance Co., 31 Jul 1927, *Braxton s.n.* (NCU); Buncombe Co., *Biltmore Herb. 1188B* (A, GH, MO, NCU, NY, PH, US); Chatham Co., 22 Sep 1964, *Tucker s.n.* (APCR); Cherokee Co., 8 Aug 1935, *Wells s.n.* (NCSC); Durham Co., *Davis 2018* (WVA); Granville Co., *Ahles 11537* (NCU, WVA); Guilford Co., *Radford 8409* (NCU); Heywood Co., *Shanks 11339* (FLAS); Henderson Co., *Coker 4982* (NCU); Jackson Co., *Horton 2111* (FUGR); Macon Co., *Godfrey 51874* (FSU, NA, NCSC); Northampton Co., *Fox 2771* (NA, NCSC); Orange Co., *Batson 753* (DUKE); Rockingham Co., *Ahles 10759* (NCU, US); Rowan Co., *Small 109* (NY); Swain Co., *Bruhn 199* (TENN); Transylvania Co., *House 4232* (US).—SOUTH CAROLINA: Darlington Co., 17 Apr 1933, *Coker s.n.* (NCU); Florence Co., *Coker 2* (NCU, UARK).—TENNESSEE: Anderson Co., *DeSelm 1873* (TENN); Bledsoe Co., *Shanks 3589* (TENN); Blount Co., *Wilson 1945* (MUHW, TENN); Cannon Co., *Kral 53755* (VDB); Coffee Co., *Kral 34354* (US, VDB, WILLI); Cumberland Co., *Weatherby 6253* (A, GH, NY, TENN, US); Davidson Co., 13 Jul 1897, *Eggert s.n.* (MO); DeKalb Co., *DeSelm 144* (TENN); Dickson Co., *Kral 48297* (VDB); Fentress Co., *Shanks 3040* (TENN); Franklin Co., *Argus 10853* (CAN); Grundy Co., *Kral 43758* (VDB); Hickman Co., *Weatherby 452* (TENN); Lawrence Co., *Kral 47806* (VDB); Lewis Co., *King 353* (US); Putnam Co., *DeSelm 1135* (TENN); Stewart Co., *Kral 48346* (VDB); Sumner Co., *Clebsch 14277* (TENN); Warren Co., *Kral 54854* (VDB); Wayne Co., *Sharp 9629* (NCSC, TENN).—VIRGINIA: Appomattox Co., *Reed 83697* (CFR); Arlington Co., *Blake 10827* (MTJB, NA); Augusta Co., *Argus 10972* (CAN); Dinwiddie Co., *Jones 2448* (NCU); Fairfax Co., *Fosberg 39810* (NY); Faquier Co., *Allard 7608* (GH, NY); Floyd Co., *Argus 10940* (CAN); Giles Co., *Fogg 12864* (GH, PH); Louisa Co., *Reed 89517* (CFR); Montgomery Co., 27 Mar 1945, *Massey s.n.* (VPI); Northampton Co., *Fernald 5287* (CFR, GH); Page Co., *Artz 624* (US); Rockbridge Co., 2 May 1915, *Churchill s.n.* (MO, WVA); Shenandoah Co., 2 Aug 1949, *Artz s.n.* (TENN); Surry Co., *Lord 163* (LYN); York Co., 12 Apr 1964, *Knorr s.n.* (WVA).—WEST VIRGINIA: Calhoun Co., 15 May 1933, *Harris s.n.* (WVA); Grant Co., 15 Jun 1959, *Bartholomew s.n.* (WVA); Monongalia Co., 10 May 1932, *Sharp s.n.* (MUHW); Pocahontas Co., 13 Jul 1934, *Strausbaugh s.n.* (WVA); Summers Co., *Boone 1158* (WVA); Tyler Co., *Bartholomew 52* (WVA); Webster Co., *Davis 3746* (WVA); Wirt Co., *Bartholomew 4142* (WVA).

Recognition. Salix humilis var. *microphylla* is a low, slender-stemmed shrub that forms colonies by layering. Its leaves are small, narrow, densely tomentose, and lack stipules. It may be distinguished from *S. sericea* in the same way as described for var. *humilis* (see above).

Pine cone galls are produced by the gall midge *Rhabdophaga strobiloides* Walsh (Felt 1940).

Nomenclature. I have discussed the nomenclature and typification of *S. occidentalis* and *S. alpina* elsewhere (Argus 1984b). Both names apply to what generally has been called *S. tristis* Aiton. The name *S. alpina* Walter is a later homonym of *S. alpina* Scopoli and is therefore rejected. A photograph of the holotype of *S. occidentalis* Walter shows a plant that is typical of *S. tristis,* including the absence of stipules or stipule scars (R. Vickery, pers. comm.). It is the earliest specific name for this taxon and takes precedence over *S. tristis* at this rank.

Aiton (1789: 393) described *Salix tristis* as "S. exstipulaceae, foliis lineari-lanceolatis petiolatis rugosis tomentosis. Narrow-leav'd American willow. Nat. of Pennsylvania. M. William Young. Intro. 1765. Fl., April." The probable holotype is in the British Museum. It is labelled as "s. Hort Kew. 1779. Short specimen without flowers." The specimen has leaves that are larger than usual for the species, but this may be due to its having been in cultivation. The specimen lacks stipules, a characteristic noted by Aiton.

A photograph of the holotype of *S. muhlenbergiana* Willdenow reveals leaves that are typical of *S. occidentalis;* it lacks stipules and stipule scars (F. Butzin, pers. comm.).

The identity of *S. recurvata* Pursh is problematical; type material could not be located in the Philadelphia Academy of Natural Sciences or in the Darlington Herbarium. Some of the elements in Pursh's description could apply equally well to *S. humilis* var. *humilis.*

Salix tristis var. *microphylla* Andersson is not a distinct taxon. Andersson recognized it in his first publication on North American willows (1858a, 1858b), but did not mention it in his later papers (1867, 1868) and evidently also questioned its distinctness. Specimens bearing this name could not be found in Andersson's herbarium (at S), and a type could not be selected from the available fragments. The original description has been used to fix the application of this name.

Fernald (1946) described three forms of *S. humilis* var. *microphylla,* namely f. *curtifolia,* f. *festiva,* and f. *tortifolia.* The first two fall within the range of variation of the variety, and the latter is a monstrous form with twisted leaves. None of them is taxonomically significant.

Hybridization. Salix humilis var. *microphylla* (often as *S. tristis*) is reported to hybridize with *S. eriocephala* (as *S. rigida*), *S. petiolaris* (as *S. gracilis*), and *S. humilis* var. *humilis* (Schneider 1921b; Fernald 1950; Steyermark 1963). Hybrids with these species have not been recognized in this study (see also *S. humilis* var. *humilis*).

14. Salix caprea Linnaeus, *Sp. pl.* 2: 1020. 1753.—Goat willow, Pussy willow.

Trees to 12–15 m tall, or large shrubs up to 8 m tall. Branches light brown, glabrous to sparsely pubescent; branchlets chestnut-brown, velutinous to villous; decorticated wood of 1–4 year old stems smooth or with a few short ridges on older wood; buds reddish brown, coarsely tomentose with white and ferruginous hairs to glabrate, ribbed laterally, short-beaked. Leaf blades narrowly to broadly elliptical to obovate or narrowly so, the larger ones 6.5–13 cm long, 2.5–5.5 cm wide, length/width 2–3, apex abruptly short-acuminate, sometimes conduplicate when pressed, base rounded to acute, margin crenulate to subentire, revolute; immature leaf blades densely white-tomentose beneath, less densely so above; mature leaf blades sparsely pubescent to glabrate above, sparsely tomentose and glaucous beneath with white or rarely ferruginous, erect and wavy hairs, hypostomatous, midrib yellowish; petioles tawny, 7–18 mm long; tomentose adaxially becoming glabrate; stipules acute to rounded, glandular adaxially, up to 14 mm long. Aments precocious, sessile, sometimes with 3–4 leaflike bracts at base. Staminate aments 2–2.5 cm long; stamens 2, filaments glabrous, dry anthers 0.8–1 mm long; nectary 1, adaxial; bracts dark brown, 2–4 mm long, narrowly oblong to oblong, apex acute to rounded, straight hairs on both sides 2 to 3 times the length of bracts. Pistillate aments densely flowered, 3.5–6 cm long; pistils pyriform, long-beaked, sericeous with refractive hairs, capsules brownish, about 6 mm long,

FIG. 41. *Salix caprea*. a. Vegetative branchlet (×0.75). b. Branch with staminate aments in anthesis and vegetative branchlets just starting to emerge (×1). c. Staminate flower (×7.5). d. Young pistillate flowers (×7.5).

FIG. 42. Distribution of *Salix caprea* in the Southeastern States; no distributional data are available for peripheral states. Solid dots, naturalized; open diamonds, cultivated; solid diamonds, naturalization or cultivation status unknown.

styles about 0.4 mm long, stigmas 2 slender lobes 0.6–0.8 mm long up to 2 times the length of style, stipes 2–2.5 mm long; nectary 1, adaxial, about ⅕ as long as stipe; ovules 6–9 per carpel; bracts as in staminate flowers. Chromosome number 2n = 38. Fig. 41.

Distribution. (Fig. 42). *Salix caprea* is a Eurasian species that has been introduced into the United States. It rarely escapes in the Southeastern States; only two naturalized occurrences are known: Baltimore Co., Maryland, and Macon Co., North Carolina.

REPRESENTATIVE SPECIMENS. U.S.A. ALABAMA: Madison Co., *Heller 91* (VDB).—DISTRICT OF COLUMBIA: *Ball 1609* (CAN, GH, NA).—MARYLAND: Baltimore Co., *Reed 64070, 64071* (CFR); Carroll Co., *Smith 22981* (CAN); Wicomico Co., *Smith 19385* (CAN).—NORTH CAROLINA: Macon Co., 29 Aug 1937, *Coker s.n.* (NCU).—WEST VIRGINIA: Monongalia Co., 13 Oct 1891, *Ramsey s.n.* (WVA).

Recognition. This species is cultivated as a "pussy willow," but its aments are often smaller than those of *S. cinerea* and some other precocious flowering species, and it may not be the best species for this purpose. In this study I found that many specimens originally identified as *S. caprea* were misidentified specimens of *S. cinerea* and that *S. caprea* is cultivated much less in North America than is suggested both by Floras and the horticultural literature.

Salix caprea and *S. cinerea* are similar and can easily be confused. *Salix caprea* can be recognized by the absence of prominent ridges on decorticated wood and to a cerain extent by its shorter staminate aments and other characters (see Tables 5 and 6, and *S. cinerea,* no. 15). Growth form was used as a diagnostic characteristic by Skvortsov (1968) in his key to the species in the USSR. He stated that *S. caprea* is a tree 12–15 m tall and grows as a shrub only when damaged in some way. *Salix cinerea,* on the other hand, is a tall shrub growing only to 4–5 m. The data available to me on habit and stature of North American specimens are very scanty, and the heights given in my descriptions are derived mostly from the

Table 5. Comparison of *Salix cinerea, S. caprea,* and *S. discolor.*

	S. cinerea	*S. caprea*	*S. discolor*
Habit	shrubs 3–7 m tall	trees or large shrubs up to 8 m tall	shrubs 2–6 m tall
Immature leaves	densely tomentose with white or gray hairs; ferruginous in subsp. *oleifolia*	densely tomentose with white hairs	sparsely to densely sericeous with white and/or ferruginous hairs
Mature leaves	pubescent to glabrate above, tomentose beneath	pubescent to glabrate above, sparsely tomentose beneath	glabrate on both sides or with scattered ferruginous hairs beneath
Leaf margin	crenulate; irregularly revolute	crenulate to subentire; irregularly revolute	crenate to coarsely toothed; flat
Pistillate aments	2–5 cm long	2–2.5 cm long	2–3 cm long
Decorticated wood	with long, prominent ridges	lacking ridges	lacking ridges, or short, scattered, and indistinct

literature. Roer and Elven (1975) use differences in the position of the stigma to separate *S. caprea* and *S. cinerea:* stigmas commissural in *S. caprea* and carinal in *S. cinerea;* but this characteristic has not been carefully studied in our material.

15. Salix cinerea Linnaeus, *Sp. pl.* 2: 1021. 1753.—Gray willow, Pussy willow.

Branches light brown, glabrous to sparsely pubescent; branchlets chestnut-brown, villous to pubescent with gray or white hairs; decorticated wood of 1–4 year old stems with many long, prominent ridges; buds yellow-brown, glossy, coarsely tomentose to glabrate with a few hairs at tip, ribbed laterally, apex blunt. Leaf blades elliptical, oblong to obovate, the larger ones 6.5–10.5 cm long, 2.2–5.2 cm wide, length/width 2–3, apex acute, imbricate when pressed, base obtuse to rounded, margin crenulate with irregular teeth, revolute; immature leaf blades sparsely tomentose above, densely tomentose beneath; mature leaf blades pubescent to glabrate above, glaucous and tomentose beneath with erect, wavy hairs, hypostomatous; petioles tawny, 4–15 mm long, tomentose abaxially; stipules acute, palmately veined, with 3–4 acute teeth. Aments precocious, sessile, sometimes with 4–5 small, leaflike bracts at base. Staminate aments 2–5 cm long; stamens 2, filaments pilose or glabrous at base, dry anthers 0.7–1 mm long; nectary 1, adaxial; bracts dark brown, pale at base, 2–3 mm long, oblong, apex acute or obtuse, long-hairy on both sides. Pistillate aments densely flowered, 3–7 cm long; pistils greenish tawny, pyriform, long-beaked, sericeous with refractive hairs, capsules tawny, 5–6.5 mm long, styles 0.2–0.5 mm long, stigmas 0.3–0.6 mm long, stipes about 2 mm long; nectary 1, adaxial, square, about ⅕ as long as stipe; ovules 6 per carpel; bracts as in staminate flowers. Fig. 43.

This species is cultivated for its spring display of soft, gray aments and is the species usually referred to in North America as the "pussy willow."

It is related to *S. caprea* and *S. discolor* from which it can be distinguished by the characters listed in Table 5. An important diagnostic feature, and the one most commonly used in European Floras to separate *S. cinerea* and *S. caprea,* is the presence of ridges on the decorticated wood of *S. cinerea.*

FIG. 43. *Salix cinerea.* a. Vegetative branchlet (×0.75). b. Portion of a decorticated branch showing ridges on wood. c. Branch with staminate aments in anthesis (×1). d. Staminate flower (×7.5). e. Young pistillate flowers (×7.5).

Table 6. Frequency of ridges on decorticated wood of *Salix cinerea, S. caprea,* and *S. discolor.*

	S. cinerea subsp. *cinerea*	*S. cinerea* subsp. *oleifolia*	*S. caprea*	*S. discolor*
Wood smooth	3	0	36	68
Wood with short ridges*				
ridges indistinct	2	0	2	20
ridges distinct	2	0	1	7
Wood with long ridges**				
ridges indistinct	14	0	0	0
ridges distinct	28	5	0	0
Total number of specimens	49	5	39	95

*Ridges usually less than 5 mm long.
**Ridges several cm long.

Three similar species in the Southeastern States, *S. cinerea, S. caprea,* and *S. discolor,* were studied for the presence of ridges on wood. Specimens were examined by soaking sections of branches in water and peeling the bark. All specimens of species introduced into the Southeastern States were examined and supplemented by specimens from Eurasia that had been identified by either A. K. Skvortsov, Moscow, USSR, or J. Chmelar, Brno, Czechoslovakia. *Salix discolor* was sampled from throughout its North American range. The study was based mainly on 1–2 year old wood, because older wood was only rarely available on herbarium specimens. Variation in ridges was recorded in five classes: (1) smooth (complete absence of ridges even on old wood); (2) short, indistinct ridges (less than 5 mm long, low, and rounded); (3) short, distinct ridges (short but sharply raised); (4) long, indistinct ridges (continuous, often several centimeters long, but low and rounded); (5) long, distinct ridges (prominent and sharply raised, sometimes uniformly so or with irregular peaks and valleys). The results are tabulated in Table 6.

In *Salix cinerea,* 87% of plants examined had long ridges most of which were prominently raised. In a few specimens from the USSR and Czechoslovakia ridges were absent or were very indistinctly developed on young wood. In *S. caprea* 92% of plants had smooth wood, and only three showed any ridge development at all. In the native *S. discolor,* the decorticated wood varied from smooth (72%) to short, indistinct ridges (21%) to those with short, distinct ridges (7%). Long ridges did not occur on the wood of *S. discolor.*

Key to the Subspecies of Salix cinerea

Leaves tomentose beneath with white or gray hairs. 15a. *S. cinerea* subsp. *cinerea.*
Leaves tomentose beneath with a mixture white and ferruginous hairs. 15b. *S. cinerea* subsp. *oleifolia.*

15a. **Salix cinerea** subsp. **cinerea.**

Shrubs 3–7 m tall. Leaf blades white- or gray-tomentose beneath. Chromosome number 2n = 76.

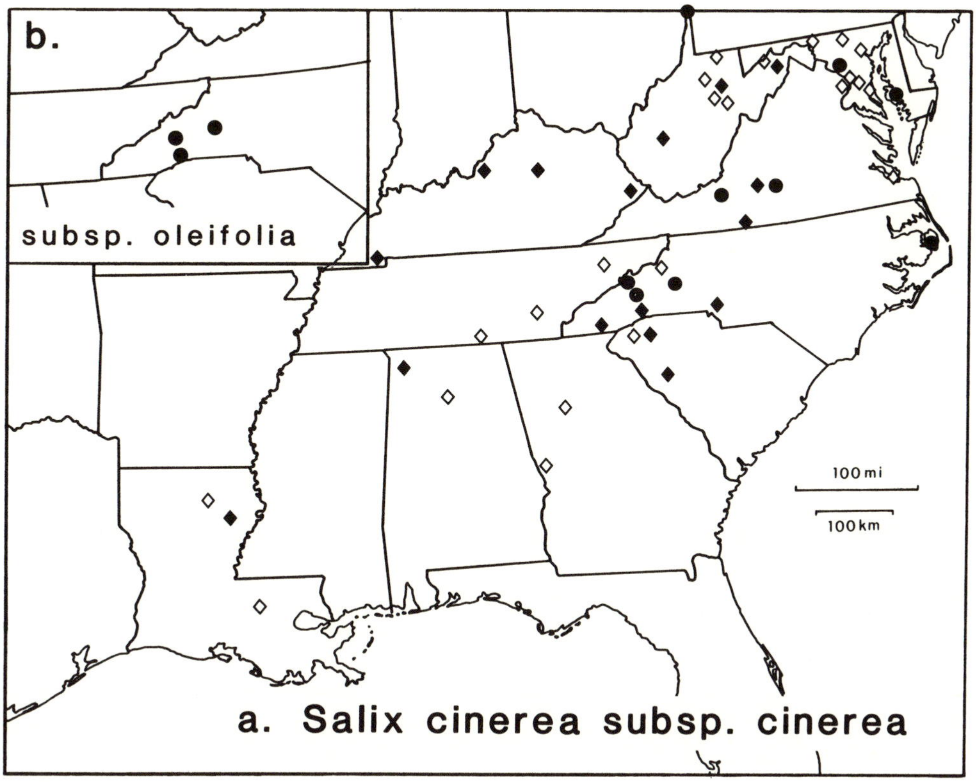

FIG. 44. Distribution of *Salix cinerea* in the Southeastern States. a. *S. cinerea* subsp. *cinerea*. b. *S. cinerea* subsp. *oleifolia*. Solid dots, naturalized; open diamonds, cultivated; solid diamonds, naturalization or cultivation status unknown. No distributional data are available for either subspecies in the peripheral states.

Distribution. (Fig. 44a). *Salix cinerea* subsp. *cinerea* is a Eurasian species that is cultivated and naturalized in the Southeastern States.

Representative Specimens. U.S.A. Alabama: Colbert Co., 18 Feb 1950, *T.F.H. s.n.* (MUR); Cullman Co., *Gould 101* (AUA).—District of Columbia: *Ball 72* (MO, NA).—Georgia: DeKalb Co., 31 May 1937, *Brinkley s.n.* (GEO); Harris Co., *Mazzeo 10313* (NA).—Kentucky: Calloway Co., *Easley 4* (MUR); Fayette Co., *Northcut 48* (KY); Jefferson Co., 4 May 1940, *Huber s.n.* (DHL); Pike Co., 16 Oct 1960, *Miniard s.n.* (KY).—Louisiana: East Baton Rouge Par., *Tollett 13* (NLU); Franklin Par., *Mason 41* (NLU); Ouachita Par., *Thomas 33411* (NLU).—Maryland: Baltimore Co., *Beach 107* (MARY); Calvert Co., *Meyer 9796* (NA); Carroll Co., 1 Apr 1929, *Ditman s.n.* (MARY); Dorchester Co., *Stibolt 70* (MARY); Montgomery Co., *Benedict 4846* (VPI); Prince Georges Co., *Dudley 12362* (CAN, NA); Washington Co., *Downs 5324* (NCU).—North Carolina: Avery Co., 7 Jun 1918, *Holmes s.n.* (NCU); Buncombe Co., *Palmer 39933* (A, MO); Burke Co., *Harbison 1041* (NCU); Dare Co., *Crutchfield 2848* (NCU); Henderson Co., *Pittillo 6531* (WCUH); Macon Co., 29 Aug 1937, *Coker s.n.* (NCU); Madison Co., *Thomas 21094* (NLU); Mecklenburg Co., 13 Apr 1970, *DeLany s.n.* (UNCC).—South Carolina: Greenville Co., *Elliott 51* (FUGR); Greenwood Co., 29 Jul 1972, *Ridlehoover s.n.* (FUGR); Pickens Co., *Hanna 58* (CLEMS).—Tennessee: Franklin Co., *Mazzeo 13032* (NA); Hamblen Co., *Thomas 12996* (NLU, TENN); Meigs Co., 7 Mar 1974, *DeSelm s.n.* (TENN).—Virginia: Bedford Co., *Elmore 25* (WILLI); Campbell Co., *Clemenson 40* (LYN); Fairfax Co., 5 Oct 1941, *Mason s.n.* (NA, VPI); Henry Co., *Heck 21* (LYN); Montgomery Co., *Massey 2406*

(NA, VPI); Warrick Co., 25 Mar 1978, *Ernsberger s.n.* (LYN).—West Virginia: Barbour Co., 18 Jun 1951, *Haller s.n.* (WVA); Hampshire Co., 26 May 1940, *Frye s.n.* (WVA); Hancock Co., *Bonar 1* (WVA); Harrison Co., 29 May 1947, *Fisher s.n.* (CM); Kanawha Co., 22 Mar 1969, *Denison s.n.* (WVA); Mineral Co., 4 Aug 1950, *Brown s.n.* (WVA); Monongalia Co., 20 Nov 1938, *Core s.n.* (WVA); Randolph Co., *Clarkson 14* (WVA); Upshur Co., 24 Mar 1953, *Rossbach s.n.* (WVW).

15b. Salix cinerea subsp. **oleifolia** (Smith) Macreight, *Man. Brit. bot.* 212. 1837. *Salix oleifolia* Smith, *Fl. brit.* 1065. 1804, non Villars, 1789.
Salix atrocinerea Brotero, *Fl. lusit.* 1: 31. 1804. *Salix cinerea* subsp. *atrocinerea* (Brotero) Guinier, *Atlas des Arbres* 9(29): 2. 1912.

Shrubs up to 12 m tall, branches and buds glabrate or almost so (fide Bean 1980: 267). Leaves tomentose beneath with a mixture of white and ferruginous hairs. Chromosome number 2n = 76.

Distribution. (Fig. 44b). A western European subspecies common in the British Isles (Bean 1908); rarely cultivated and naturalized in North America.

Salix cinerea subsp. *oleifolia* was first recognized in the flora of the Southeastern States on the basis of a widely distributed North Carolina collection identified as *S. caprea* (*Pittillo 4389*). The specimens had ferruginous hairs on the leaves and therefore could not have been *S. caprea,* and, although they resembled *S. discolor,* the locality was disjunct from the main range of the species. An examination of decorticated wood showed that even one-year-old wood bore long, distinct ridges, a condition that does not occur in *S. discolor.* The possibility that this specimen was introduced and naturalized was later confirmed by the collector (D. Pittillo, pers. comm.), who commented that it is now naturalized "downstream of Asheville at Marshall," but not outside of the Asheville-Hendersonville basin or on the French Broad River. It is now also known from Burke Co. (*Harbison 1178*) and may prove to be common in the region around the site of the original introduction.

Specimens Examined. U.S.A. North Carolina: Buncombe Co., *Palmer 35441* (A, MO), *Pittillo 4389* (AUA, CLEMS, FSU, GH, LYN, MO, NCU, NLU, PAC, TENN, UNA, UNCC, USF, VSC), *3353* (WCUH), 9 Jul 1939, *Totten s.n.* (NCU); Burke Co., *Harbison 1178* (NCU); Henderson Co., *Pittillo 4526* (WCUH), *Radford 7098* (NCU).

VII. Salix subgenus **Vetrix** section **Vimen** Dumortier, Bijdr. Natuurk. Wetensch. 1(1): 56. 1825.—Type: *Salix viminalis* Linnaeus.

Salix subgenus **Vetrix** section **Vimen** subsection **Arbusculae** Hayek, *Fl. Steiermark* 1: 162. 1908.—Type: *Salix arbuscula* Linnaeus.

16. Salix sericea Marshall, *Arbust. amer.* 140. 1785. *Salix grisea* Willdenow, *Sp. pl.* 4: 699. 1806. *Salix petiolaris* var. *grisea* (Willdenow) Torrey, *Fl. New York* 2: 207. 1847. *Salix petiolaris* [var.] ϵ *sericea* (Marshall) Andersson in DeCandolle, Prodr. 16(2): 235. 1868.—Type: unknown.—Silky willow.
? *Salix pensylvanica* Forbes, *Salict. woburn.* 189. 1829.—Type: unknown.
Salix sericea f. *glabra* E. J. Palmer & Steyermark, Ann. Missouri Bot. Gard. 25: 769. 1938.—Type: Missouri. Dent Co., "north prong (Hutchins Creek) of Meramec river, between Stone Hill and Indian Trail State Park," 4 Aug 1936, *Steyermark 12548* (holotype: MO!).

Shrubs 0.5–4 m tall. Branches reddish brown, densely gray- to brown-tomentose, becoming glabrate, but hairs often persistent at nodes; branchlets greenish to mottled yellow-brown, velutinous, brittle at base; buds brown, velutinous to puberulent or glabrate, floral buds navicular, vegetative buds flat adaxially, convex abaxially, strongly ribbed laterally, apex blunt. Leaf blades often drying dark, narrowly lanceolate to narrowly oblong, the larger ones (4.8–) 6.5–10 (–12.5) cm long, (0.7–) 1.3–2.5 cm wide, length/width (3.5–) 4–5–6 (–11), apex acute to short acuminate, base cuneate rarely obtuse, margin glandular-serrate to -serrulate; immature leaf blades reddish green or yellowish green, sparsely villous to glabrate above, densely short-sericeous beneath with white hairs intermixed with ferruginous hairs; mature leaf blades sparsely pubescent to glabrate and dark green above, densely silvery short-sericeous and glaucous beneath, sometimes becoming glabrate, midrib velutinous, hypostomatous; petioles yellowish, 3.5–12 (–21) mm long, velutinous, sometimes dark, glandular dots at base of blade; stipules usually small, glandular lobes or apparently absent, on vigorous shoots oblong or half-ovate, 1.2–4 mm long, margin glandular. Aments precocious or subprecocious, sessile, with 2 or 3 leafy bracts at base. Staminate aments 1.5–3.5 cm long; stamens 2, filaments pilose at base, rarely glabrous, dry anthers yellow or purplish, 0.4–0.6 mm long; nectary 1, adaxial, 0.8 mm long; bracts dark brown or blackish, often pale at base, 0.8–1.5 mm long, ovate, apex rounded, with long, straight or wavy hairs. Pistillate aments densely flowered, 2–4.5 cm long; pistils tawny to greenish brown, ovate, unbeaked, short-sericeous, capsules tawny, 2.5–4 mm long, styles 0.2–0.4 mm long, may be obscured by indumentum, stigmas 2-lobed, stipes 0.6–1.5 mm long; nectary 1, adaxial, oblong, ⅓ to ½ as long as stipe; ovules 3 per carpel; bracts as in staminate flowers. Chromosome number unknown. Fig. 45.

Phenology. Flowering from early March to mid-May with a mean in mid-April; based on 94 records.

Distribution. (Fig. 45-inset, 46). *Salix sericea* is a northeastern species whose range extends into the Southeastern States along the Appalachian System and reaches as far south as northeastern Georgia. It occurs on the Coastal Plain from Delaware to North Carolina with an apparently disjunct locality in South Carolina.

This species forms thickets in marshes, sedge meadows, acid bogs, on rocky, silty, or sandy streambanks, and on open seepages. It is a pioneer species on floodplains along with *S. exigua* (Core 1966). In the Ozarks of southeastern Missouri it occurs in calcareous, swampy, spring-fed meadows in association with northern herbaceous species. Steyermark (1963) thought that the Missouri populations may be Pleistocene relicts; this explanation may also apply to the Arkansas population that was collected near Mammoth Springs, Fulton Co. (*Demaree 25872*).

Salix sericea is rare in Alabama where it is known from one collection made in Colbert Co., in 1972. It is listed by Kartesz and Kartesz (1977) as a species of special concern in the state, and it was reported by Clark (1971) to be rare. In Arkansas it was collected only once in Fulton Co., in 1947. It is listed by Kartesz and Kartesz (1977) as rare and endangered in Georgia, where it is restricted to the Blue Ridge Province (Duncan & Kartesz 1981). I have seen specimens from five counties, but in only one of them was it collected more than once: Fannin in

FIG. 45. *Salix sericea*. a. Vegetative branchlet (×0.75). b. Staminate flower (×10). c. Branch with young pistillate aments and a young vegetative branchlet (×1). d. Pistillate ament in late flower (×1). e. Young pistillate flower (×10). Inset: map of total distribution, based on Argus (1964), Little (1977), Mohlenbrock (1980), Steyermark (1963), and herbarium specimens.

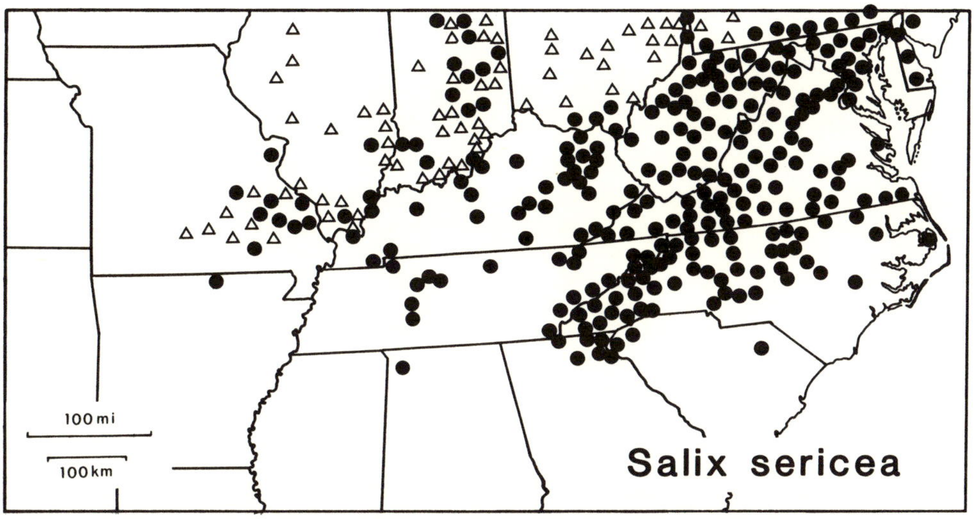

FIG. 46. Distribution of *Salix sericea* in the Southeastern and peripheral states. Solid dots, herbarium specimens; open triangles, literature reports.

1946, Habersham in 1931, Stephens in 1978, Towns in 1966, and Rabun (12 collections, the most recent in 1953). In South Carolina it is known only from three counties, but the collections are recent and the populations, at least in the Blue Ridge Province, may be secure.

REPRESENTATIVE SPECIMENS. U.S.A. ALABAMA: Colbert Co., *Kral 46734* (VDB).—ARKANSAS: Fulton Co., *Demaree 25872* (NA).—DELAWARE: Kent Co., 10 May 1824, *Commons s.n.* (PH); New Castle Co., *Tatnall 2142* (DOV); Sussex Co., 25 Apr 1957, *Buker s.n.* (CM).—DISTRICT OF COLUMBIA: *Ball 106* (CAN, GH, MO, NA).—GEORGIA: Fannin Co., *Duncan 7340* (GA); Habersham Co., *Eyles 998* (NA); Lumpkin Co., *Kral 57580* (VDB); Rabun Co., *Hardin 16480* (GA), *Miller 192* (LSU, PH, US); Stephens Co., 1 Apr 1978, *Graves s.n.* (GA); Towns Co., 8 May 1966, *Park s.n.* (NCU).—KENTUCKY: Bath Co., *Wharton 5909* (NA); Bell Co., *Reed 8393* (CFR); Breathit Co., *Lassetter 2404A* (EKY); Calloway Co., *McFarland 52* (NA); Carter Co., *McFarland 3337* (NA); Casey Co., *Murphy 5135* (KY); Elliot Co., *Reed 18394* (CFR); Fleming Co., *McInteer 1217* (KY); Franklin Co., *McFarland 72* (NA); Hardin Co., *Cranfill 127* (KY); Harlan Co., *McFarland 17* (DUKE, GH, MO, MUHW, NA, NY, PH, TENN, WVA); Hart Co., *Wharton 49* (KY, NA); Jefferson Co., *Davies 203* (DHL); Letcher Co., *Sale 677* (EKY); Lewis Co., *Braun 323* (US); Lincoln Co., 14 Aug 1974, *Meijer s.n.* (KY); Madison Co., *Lassetter 2502* (EKY, NLU); McCracken Co. *McFarland 33* (MO, NA); McLean Co., *Palmer 17703* (A, MO, US); Meade Co., *Haskinson 150* (DHL); Menifee Co., *Higgins 1178* (DHL, NCU); Morgan Co., 7 May 1950, *Perkins s.n.* (CFR); Pike Co., *Braun 3819* (A, US); Powell Co., *McFarland 3323* (NA); Rockcastle Co., *McFarland 3590* (NA); Rowan Co., *Netherly 32* (CFR); Trigg Co., *Ellis 1640* (APSC); Wayne Co., *Braun 2850* (A, US); Wolfe Co., *Higgins 1073* (NCU).—MARYLAND: Allegany Co., *Ball 3024* (CAN, MARY, NA, NY, US); Ann Arundel Co., 27 Apr 1941, *Waite s.n.* (NA, PH); Baltimore Co., *Norton 3013* (MARY); Carroll Co., *Reed 3238* (CFR); Cecil Co., *Long 2552* (MOAR, PH); Charles Co., *Reed 27758* (CFR); Frederick Co., *Balton 528* (MARY); Garrett Co., *Ball 3012* (CAN, MARY, NA, NY, PH, US); Harford Co., *Reed 3554* (CFR); Howard Co., Apr 1966, *Engh s.n.* (MARY); Montgomery Co., *Maxon 6083* (NA); Prince Georges Co., 3 May 1967, *Eaton s.n.* (NCU); Washington Co., 16 Aug 1907, *Norton s.n.* (MARY).—NORTH CAROLINA: Alamance Co., *Radford 9669* (NCU); Alexander Co., *Keever 223* (DUKE); Alleghany Co., *Correll 14631* (DUKE, FSU, NA); Ashe Co., *Radford 32887* (FSU, TENN); Avery Co., *Rohrer 1759* (NCU); Bertie Co., *Ahles 52002* (NCU); Buncombe Co., *Palmer 35424* (A, MO, NY,

US); Burke Co., *Smith 436* (NCSC); Cabarrus Co., *Ahles 15947* (NY); Caswell Co., *Fogg 5435* (GH, MOAR, PH); Chatham Co., *Ahles 53182* (CM, FSU, GH, NCU, NY, USF, VDB, WVA); Clay Co., *Solomon 1824* (CAN); Dare Co., 14 Jul 1959, *Burke s.n.* (NCU); Davidson Co., 9 Jun 1915, *Totten s.n.* (NCU); Durham Co., *Blomquist 2925* (DUKE, PH); Forsyth Co., *Correll 14457* (DUKE); Graham Co., *Radford 11822* (NCU); Granville Co., *Oosting 3458* (DUKE); Haywood Co., *Bozeman 9111* (AUA, DOV, NO, NCU, PAC, WIS); Henderson Co., *Leonard 2378* (DHL, DUKE, FUGR, LYN, MARY, MISS, NCU, UNA, UNCC, VSC, WILLI); Iredell Co., *Palmer 39944* (A, MO, NY); Jackson Co., *Palmer 42482* (A, PH); Johnston Co., *Ahles 10055* (NCU); Lee Co., *Argus 11383* (CAN); Lenoir Co., *Burk 28* (NCU); Macon Co., *Godfrey 59977* (FSU, NA); Madison Co., 9 Jul 1946, *Radford s.n.* (VPI); McDowell Co., *Fox 2168* (GH, NA, NCSC); Mecklenburg Co., 13 May 1916, *Coker s.n.* (NCU); Mitchell Co., 18 Apr 1960, *Harrison s.n.* (NCSC); Montgomery Co., *Radford 43443* (MUR, NCU); Northampton Co., *Carroll 371* (NCU); Orange Co., *Ahles 53227* (CM, FSU, GH, NY, NCU, PAC, TENN, UNA, USF, WVA); Person Co., *Bell 14618* (NCU, USF); Polk Co., *Fosberg 18939* (NA, PH); Randolph Co., *Radford 8867* (NCU); Rowan Co., *Wickland 1531* (NCU); Stanley Co., 30 Jul 1917, *Holmes s.n.* (NCU); Stokes Co., *McCurdy 333* (NCU); Surry Co., 14 Aug 1937, *Totten s.n.* (NCU); Swain Co., *Argus 10862* (CAN); Transylvania Co., *Ware 3525* (NCU, WILLI); Wake Co., *Fox 1326* (GH, NA, NCSC, TENN); Warren Co., *Argus 11378* (CAN); Watauga Co., *Kologiski 160* (NCSC, UNCC); Wilkes Co. *Downs 12704* (VDB); Yancy Co., *DuMond 921* (LTU, NCSC).—South Carolina: Darlington Co., *Williamson 1412* (NCU); Oconee Co., *Rodgers 73053* (CITA, FUGR, NLU, WILLI); Pickens Co., *Kral 61556* (VDB).—Tennessee: Blount Co., *Thomas 57899* (NLU); Campbell Co., *Morton 47673* (NA, NCU, NY, TENN, VPI, WIS); Carter Co., *Palmer 42559* (A, MO, NY); Cheatham Co., *Jones 359* (VDB); Claiborne Co., *Sharp 26984* (TENN); Cocke Co., *Thomas 73584* (NLU); Davidson Co., *Hyman 11* (VDB); Dickson Co., *Svenson 10320* (GH); Hickman Co., *Kral 34102* (GH, NCU, TENN, VDB); Jackson Co., *Jennison 3880* (TENN); Johnson Co., *Gonsoulin 3610* (VDB, WVA); Lewis Co., *Kral 36375* (FSU, VDB); Monroe Co., *Malter 53932* (TENN); Polk Co., *Sharp 45621* (NA, NCU, TENN, VDB); Sevier Co., *Ball 2713* (MARY, MUHW, NA, TENN, WVA); Stewart Co., *Stack 125* (APSC); Unicol Co., *Lyle 19266* (TENN); Union Co., 23 Jun 1936, *Kalter s.n.* (TENN); Washington Co., *James 16880* (TENN).—Virginia: Albemarle Co., *Ball 2741* (DUKE, NA, NCSC, UNA, VPI); Amelia Co., *Lewis 366* (NA, VPI); Amherst Co., *James 12233* (NCU); Appomattox Co., *Reed 83605* (CFR); Arlington Co., *Blake 10560* (GH); Augusta Co., *Wood 4195* (CM); Bath Co., *Johnson 4221* (NCU); Bedford Co., *Freer 1694* (GH, LYN, VA, WILLI); Bland Co., *James 8429* (NCU); Botetourt Co., *Freer 1693* (LYN, NCU, NLU); Brunswick Co., *James 13282* (NCU); Buchanan Co., 18 Apr 1945, *Ramey s.n.* (VPI); Buckingham Co., *Little 14604* (US, VPI); Campbell Co., *White 9* (LYN); Carroll Co., *James 8140* (NCU); Chesterfield Co., 22 Apr 1915, *Churchill s.n.* (NCSC); Craig Co., *James 7768* (NCU); Culpeper Co., *Reed 99239* (CFR); Dickenson Co., *James 9871* (NCU); Dinwiddie Co., *Fernald 10607* (GH, PH); Fairfax Co., *Erlanson 53* (NA); Faquier Co., *Allard 9760* (CM, GH, MO, NA, NY, TENN); Floyd Co., *Argus 10947* (CAN); Fluvanna Co., *Diggs 1352* (WILLI); Franklin Co., *James 10775* (NCU); Frederick Co., *James 10878* (NCU); Giles Co., *Fogg 16991* (PH, VA); Grayson Co., *James 9264* (NCU); Greene Co., *James 12052* (NCU); Halifax Co., *Argus 10960* (CAN); Henrico Co., *Fernald 6977* (GH, MO, PH); Henry Co., *James 10506* (NCU); Highland Co., *James 7689* (NCU); Lee Co., *James 9733* (VPI); Lunenburg Co., *James 3068* (NCU); Madison Co., *Fosberg 16916* (PAC); Mecklenburg Co., *Seaman 3624* (NCU); Middlesex Co., *Shull 419* (US); Montgomery Co., *James 10054* (NCU); Nansemond Co., *Ahles 58300* (NCU); Nelson Co., *Freer 1879* (GH, LYN, NCU); Norfolk Co., *Fernald 3889* (GH, PH); Patrick Co., *Kral 9842* (FSU, NCSC, NCU, VPI); Pittsylvania Co., *Wherry 2757* (A, GH); Prince William Co., *Allard 9816* (GH, VPI); Pulaski Co., *Reed 98077* (CFR); Rappahannock Co., 26 Sep 1978, *Frey s.n.* (LFCC); Roanoke Co., *Argus 10949* (CAN); Rockbridge Co., 29 Apr 1915, *Churchill s.n.* (GH, MO, WVA); Rockingham Co., *Ball 2705* (MUHW, NA, TENN, US, WVA); Scott Co., *Peake 793* (WILLI); Shenandoah Co., *Artz 1201* (GH, NA, VPI); Smythe Co., *James 8959* (NCU); Southampton Co., *Fernald 6580* (PH); Washington Co., *Nicely 831* (VPI, WVA); Wythe Co., *Ball 2746* (see text).—West Virginia: Barbour Co., 13 Jun 1951, *Haller s.n.* (WVA); Braxton Co., *Gilbert 1001* (MUHW, NA, WVA); Cabell Co., *Williams 674* (CLEMS, CM, DUKE, GH, MO, MTL, MUHW, NA, NY, PH, TENN, US, WIS, WVA); Calhoun Co., 29 Apr 1933, *Harris s.n.* (MUHW); Clay Co., *Rossbach 875* (WVA, WVW); Doddridge Co., *Bartholomew 266* (WVA); Fayette Co., 30 Mar 1968, *Phillips s.n.* (WVA); Gilmer Co., 31 Jul 1965, *Bartholomew s.n.* (WVA); Grant Co., *Allard 10201* (CM, NA); Greenbrier Co., *Martin 1013* (DUKE, MUHW, NA); Hampshire Co., 11 May 1940, *Frye s.n.* (WVA); Hancock Co., Jul 1938, *Sumptine s.n.*

(WVA); Hardy Co., *Allard 9391* (US, WVA); Harrison Co., 17 Sep 1965, *Stewart* (WVW); Jackson Co., *Bartholomew 118* (WVA); Kanawha Co., 25 Jun 1966, *Denison s.n.* (WVA); Logan Co., 24 Aug 1954, *Ross s.n.* (WVA); Marion Co., 29 Jul 1930, *Miller s.n.* (WVA); Mineral Co., *Downs 10670* (LYN, UNCC); Monongalia Co., *Millspaugh 89* (NY, WVA); Monroe Co., 1 Aug 1940, *Fox s.n.* (WVA); Morgan Co., *Downs 10196* (NCU); Nicholas Co., 9 Aug 1940, *Fox s.n.* (WVA); Ohio Co., 19 Apr 1877, *Guttenberg s.n.* (CM); Pendleton Co., *Rossbach 4117* (USF, WVA, WVW); Pocahontas Co., *Moore 2398* (GH, PH); Preston Co., *Downs 6837* (NCU); Raleigh Co., *Tosh 920* (US, WVA); Randolph Co., *Rossbach 2500* (WVA, WVW); Ritchie Co., *Bartholomew 49* (CM, WVA); Summers Co., *Boone 271* (WVA); Taylor Co., 21 Jul 1959, *Constable s.n.* (TENN, WVA); Tucker Co., *Allard 11462* (PH, US, WVA); Upshur Co., 9 Apr 1953, *Rossbach s.n.* (WVW); Webster Co., *Argus 10831* (CAN); Wetzel Co., Nov 1940, *Dillaman s.n.* (WVA); Wirt Co., *Bartholomew 1676* (CM, WVA); Wyoming Co., Sep 1964, *Evans s.n.* (NLU, WVA).

Recognition. Salix sericea is a subprecocious willow with sessile aments that emerge just before the leaves appear. Its leaves are conspicuously sericeous beneath with short, appressed hairs; the upper leaf surface is glabrate and dark green, often drying blackish.

Immature flowering specimens of *S. sericea* resemble *S. humilis* s.l. They can be recognized by their very white pistil indumentum and obscure styles. Their leaves begin to emerge very soon after the aments, and even at this stage of development their glandular-serrate margins are recognizable (see also *S. humilis,* no. 13).

The mature foliage of *S. sericea* can sometimes be confused with that of *S. eriocephala. Salix sericea* differs from *S. eriocephala* in its brittle branches, which are velutinous, especially at the nodes between the bud and the branchlet, and by buds that lack the loose inner membrane characteristic of *S. eriocephala.* Its leaves are typically sericeous beneath, have a cuneate base, and usually lack stipules. Both species have pine cone galls produced by gall midges.

Salix sericea is morphologically similar to *S. petiolaris,* and in the past the two have been variously treated as varieties or subspecies. There are sufficient differences between these taxa, however, especially in floral characters, to warrant specific rank. Problems in recognition occur not because the species are confluent but because of seasonal variations in the foliage of both species. The leaves of *S. petiolaris,* as they unfold from the bud, are densely sericeous on both sides, and this indumentum may be partially retained in some plants. Such specimens have been confused with *S. sericea* or with putative hybrids of the two (see below). On the other hand, the mature foliage of *S. sericea* may sometimes become completely glabrous, and such specimens are difficult to separate from *S. petiolaris.* Fortunately, completely glabrous leaves are infrequent in *S. sericea;* I have seen only six such specimens in this study: GEORGIA: Rabun Co., *Ball 2602* (GH, NA, NY, US); NORTH CAROLINA: Iredell Co., *Ahles 51981* (USF); Wake Co., *Pultorak 527* (NCSC); VIRGINIA: Floyd Co., *Morf 12* (VPI); Wythe Co., *Ball 2746* (CAN, DUKE, GH, MARY, MO, MUHW, NA, NCSC, NY, UNA, WVA). The two species can be separated by the presence of ferruginous hairs in *S. petiolaris* and a tendency in that species to have subentire or irregularly serrate leaf margins and narrower leaves. These characters, however, are variable and some specimens cannot be positively identified.

The greatest difficulty in this regard is posed by two collections made by C. R. Ball. One of them (*Ball 2746*) was collected in Wythe Co., Virginia (grassy margin of Reed Cr., 6.8 mi E of Wytheville on Highway 11); duplicates were distributed to more than 13 herbaria. This collection has formed the basis of reports and entries on distribution maps of *S. petiolaris* in Virginia (Massey & Ball 1944; Little 1976a, 1977; and probably others). A second collection (*Ball 2602*) was made in Rabun Co., Georgia (Mountain City near Clayton), but was not as widely distributed. Both specimens are vegetative and consist of mature, glabrate leaves and young unfolding leaves that are densely sericeous above and glabrous beneath. *Salix sericea* has since been collected near Clayton, Georgia, but no other specimens that resemble *S. petiolaris* have been collected there. I searched unsuccessfully for *Salix* at Ball's Virginia locality; further exploration is necessary. I have identified these and similar problematical specimens as *S. sericea,* because they are located far outside of the range of *S. petiolaris* and because positively identified flowering or fruiting material of *S. petiolaris* is not known from the Southeastern States.

Salix sericea and *S. petiolaris* not only are morphologically similar but occupy the same ecological niche; however, just how closely related they are is uncertain. Schneider (1920), Fernald (1950), and others, placed them in sect. *Griseae,* but Dorn (1976) placed them in different sections. He included *S. sericea* with *S. arbusculoides* Andersson and *S. argyrocarpa* Andersson in sect. *Vimen,* subsect. *Arbusculae;* and *S. petiolaris* with *S. bebbiana, S. geyeriana* Andersson, and *S. lemmonii* Bebb in sect. *Vetrix,* subsect. *Substriatae.* I have no evidence to support or refute either hypothesis, but I am inclined to follow Dorn's disposition. A comparative study of these species where they are sympatric would help to solve this problem.

Nomenclature. There seems to be no type material of *S. sericea* in existence. According to Ewan (1967), Marshall did not preserve specimens, and a search of the Darlington Herbarium (DWC), where some specimens from his garden are located, and at the Philadelphia Academy of Natural Sciences Herbarium did not turn up any Marshall specimens of *Salix.*

The plant was described by Marshall (1785: 207) as one that "generally rises to the height of eight to ten feet, with many shrubby stalks covered with pretty smooth, dark, greenish bark. The leaves are shorter and somewhat broader than the other kind [*S. nigra*], lance-shaped, silky underneath, and very slightly serrated on the edges." This description fits *S. sericea* as it is commonly understood very well. There is no problem identifying Marshall's species, but the name is not without nomenclatural problems.

The North American *S. sericea* (1785) was treated incorrectly by Willdenow in *Species Plantarum* (1806) as a later homonym of *S. sericea* Villars (1786), only to be reinstated by Carey in 1848 (Schneider 1920). Now Edward G. Voss (pers. comm.) has called to my attention another possible earlier homonym of *S. sericea.*

Salix sericea Marshall was published in *Arbustum Americanum* in 1785; this work was advertised for sale in two newspapers by the publisher on 24 December 1785, on which date he also sent an invoice for printing to the author (Ewan 1967). The new, possibly earlier homonym, *S. sericea* Chaix, was also published in 1785 in *Plantae Vapincenses* (Perret & Burdet 1981). One copy of this volume bears a dedication dated 24 December 1785, inscribed by the author to Gaude, Curé d'Oze. In the past Chaix's *Plantae Vapincenses* was generally ignored by botanists (Perret & Burdet 1981), and his names were either attributed to Villars, as was the case with *S. sericea* and *S. virescens* (cf. *Index Kewensis,* Pt. 4, 1885), or to Chaix in Villars, as was done with *S. dubia* and *S. spadicea* (cf. *Index Kewensis,* Pt. 4, 1885, and Supplement 7. 1929). The exact date of effective publication, that is, the date on which the work was distributed to the general public "through sale, exchange, or gift," as required to establish priority by the International Code of Botanical Nomenclature (Voss 1983, Art. 29), cannot be definitely established for either of these floras. It is probable that *Arbustum Americanum* was available to the public by the date that it was advertised, but there is no way of knowing when *Plantae Vapincenses* was available. It could either have been prior to the date of dedication, as was assumed by Perret and Burdet, or, if the dedication copy had been sent to Gaude before its general distribution, it would not have been available to the "general public" until after that date. A similar argument, however, could also be made for *Arbustum Americanum,* which emphasizes the difficulty in establishing a date of effective publication for these two works. There are other considerations that must come into play in solving this question of nomenclatural priority.

In 1786 *S. sericea* appeared in Villars's *Histoire de plantes de dauphine,* and the name, as it applies to a European species, has been traditionally, but apparently incorrectly, attributed to Villars. As a presumed later homonym of *S. sericea* Marshall the Villars name was replaced by *S. glaucosericea* Floderus (1943), a species related to *S. glauca* Linnaeus (Rechinger 1964) or a synonym of it (Skvortsov 1968). If *S. sericea* Chaix is proved to be earlier than *S. sericea* Marshall, two name changes would result: *S. sericea* Chaix would replace *S. glaucosericea* Floderus and a new name would be required for *S. sericea* Marshall.

In the interest of nomenclatural stability and in view of the impossibility of definitely establishing dates of effective publication for these names, I recommend that we continue to regard the Marshall name as having priority. If, however, the name *S. sericea* Marshall is to be replaced, the next available name is *S. grisea* Willdenow.

Salix grisea was proposed by Willdenow in 1806 as a nomen novum for the North American species after he had made the double "error" of attributing Chaix's *S. sericea* (1785) to Villars (1786) and Marshall's *S. sericea* (1785) to Muhlenberg (1803), thus incorrectly establishing priority for the European name. Willdenow's authority was such that his opinion was followed for the next 45 years by most authors, including Pursh (1814), Koch (1828), and Barratt (1840); it was not until 1848 that J. Carey reestablished Marshall's name. From that time on, with a minor lapsus when Andersson (1858a) used *S. grisea* Marshall (corrected in 1858b), the name *S. sericea* Marshall has been in use for the North American species. It would be regrettable to replace this name without conclusively proving the priority of *S. sericea* Chaix.

Salix sericea f. *glabra* E. J. Palmer & Steyermark is of no taxonomic significance, even as a forma. Glabrescent leaves in *S. sericea* are a reflection of stage of development and not of genetic variation. Late in the season the older leaves on some plants may become glabrate, while younger leaves on the same plant are still sericeous. The holotype is not even conspicuously glabrate, although the paratype does have some glabrate leaves.

Hybridization. Salix sericea is said to hybridize with *S. eriocephala* (as *S. cordata* Muhlenberg and *S. rigida*), *S. humilis* var. *humilis,* and *S. petiolaris* (as *S.*

gracilis) (Schneider 1921b; Fernald 1950). Only *S. eriocephala* × *sericea* is recognized by me in the Southeastern States (see *S. eriocephala,* no. 17). Hybrids with *S. humilis* var. *humilis* and *S. petiolaris* have not been recognized by me, and specimens so identified may be in error (see *S. humilis,* no. 13).

The binomial *S.* ×*subsericea* (Andersson) C. Schneider (*Ill. Handb. Laubholzk.* 65. 1904), non Doell, 1858, is sometimes applied to *S. petiolaris* × *sericea* (Schneider 1921b). I have not seen the "type" of Andersson's (1868) name, *S. petiolaris* var. *subsericea,* as reported by Forbes (1909), but all "*subsericea*" specimens that I have seen are merely sericeous-leaved plants of *S. petiolaris.* This variant occurs throughout the range of *S. petiolaris,* even well outside of the range of *S. sericea,* and is almost certainly not due to hybridization.

VIII. Salix subgenus **Vetrix** section **Cordatae** J. Barratt ex Hooker, *Fl. bor.-amer.* 2: 149. 1838.—Type: *Salix eriocephala* Michaux.

Salix subgenus **Vetrix** section **Cordatae** subsection **Lutea** Dorn, Canad. J. Bot. 54: 2782. 1976.—Type: *Salix lutea* Nuttall.

17. Salix eriocephala Michaux, *Fl. bor.-amer.* 2: 225. 1803. *Salix discolor* [var.] *S. eriocephala* (Michaux) Andersson, *Monogr. Salicum* 85. 1867.—Type: Illinois. "in regione Illinoensi," 1795, *Michaux s.n.* (lectotype and isolectotypes, designated by Argus, 1980: P–herb. Michaux!).—Heart-leaved willow.

Salix cordata Muhlenberg, Ges. Naturf. Freunde Berlin II. 4: 236. 1803, non Michaux, 1803. *Salix myricoides* var. *cordata* (Muhlenberg) Dippel, *Handb. Laubholzk.* 2: 134. 1891.—Type: Pennsylvania. Lancaster, *Muhlenberg 569* (holotype: PH!).

Salix rigida Muhlenberg, Ges. Naturf. Freunde Berlin II. 4: 237. 1803. *Salix cordata* var. *rigida* (Muhlenberg) J. Carey in A. Gray, *Manual* 427. 1848. *Salix cordata* var. *rigida* (Muhlenberg) J. Boivin, Naturaliste Canad. 93: 436. 1966.—Type: Pennsylvania. Lancaster, *Muhlenberg 570* (holotype: PH!).

Salix angustata Pursh, *Fl. Amer. sept.* 1: 613. 1814. *Salix cordata* [var.] 2. *Salix angustata* (Pursh) Andersson, *Monogr. Salicum* 159. 1867. *Salix rigida* var. *angustata* (Pursh) Fernald, Rhodora 48: 38. 1946.—Type: unknown.

Salix torreyana J. Barratt, *Salices amer.,* Sp. 29. 1840.—Type: Connecticut. Middletown, *J. Barratt s.n.* (holotype: NY!).

Salix cordata 1. [f.] d. *vestita* Andersson, *Monogr. Salicum* 159. 1867. *Salix missouriensis* Bebb, Gard. & Forest 8: 373. 1895. *S. cordata* var. *vestita* (Andersson) Bebb, Gard. & Forest 8: 373. 1895, pro syn. *Salix cordata* var. *missouriensis* (Bebb) Mackenzie & Bush, *Man. fl. Jackson County* 69. 1902. *Salix rigida* var. *vestita* (Andersson) C. Ball in Gleason, *Illustr. fl. northeastern U.S.* 2: 10. 1952.—Type: Missouri. Fort Osage, 1833, *Prince Wied-Neuwied s.n.* (holotype: BR!).

Salix acutidens Rydberg in Britton, *Man. fl. n. states.* 315. 1901.—Type: Pennsylvania. Lancaster Co., Dillerville Swamp, 10 July 1901, *Heller 4917* (lectotype, here designated: NY!).

Salix cordata f. *mollis* E. J. Palmer & Steyermark, Ann. Missouri Bot. Gard. 25: 770. 1938.—TYPE: MISSOURI. Dent Co., spring branch of Twin Springs, 4 Aug 1936, *Steyermark 12491* (holotype: MO!).

? *Salix cordata* f. *subintegra* E. J. Palmer & Steyermark, Ann. Missouri Bot. Gard. 25: 770. 1938.—TYPE: MISSOURI. Phelps Co., lake at Yancy Mills Spring, 28 Nov 1936, *Steyermark 20910* (holotype: MO, not located).

Salix cordata var. *abrasa* Fernald, Rhodora 48: 34. 1946.—TYPE: QUÉBEC. Gaspé Co., R. Ste. Annes des Monts, 14–17 July 1906, *Fernald & Collins 484* (holotype: GH!; isotype: CAN!).

Shrubs 0.2–6 m tall. Branches reddish brown, glabrate; branchlets yellowish brown, velutinous or tomentose, becoming villous or glabrate, tenacious or sometimes brittle at base; buds yellowish brown, pubescent or thinly tomentose to glabrate, obclavate, beak flattened, apex retuse, inner membrane loose. Leaf blades commonly narrowly oblong, but varying from elliptic to very narrowly oblong, the larger ones (5.8–) 7–10.5 (–13.5) cm long, (0.9–) 1.5–3.6 cm wide, length/width 2.5–5, apex acute to subacuminate, base cordate or rounded, cuneate in smaller leaves, margin glandular-serrate; immature leaf blades reddish or yellowish green, translucent, glabrous or sparsely villous above, sparsely to densely villous beneath, midrib velutinous or tomentose on both sides; mature leaf blades glabrous or glabrate, glaucous beneath, wax sometimes very thin, midrib yellowish, tomentose or glabrate, hypostomatous; petioles yellowish, 3–14 (–18) mm long, tomentose; stipules half-ovate, base half-cordate, apex rounded or acute, margin glandular-serrulate or glandular-denticulate, glabrous to sparsely villous, 5–15 mm long. Aments coetaneous, sessile or on flowering branchlets. Staminate aments 2–4 cm long, sessile, subtended by 2 or 3 small leaves; stamens 2, filaments glabrous, dry anthers 0.4–0.6 mm long; nectary 1, adaxial; bracts dark brown, paler near base, 0.8–1.6 mm long, oblong, apex rounded, with wavy hairs on both sides. Pistillate aments densely flowered, 2.5–4.5 cm long, borne on leafy branchlets 0.4–1.5 cm long; pistils greenish, sometimes suffused with red, pyriform, short-beaked, glabrous, capsules tawny, 3.5–7 mm long, styles 0.3–0.6 mm long, stigmas 2 short lobes 0.1–0.2 mm long, stipes 1.2–1.6 (–2.8) mm long; nectary 1, adaxial, oblong or obclavate, ¼ to ½ length of stipe; ovules 6–8 per carpel; bracts as in staminate flowers. Chromosome number 2n = 38. Fig. 47.

Phenology. Flowering in the South (Georgia to Florida and Louisiana) in about late March, in the North from early April to early June, with a mean in late April; based on 17 records.

Distribution. (Figs. 47-inset, 48). *Salix eriocephala* is a northern and eastern Great Plains species that enters the Southeastern States along the floodplain of the Mississippi River in Kentucky and Tennessee, the Blue Grass Basin of Kentucky, and the Cumberland Plateau, Appalachian Mountains, and the Piedmont Plateau in West Virginia, Maryland, and Virginia. It occurs at scattered localities on the Piedmont Plateau in Maryland and Virginia, and it sometimes extends onto the Coastal Plain in southern Virginia and Delaware. A remarkable disjunct population of *S. eriocephala* in Georgia, Florida, and Alabama occurs over several physiographic provinces.

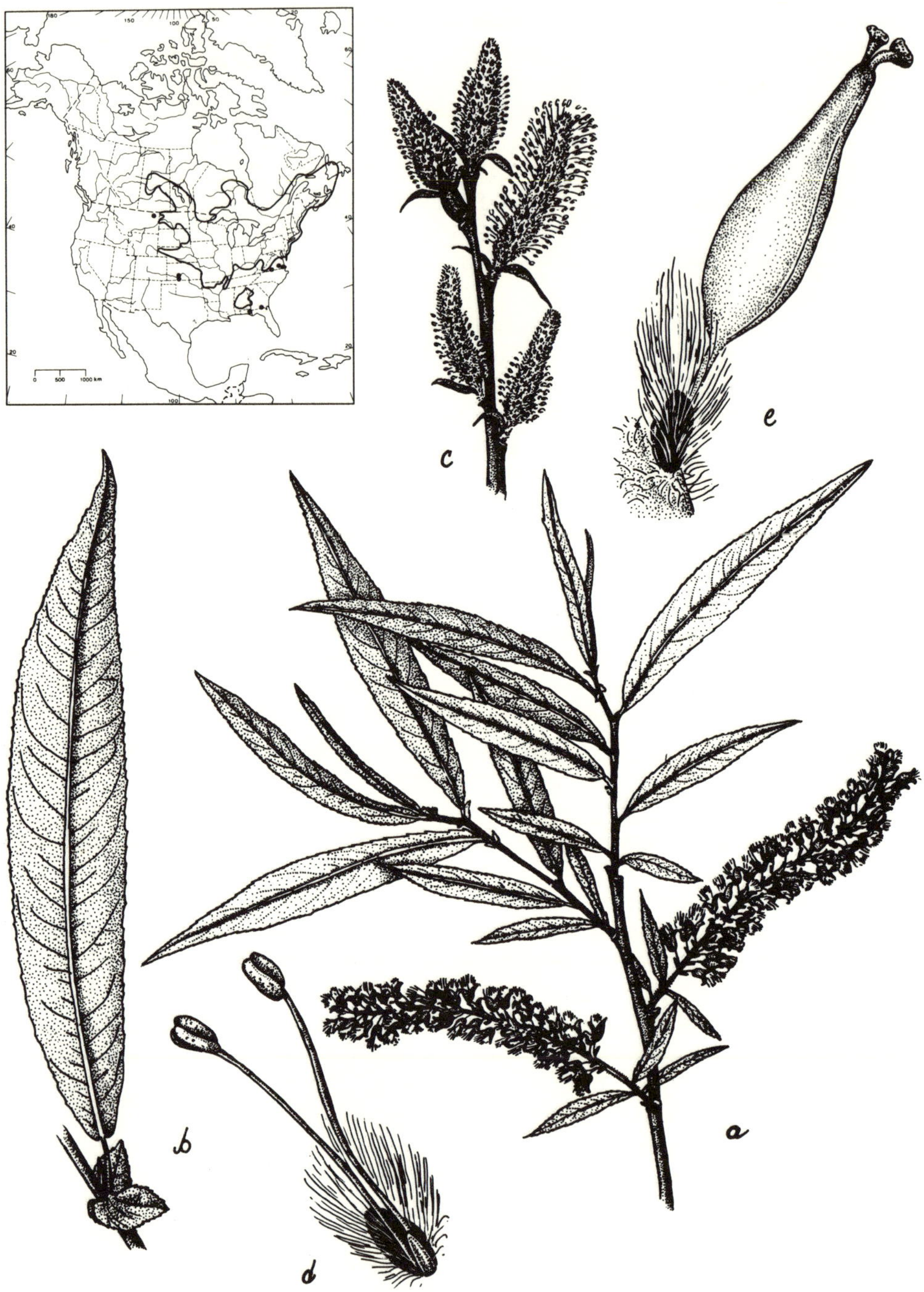

FIG. 47. *Salix eriocephala.* a. Branch bearing vegetative branchlets and two pistillate aments in fruit (×0.75). b. Mature leaf (×0.75). c. Branch with staminate aments in anthesis (×1). d. Staminate flower (×10). e. Young pistillate flower (×10). Inset: map of total distribution, based on Dorn (1970), McGregor and Barkley (1977), and herbarium specimens.

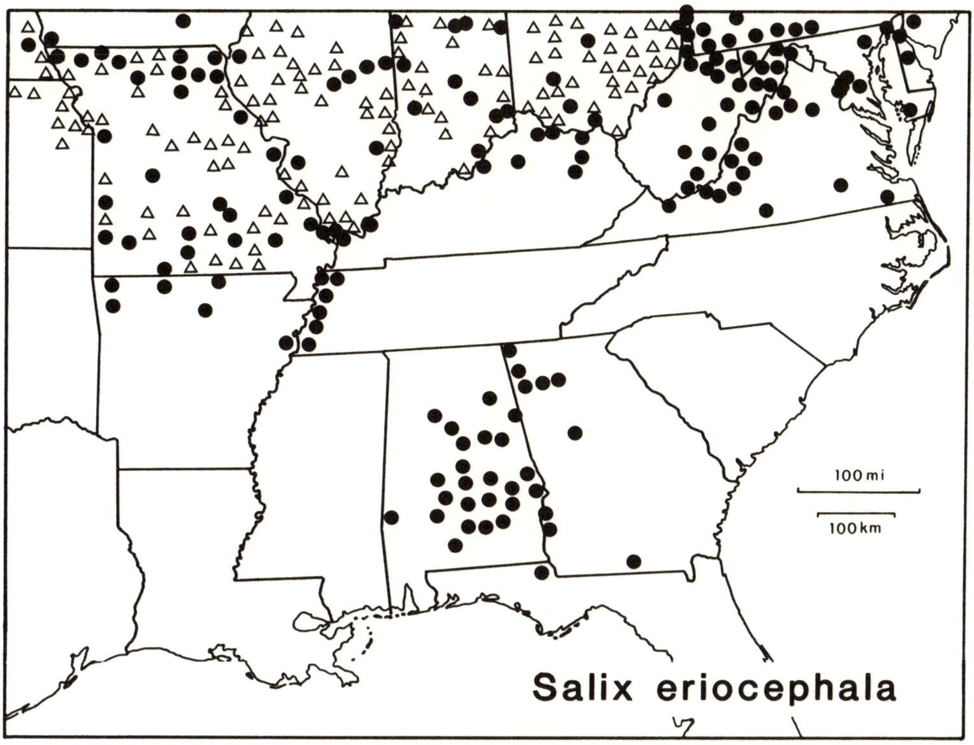

FIG. 48. Distribution of *Salix eriocephala* in the Southeastern and peripheral states. Solid dots, herbarium specimens; open triangles, literature reports.

The species occurs on gravelly or rocky river and stream banks, in marshy fields, and in mixed mesophytic woods on alluvium.

Salix eriocephala is rare in Florida, where it is known only from Jackson County. It was first collected there by R. K. Godfrey in 1981.

REPRESENTATIVE SPECIMENS. U.S.A. ALABAMA: Autauga Co., *Harper 3025* (GH, MO, NY, PH); Bullock Co., *Clark 16346* (NCU); Butler Co., *Clark 14597* (NCU); Chilton Co., *Clark 17619* (NCU); Choctaw Co., 14 May 1898, *Mohr s.n.* (US); Clay Co., *Clark 9547* (NCU); Cleburne Co., *Clark 2569* (NCU); Conecuh Co., *Clark 10282* (NCU); Crenshaw Co., *Clark 14776* (NCU); Dallas Co., *Clark 18149* (NCU); Elmore Co., *Clark 17107* (NCU); Etowah Co., *Clark 14999* (NCU); Jefferson Co., *Clark 15992* (NCU); Lee Co., 29 Jul 1959, *Carothers s.n.* (UNA); Lowndes Co., *Palmer 27217* (A, MO); Macon Co., *Clark 16614* (NCU); Montgomery Co., *Clark 16957* (NCU); Perry Co., *Clark 17978* (NCU); Pike Co., *Kral 299667* (GH, NCU, US, VDB); Russell Co., *Clark 16105* (NCU); Shelby Co., *Argus 11513* (CAN); Talladega Co., *Argus 11511* (CAN); Walker Co., *Clark 12559* (NCU); Wilcox Co., *Clark 14002* (NCU).—ARKANSAS: Benton Co., *Moore 39* (UARK, WIS); Boone Co., 1929, *Demaree s.n.* (UARK); Crittenden Co., *Wilcox 1505* (MEM); Fulton Co., *Demaree 25873* (NCSC); Stone Co., *Moore 632* (UARK); Washington Co., *Palmer 29651* (A, MO).—DELAWARE: Kent Co., 1 Oct 1893, *Tatnall s.n.* (DOV); New Castle Co., *Tatnall 2141* (DOV, PH).—DISTRICT OF COLUMBIA: *Ball 2594* (DUKE, GH, MO, NY).—FLORIDA: Jackson Co., *Godfrey 79419* (FLAS).—GEORGIA: Bartow Co., *Christensen 34* (FSU); Chatooga Co., *Mobbs 19* (GA); Cherokee Co., 18 Jun 1958, *Forman s.n.* (MOAR); Clay Co., *Thorne 3799* (GEO, NA); Dade Co., *Whetstone 29* (GA); Floyd Co., 10 Apr 1964, *Lipps s.n.* (GA); Lowndes Co., *Faircloth 8222* (VSC); Quitman Co., *Thorne 3147* (GA, GEO, NA); Spalding Co., 1 Sep 1925, *Woodroog s.n.* (GA).—KENTUCKY: Ballard Co., *McFarland 38* (MO,

NA); Bracken Co., *Thieret 52267* (KNK); Franklin Co., *McFarland 73* (NA); Jefferson Co., *McFarland 7* (MO, NA); Lewis Co., *Thieret 50633* (KNK); Menifee Co., *Braun 2744* (A, CFR); Pendleton Co., *Braun 4411* (A, CFR); Rowan Co., *Reed 16564* (CFR).—MARYLAND: Allegany Co., *Ball 3026* (CAN, DAO, NA, NY, PH); Cecil Co., *Tanger 3030* (PH); Garrett Co., *Downs 7937* (NCU); Harford Co., *Reed 26505* (CFR); Prince Georges Co., *Meyer 13697* (NA); Somerset Co., *Ball 3019* (PH).—TENNESSEE: Dyer Co., *Shanks 15837* (TENN); Lake Co., *Shanks 15898* (TENN); Lauderdale Co., *Sharp 7958* (NCU, TENN); Obion Co., *Eyles 8348* (DUKE, NA); Shelby Co., *Sharp 16334* (TENN); Tipton Co., *Warrington 172* (FUGR).—VIRGINIA: Alleghany Co., *James 7893* (NCU); Arlington Co., *Ball 2594* (DAO, MARY, MUHW, NA, NCSC, UNA, VPI, WVA); Bath Co., *James 8042* (NCU); Botetourt Co., *Crooks 450A* (VDB); Culpeper Co., *Allard 20829* (WVA); Dinwiddie Co., *Kiltz 187* (VPI); Fairfax Co., *Bradley 6359* (CM); Giles Co., *Kral 12185A* (FSU, VDB, VPI); Montgomery Co., *Ball 2745* (DUKE, MARY, MO, MUHW, NA, NCSC, NY, TENN, UNA, US, VPI, WVA); Nansemond Co., *Ahles 58299* (NCU); Page Co., *Argus 10974* (CAN); Pittsylvania Co., *Fauntleroy 3* (NCU); Roanoke Co., *Wood 6018* (GH, US); Rockbridge Co., *Gupton 3334* (NCU, UNCC); Rockingham Co., *Ball 2710* (DUKE, MARY, MUHW, NA, NCSC, TENN, UNA, US, VPI, WVA); Shenandoah Co., *Allard 9359B* (GH, NA, US, VPI, WVA); Tazewell Co., *Massey 2324* (VPI).—WEST VIRGINIA: Brooke Co., *Bartholomew 13* (WVA); Fayette Co., 13 Apr 1892, *Nuttall s.n.* (WVA); Grant Co., *Davis 6623* (NA, WVA); Greenbrier Co., 31 Aug 1940, *Fox s.n.* (WVA); Hampshire Co., *Ball 2604* (GH, MO, NA, NY, US, WVA); Hancock Co., *Bonar 2* (FLAS); Hardy Co., *Argus 10821* (CAN); Marion Co., *Bartholomew 181* (WVA); Marshall Co., *Bartholomew 137* (WVA); Mercer Co., 6 Sep 1940, *Fox s.n.* (NA, WVA); Mineral Co., Aug 1956, *Brown s.n.* (NA, NCU, WVA); Monongalia Co., 25 Apr 1971, *Six s.n.* (WVA); Morgan Co., *Downs 2821* (NCU); Ohio Co., *Guttenberg 257* (CM); Pendleton Co., *Reed 103571* (CFR); Preston Co., 15 Aug 1953, *Bartholomew s.n.* (WVA); Randolph Co., *Rossbach 4154* (WVA, WVW); Summers Co., *Boone 587* (WVA); Taylor Co., 8 Aug 1959, *Constable s.n.* (WVA); Tucker Co., *Allard 11577* (PH, US, WVA); Webster Co., *Argus 10832* (CAN); Wetzel Co., *Haught 7017* (NLU); Wirt Co., *Bartholomew 3206* (WVA).

Recognition. *Salix eriocephala* is characterized by leaves that are usually cordate or at least rounded at the base, by prominent, persistent stipules, and by glabrous pistils borne on distinct stipes and subtended by bracts clothed with wavy hairs.

The mature foliage of the species can be confused with that of *S. sericea, S. caroliniana,* or *S. nigra.* From *S. sericea* it can be distinguished by branchlets that are tenacious at the base and loosely tomentose becoming glabrate, and buds with a loose inner membrane. Its leaves are glabrous or glabrate beneath, cordate or rounded at the base and subtended by prominent stipules. *Salix eriocephala* can be easily distinguished from *S. caroliniana* and *S. nigra* by buds with a single, calyptrate bud scale, petioles that lack glands at the base of the leaf blade, and in rarely attaining tree stature.

Nomenclature. The identity and typification of *S. eriocephala* and its priority over *S. rigida* were discussed by Argus (1980).

Salix acutidens was described by Rydberg (in Britton) in October 1901. It was compared with *S. cordata* and *S. eriocephala* and thought to be the same as Torrey's *S. myricoides* (non Muhlenberg). No specimens were cited, but six sheets at the New York Botanical Garden were annotated by Rydberg as this species and constitute the type material. A lectotype was selected from this material (see above). Five of the specimens are *S. eriocephala* and one is *S. sericea.*

In 1938 Palmer and Steyermark published two forms of *S. cordata:* f. *mollis* and f. *subintegra.* The type of f. *mollis,* characterized by glabrous leaves, actually has mature leaves that are very sparsely tomentose beneath, especially on the veins; it does not differ significantly from plants typical of the species. The type of f. *subintegra* could not be located in the herbarium of the Missouri Botanical Garden (N. Morin, pers. comm.); the identity of this form, characterized by subentire leaves, cannot be determined with certainty without seeing the type.

Taxonomy. I have shown that the name *S. eriocephala* applies to the same plant as the two Muhlenberg names *S. cordata* and *S. rigida,* and that it has priority over them (Argus 1980). The close relationship of these taxa was recognized also in Bailey et al. (1976), where it was stated that *S. eriocephala* (including *S. missouriensis*) was "closely related to *S. rigida* and perhaps only a form of it." When Bebb (1895b) proposed the name *S. missouriensis* for the already occupied name *S. cordata* var. *vestita,* he did so not because he was confident of its specific rank, but because he felt that insufficient attention would be drawn to it if it were merely a variety. Some authors have treated *S. eriocephala* (including *S. missouriensis*) as distinct from *S. rigida* (Ball 1938; Fernald 1950; Steyermark 1963) on the basis of its tree habit and permanently villous branchlets and leaf veins. Specimens from western Tennessee, where *S. eriocephala* in the sense of these authors is to be expected, are not significantly different from the species elsewhere.

The disjunct occurrence of *S. eriocephala* in Georgia and Alabama was recognized by Thorne (1949, 1951), but the Florida occurrence was not discovered until 1981 [*Godfrey 79125* (FLAS)]. The magnitude of the disjunction, however, was previously not appreciated, because most specimens of the species were misidentified as *S. caroliniana* or *S. nigra.* The distributional gap in eastern Tennessee and North Carolina seems to be real since other species of *Salix* are known from the gap region, and it is unlikely that a species as conspicuous as *S. eriocephala* could have been overlooked. Specimens from the disjunct area do not appear to be morphologically separable from those of the continuous range, but field study may reveal differences.

It is probable that the disjunction is the result of Pleistocene events (see above). During Late-Glacial and Full-Glacial times *S. eriocephala* may have entered the Gulf watershed by way of the Chattahoochee-Appalachicola river system in Georgia and Florida (Thorne 1949) and the Alabama-Tombighee river system in Alabama. It was not eliminated from this region during the climatic amelioration, but it became isolated as populations in the southern Appalachians disappeared.

Hybridization. Salix eriocephala (as *S. cordata* and *S. rigida*) is reported to hybridize with a large number of species, including *S. amygdaloides, S. bebbiana, S. caroliniana, S. discolor, S. exigua* (as *S. interior*), *S. humilis* var. *humilis, S. lucida, S. nigra, S. petiolaris* (as *S. gracilis*), and *S. sericea* (Hitchcock & Standley 1919; Schneider 1921b; Fernald 1950; Braun 1961; Steyermark 1963). All of these supposed hybrids, except for the last, are either unsubstantiated, doubtful, or unrecognized in the southeastern flora.

The hybrid *S. eriocephala* × *sericea* occurs in the Southeastern States as individual plants usually in association with the parental species (Fig. 49). They are morphologically intermediate. They resemble *S. sericea* in their brittle branches (*Argus 10830*) and leaves that are sparsely sericeous beneath and acute

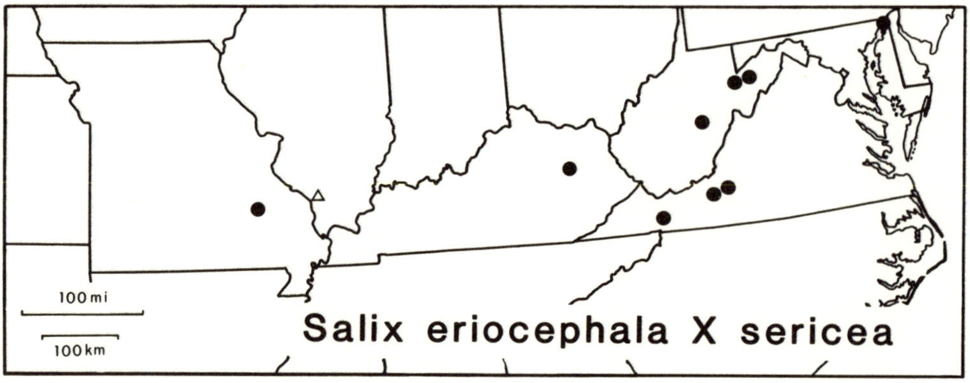

FIG. 49. Distribution of *Salix eriocephala* × *sericea* in the Southeastern States. Only incidental distributional records are available for peripheral states. Solid dots, herbarium specimens; open triangles, literature reports.

to cuneate at the base, and they resemble *S. eriocephala* in their prominent stipules associated with even the earliest leaves, distinct styles, and sparsely villous upper leaf surfaces. The capsules are mostly glabrous, as in *S. eriocephala,* but they have patches of pubescence near the base. The name *S. myricoides* Muhlenberg (1803) is often applied to this hybrid (Schneider 1921a, 1921b; Fernald 1950), but this name is not available because its type is the same as for *S. glaucophylloides* Fernald, which name it supersedes (Dorn 1976). The correct binomial for this hybrid may be *S.* ×*bebbii* Gandoger (*Fl. Eur.* 21. 167. 1892) (Schneider 1921a, 1921b), but I have not seen the type.

REPRESENTATIVE SPECIMENS. U.S.A. KENTUCKY: Menifee Co., *Higgins 1257* (DHL).—MARYLAND: Cecil Co., *Long 1536* (PH).—VIRGINIA: Montgomery Co., *Smyth 5087* (VPI); Roanoke Co., *Wood 5980* (GH, WVA); Smythe Co., *James 9024* (NCU).—WEST VIRGINIA: Grant Co., 29 Apr 1939, *Appalachian Botany Club s.n.* (WVA); Tucker Co., *Allard 19692* (US, WVA); Webster Co., *Argus 10830* (CAN).

IX. Salix subgenus **Helix** Dumortier, Bijdr. Natuurk. Wetensch. 1(1): 56. 1825.—TYPE: *Salix purpurea* Linnaeus.

Salix subgenus **Helix** section **Helix** subsection **Purpurea** Linnaeus, *Fl. Steiermark* 1: 154. 1908.—TYPE: *Salix purpurea* Linnaeus.

18. Salix purpurea Linnaeus, *Sp. pl.* 2: 1017. 1753.—Basket willow.

Shrubs up to 5 m tall. Branches grayish brown to olive-brown, slender; branchlets reddish brown to olive-brown, glabrous, sometimes pruinose, tenacious or brittle at base; buds reddish brown to yellowish, glabrous, flower buds navicular, often longer than petioles, ribbed laterally, apex blunt. Leaves alternate, but some opposite or subopposite, leaf blades narrowly oblanceolate or lingulate, 3.5–6–7.7 cm long, 0.7–1.2–1.7 cm wide, length/width 4.5–6.5, apex acute, sometimes abruptly so, base narrowed and rounded, margin entire toward base, distantly serrulate at middle and closely serrulate toward apex; immature leaf blades reddish, glabrous or sparsely floccose with ferruginous and white hairs;

FIG. 50. *Salix purpurea.* a. Vegetative branchlet, the leaves are mostly subopposite (×0.75). b. Branch bearing staminate aments in anthesis and vegetative branchlets starting to emerge (×1). c. Staminate flower (×10). d. Young pistillate flowers (×10).

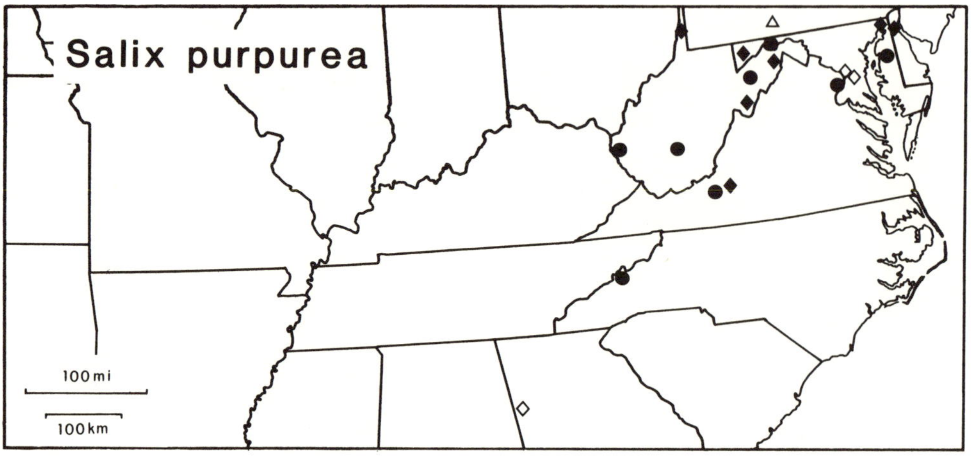

FIG. 51. Distribution of *Salix purpurea* in the Southeastern and peripheral states. Solid dots, naturalized; open diamonds, cultivated; solid diamonds, naturalization or cultivation status unknown; open triangles, literature reports.

mature leaf blades glabrate and glaucous on both sides, amphistomatous; petioles reddish brown, 2–7 mm long, glabrous; stipules absent. Aments precocious, sessile, with 2 leafy bracts at base, aments slender, often paired and recurved. Staminate aments 2–3.5 cm long; the two filaments and anthers united as one, filaments sparsely pilose at base, dry anthers 0.4 mm long; nectary 1, adaxial; bracts dark brown at tip, 0.8–1.6 mm long, obovate to oblong, apex rounded, sparsely long-hairy. Pistillate aments densely flowered, 2–3.6 cm long; pistils silvery gray, plump ovoid, unbeaked, sericeous, sessile, capsules tawny, ca 2.5 mm long, styles rudimentary or up to 0.3 mm long, stigmas very short, nectary 1, adaxial, 0.5 mm long; ovules 3 per carpel; bracts as in staminate flowers. Chromosome number 2n = 38. Fig. 50.

Distribution. (Fig. 51). This introduced European species is cultivated mainly in the northern part of our area, where it sometimes becomes naturalized.

Representative Specimens. U.S.A. Delaware: New Castle Co., *Tatnall 1440* (DOV).—District of Columbia: *Dudley 14190* (NA).—Georgia: Harris Co., *Meyer 10315* (NA).—Maryland: Allegany Co., *Westerfeld 5051* (PAC); Cecil Co., *Witman 2515* (PAC); Garrett Co., *Ball 3015* (CAN, MARY, NY, PH, US, VPI); Prince Georges Co., *Everett 13560* (NA); Queen Annes Co., *Reed 112976* (CFR).—North Carolina: Madison Co., *Ahles 42589* (LYN, NCU, USF).—Virginia: Fairfax Co., *Bradley 10456* (NLU); Montgomery Co., *Argus 10933* (CAN); Roanoke Co., *Wood 5525* (PH, VPI).—West Virginia: Fayette Co., 12 Aug 1973, *Grafton s.n.* (WVA); Grant Co., *Reed 96803* (CFR); Hampshire Co., 11 May 1940, *Frye s.n.* (WVA); Ohio Co., 28 Jul 1879, *Guttenberg s.n.* (CM); Pendleton Co., 20 Jun 1969, *Bartley s.n.* (WVA); Wayne Co., *Plymale 470* (WVA).

Cultivars. There are several cultivars and varieties in the nursery trade—cv. eugenei, cv. gracilis, cv. nana, cv. pendula, and var. *lambertiana* (Smith) Koch (Bailey et al. 1976; Chelmar 1976; Bean 1980). Undoubtedly more than one of these variants has been introduced into the Southeastern States, but because they are so difficult to identify on the basis of herbarium specimens, only the species is recognized here.

EXCLUDED SPECIES

Salix candida Fluegge. The report of this species from Smythe Co., Virginia (Small & Vail 1893), was based on *S. humilis* (Massey & Ball 1944).

Salix cordata Michaux. The reports of this species by Eiser (1961) and Massey (1961) are probably errors for *S. cordata* Muhlenberg (=*S. eriocephala* Michaux).

Salix petiolaris Smith (*S. gracilis* Andersson). Reports of this species in the Southeastern States (Ball 1938; Massey & Ball 1944; Fernald 1950) are based on *S. sericea* Marshall (see above). Peripheral populations of *S. petiolaris* occur in Pennsylvania (Bucks, Centre, Clinton, Lehigh, Montgomery, and Northampton counties), and New Jersey (Camden and Mercer counties). It is not known from Delaware or Maryland, but it is to be expected there.

LITERATURE CITED

Aiton, W. 1789. *Hortus kewensis.* Vol. 3. London: George Nicol.

Almeida, J. L. F. de. 1946. Sobre a cariologia de *Salix salviifolia* Brot. e *S. babylonica* L. Soc. Brotériana, B. (ser. 2) 20: 201–242.

Allard, H. A., and E. C. Leonard. 1952. The Canaan and the Stony River valleys of West Virginia, their former magnificent spruce forests, their vegetation and floristics today. Castanea 17: 1–60.

Andersson, N. J. 1858a. Bidrag till kannendomen om de i Nordamerika forekommande pilarter (Salices). Öfvers. Förh. Kongl. Svenska Vetensk.-Akad. 15: 109–133.

———. 1858b. Salices boreali-Americanae. A synopsis of North American Willows. Cambridge: Metcalf and Co., and New York: B. Westermann and Co.

———. 1867. *Monographia Salicum.* Pars 1. Kongl. Svenska Vetens. Akad. Handl. II.

———. 1868. Salicineae. In *Prodromus systematis naturalis regni vegetabilis,* ed. A. P. DeCandolle, 16 (2): 190–331. Paris, Strasbourg, and London: Treuttel et Würtz.

Argus, G. W. 1964. Preliminary reports on the flora of Wisconsin. No. 51. Salicaceae. The genus *Salix*–the willows. Wisconsin Acad. Sci., Arts & Lett. 53: 217–272.

———. 1965. The taxonomy of the *Salix glauca* complex in North America. Contr. Gray Herb. 196: 1–142.

———. 1973. The genus *Salix* in Alaska and the Yukon. Canad. Natl. Mus. Nat. Sci. Publ. Botany 2: 1–279.

———. 1974. An experimental study of hybridization and pollination in *Salix* (willows). Canad. J. Bot. 52: 1613–1619.

———. 1980. The typification and identity of *Salix eriocephala* Michx. (Salicaceae). Brittonia 32: 170–177.

———. 1983. *Salix.* In *Flora of Alberta,* by E. H. Moss, rev. J. G. Packer, 198–214. Toronto: University of Toronto Press.

———. 1984a. The identity of *Salix conifera.* Brittonia 36: 321–327.

———. 1984b. *Salix occidentalis,* the correct name for *S. tristis* Aiton (Salicaceae). Brittonia 36: 328–329.

———. 1985. Computerized catalogue of herbarium specimens of *Salix* of the Southeastern United States. Ottawa: National Museum of Natural Sciences.

Ashe, W. W. 1918. Additions to the arborescent flora of North Carolina. J. Elisha Mitchell Sci. Soc. 34: 130–140.

Atwood, W. W. 1940. *The physiographic provinces of North America.* Boston: Ginn and Co.

Bacone, J. A., and C. L. Hedge. 1980. A preliminary list of endangered and threatened vascular plants in Indiana. Proc. Indiana Acad. Sci. for 1979, 89: 359–371.

Bailey, L. H., E. Z. Bailey, and staff of Bailey Hortorium. 1976. *Hortus Third.* New York: MacMillan Publishing Co.

Ball, C. R. 1910. *Salix,* Varro. In *Elysium Marianum,* by I. Tidestrom, 19–37. Washington, D.C.

———. 1938. Willows of the southern states. Castanea 3: 1–9.

———. 1943. *Salix floridana* Chapman, a valid species. J. Arnold Arbor. 24: 103–106.

———. 1952. *Salix.* In *The new Britton and Brown illustrated flora of the northeastern United States and adjacent Canada,* by H. A. Gleason, 2: 6–23. New York: New York Botanical Garden.

———. 1961. *Salix.* In *Flora of Texas,* by C. L. Lundell et al., 3(6): 369–392. Renner, Texas: Texas Research Foundation.

Barratt, J. 1840. *Salices americanae.* Middletown, Connecticut: Pelton.

Bean, W. J. 1980. *Trees and shrubs hardy in the British Isles.* Vol. 4, 8th ed., rev. London: Murray.

Bebb, M. S. 1881. *Salix nigra* var. *wardii.* In The flora of Washington and vicinity, by L. F. Ward, Bull. U.S. Natl. Herb. 22: 114.

———. 1893. *Salix.* In Botany of the Death Valley Expedition, by F. V. Coville, Contr. U.S. Natl. Herb. 4: 198–199.

———. 1895a. Notes on some arborescent willows of North America—I. Gard. & Forest 8: 363–364.

———. 1895b. Notes on some arborescent willows of North America—II. Gard. & Forest 8: 372–373.

———. 1895c. Notes on some arborescent willows of North America—III. Gard. & Forest 8: 423–425.

Benson, L., and R. A. Darrow. 1981. *Trees and shrubs of the southwestern deserts.* 3d ed. Tucson: University of Arizona Press.

Blackburn, K. B., and J. W. H. Harrison. 1924. A preliminary account of the chromosomes and chromosome behaviour in the Salicaceae. Ann. Bot. (London) 38: 361–378.

Boivin, B. 1966. Enumération des plantes du Canada II—Lignidées. Naturaliste Canad. 93: 371–437.

Bowden, W. M. 1945. A list of chromosome numbers in higher plants. II. Menispermaceae to Verbenaceae. Amer. J. Bot. 32: 191–201.

Branson, B. A., D. F. Harker, J. M. Baskin, M. E. Medley, D. L. Batch, M. L. Warren, W. H. Davis, W. C. Houtcooper, B. Monroe, L. R. Phillippe, and P. Cupp. 1981. Endangered, threatened, and rare animals and plants of Kentucky. Trans. Kentucky Acad. Sci. 42: 77–89.

Braun, E. L. 1955. Phytogeography of unglaciated eastern United States and its interpretration. Bot. Rev. (Lancaster) 21: 297–375.

———. 1961. *The woody plants of Ohio.* Columbus: Ohio State University Press.

Brayshaw, T. C. 1976. *Catkin bearing plants (Amentiferae) of British Columbia.* British Columbia Provincial Museum, Occasional Papers No. 18.

Britton, N. L. 1901. *Manual of the flora of the northern states and Canada.* New York: Holt and Co.

Broome, C. R., J. L. Reveal, A. O. Tucker, and N. H. Dill. 1979. Rare and endangered vascular plants in Maryland. Newton Corner, Massachusetts: United States Fish and Wildlife Service.

Brown, C. A. 1945. *Louisiana trees and shrubs.* Louisiana Forestry Comm. Bull. 1.

Brown, R. G., and M. L. Brown. 1972. *Woody plants of Maryland.* College Park: Univ. of Maryland.

Brown, W. L., Jr., and E. O. Wilson. 1956. Character displacement. Syst. Zool. 5: 49–64.

Carey, J. 1848. *Salix.* In *A manual of the botany of the northern United States,* by A. Gray, 424–430. Cambridge, Boston, and London: Munroe and Co.

Chapman, A. W. 1860. *Flora of the southern United States.* New York: Ivison, Phinney & Co.

Chmelar, J. 1977. Taxonomic importance of capsule seed number in the genus *Salix* L. [Czech]. Cas. Slez. Muz. Ser. C (Dendrologie) [Journal of the Silesian Museum (Opava)]. 26: 1–7.

———. 1978. Taxonomic significance of bud scales in the genus *Salix* L. Folia Dendrol. 4: 5–19.

———. 1979. The taxonomic importance of chromosome numbers in the genus *Salix* L. [Czech]. Lesnictvi 52: 411–415.

Chmelar, J., and W. Meusel. 1976. *Die Weiden Europas. Die Gattung Salix.* Wittenberg: A. Ziemsen Verlag.

Clark, R. C. 1971. The woody plants of Alabama. Ann. Missouri Bot. Gard. 58: 99–242.

Clarkson, R. B. 1966. The vascular flora of the Monongahela National Forest, West Virginia. Castanea 31: 1–119.

Coker, W. C., and H. R. Totten. 1937. *Trees of the southeastern states.* 2d ed. Chapel Hill: University of North Carolina Press.

Cooperrider, T. S., ed. 1982. *Endangered and threatened plants of Ohio.* Ohio Biol. Surv. Bull. 16.

Core, E. L. 1932. Some aspects of the phytogeography of West Virginia. Torreya 32: 65–71.

———. 1938. Plant migrations and vegetational history of the southern Appalachian region. Lilloa 3: 5–29.

———. 1966. *Vegetation of West Virginia.* Parsons, West Virginia: McClain Printing Co.

Craighead, F. C. 1971. *The trees of south Florida.* Vol. 1. Coral Gables, Florida: University of Miami Press.
Critchfield, W. B. 1957. *Geographic variation in Pinus contorta.* Publ. Maria Moors Cabot Found. Bot. Res. No. 3.
Cronquist, A. 1964. *Salix.* In *Vascular plants of the Pacific Northwest,* by C. L. Hitchcock et al., 2: 37–71. Seattle: University of Washington Press.
Cusick, A. W., and G. M. Silberhorn. 1977. The vascular plants of unglaciated Ohio. Ohio Biol. Surv. Bull. 5(4): 1–153.
Davis, H. A., and C. R. Ball. 1947. The willows of West Virginia. Castanea 12: 94–100.
Deam, C. C. 1940. *Flora of Indiana.* Indianapolis: Department of Conservation, Division of Forestry.
Deevey, E. S., Jr. 1949. Biogeography of the Pleistocene. Geol. Soc. Amer. Bull. 60: 1315–1416.
Delcourt, H. R. 1979. Late Quaternary vegetation history of the eastern Highland Rim and adjacent Cumberland Plateau of Tennessee. Ecol. Monogr. 49: 255–280.
Delcourt, P. A., and H. R. Delcourt. 1977. The Tunica Hills, Louisiana-Mississippi: Late Glacial locality for spruce and deciduous forest species. Quaternary Res. 7: 218–237.
———. 1981. Vegetation maps for eastern North America. In *Geobotany,* ed. R. C. Romans, 2: 123–165. New York and London: Plenum Publishing Corp.
Delcourt, P. A., H. R. Delcourt, R. C. Brister, and L. E. Lackey. 1980. Quaternary vegetation history of the Mississippi Embayment. Quaternary Res. 13: 111–132.
Dorn, R. D. 1970. *The willows of Montana.* Bozeman: Herbarium, Dept. of Botany & Microbiology, Montana State University.
———. 1975. Cytological and taxonomic notes on North American *Salix.* Madroño 23: 99.
———. 1976. A synopsis of American *Salix.* Canad. J. Bot. 54: 2769–2789.
Duncan, W. H., and J. T. Kartesz. 1981. *Vascular flora of Georgia. An annotated checklist.* Athens: University of Georgia Press.
Eilers, L. J. 1971. The vascular flora of the Iowan area. Stud. Nat. Hist. Iowa Univ. 21(5): 1–137.
Eiser, A. L., Jr. 1961. Shrubs of Virginia, taxonomy, ecology and economics. Ph.D. diss., Virginia Polytechnic Institute. Ann Arbor: University Microfilms.
Enander, S. J. 1907. *Studier över Salices i Linnés herbarium.* Uppsala.
Ewan, J. 1967. Introduction. In *Arbustum americanum,* by H. Marshall, v–xii. New York: Hafner Publishing Co.
Federova-Sarkissova, O. V. 1946. On the chromosome number of certain willow and poplar species. Compt.-Rend. ⟨Dokl.⟩ Acad. Sci. URSS. 54: 353–356.
Felt, E. P. 1940. *Plant galls and gall makers.* Reprint. New York: Hafner Press, 1965.
Fernald, M. L. 1937. Local plants of the inner coastal plain of southeastern Virginia. III. Phytogeographic considerations. Rhodora 39: 465–491.
———. 1946. Technical studies on North American plants, II. Difficulties in North American *Salix.* Rhodora 48: 13–16, 27–40, 41–49.
———. 1950. *Gray's manual of botany.* 8th ed. New York: American Book Company.
Floderus, B. 1943 The continental *Salix glauca* species. Svensk Bot. Tidskr. 37: 169–172.
Forbes, F. F. 1909. *Salix subsericea* a distinct species. Rhodora 11: 9–12.
Fowles, H. A. 1965. *Silvics of forest tress of the United States.* U.S.D.A. Agric. Handb. 271.
Gillespie, J. P. 1963, 1964. Notes on the Pteridophytes of West Virginia: I. The presence of *Pellaea atropurpurea* (L.) Link in Cabell Co. Castanea 28: 179–180, 29: 75.
Glatfelter, N. M. 1894. A study of the relations of *Salix nigra* and *Salix amygdaloides,* together with the hybrids arising from them as these species exhibit themselves in the vicinity of St. Louis. Trans. St. Louis Acad. Sci. 6: 427–431.
———. 1897. Notes on *Salix longipes,* Shuttlw. and its relations to *S. nigra,* Marsh. Reprinted from the Ninth Annual Rep. Missouri Bot. Gard.
Gleason, H. A. 1952. *The new Britton and Brown illustrated flora of the northeastern United States and adjacent Canada.* Vol. 2. New York: Hafner Press.
Goodrich, S. 1983. Utah flora: Salicaceae. Great Basin Naturalist 43: 531–550.
Graham, A. 1965. Origin and evolution of the biota of southeastern North America: evidence from the fossil plant record. Evolution 18: 571–585.
Griggs, R. F. 1905. The willows of Ohio. Proc. Ohio Acad. Sci. 4: 256–314.
Grüger, E. 1972. Late Quaternary vegetation development in south-central Illinois. Quaternary Res. 2: 217–231.

Harlow, W. M., E. S. Harrar, and F. M. White. 1979. *Textbook of dendrology.* 6th ed. McGraw Hill Book Co.

Harper, R. M. 1904. Explorations in the coastal plain of Georgia during the season of 1902. Bull. Torrey Bot. Club. 31: 9–27.

Hitchcock, A. S., and P. C. Standley. 1919. Flora of the District of Columbia and vicinity. Contr. U.S. Natl. Herb. 21: 1–329.

Hjelmqvist, H. 1948. Studies on the floral morphology and phylogeny of the Amentiferae. Bot. Not. Suppl. 2: 1–171.

Holmberg, O. 1931. Skandinaviens Flora 1b(1). Stockholm: P. A. Norstedt & Söners Förlag.

Holmgren, P. K., W. Keuken, and E. K. Schofield. 1981. *Index herbariorum.* Part I. *The herbaria of the world.* 7th ed. Reg. Veg. 106.

Howden, H. F. 1963. Speculation on some beetles, barriers, and climate during the Pleistocene and pre-Pleistocene periods in some non-glaciated portions of North America. Syst. Zool. 12: 178–201.

Hubbs, C. L., and C. Hubbs. 1953. An improved graphical analysis and comparison of series of samples. Syst. Zool. 2: 49–56, 92.

Hyland, F., and F. H. Steinmetz. 1944. The woody plants of Maine. Maine Bull. 46 (11), Univ. Maine Stud., 2nd ser., 59: 1–72.

James, C. W. 1961. Endemism in Florida. Brittonia 13: 225–244.

Kartesz, J. T., and R. Kartesz. 1977. *The biota of North America.* Part 1: *Vascular plants.* Vol. 1: *Rare plants.* Pittsburgh: Biota of North America Committee.

Kearney, T. H. 1901. Botanical survey of the Dismal Swamp region. Contr. U.S. Natl. Herb. 5: 321–585.

Kimura, A. 1928. Über *Toisusu*—eine neue Salicaceen-Gattung und die systematische Stellung derselben. Bot. Mag. (Tokyo) 45: 28.

Kimura, C. 1969. The bud scale of some artificial hybrid willows. Sci. Rep. Tohoku Univ. Ser. 4 (Biology) 35: 33–37.

King, J. E., and W. H. Allen, Jr. 1977. A Holocene vegetation record from the Mississippi River valley, southeastern Missouri. Quaternary Res. 8: 307–323.

Koch, W. D. J. 1828. *De salicibus europaeis commentatio.* Erlangen.

Kral, R. 1981. Some taxonomic problems in woody plants of Alabama and middle Tennessee. In *Dendrology in the Eastern Deciduous Forest Biome,* eds. P. P. Feret and T. L. Sharik, Publ. FWS-2-80: 40–47. Blacksburg: Virginia Polytechnic Institute and State University.

———. 1983. *A report on some rare, threatened, or endangered forest-related vascular plants of the South.* Vol. I. *Isoetaceae through Euphorbiaceae.* U.S.D.A. Forest Serv., Southern Reg., Tech. Publ. R8-TP 2.

Kuprianova, L. A. 1965. *The palynology of the Amentiferae.* [Russian]. Leningrad: Acad. Sci. USSR.

Kurz, H., and R. K. Godfrey. 1962. *Trees of northern Florida.* Gainsville: University of Florida Press.

Lanjouw, J., and F. A. Stafleu. 1954. *Index herbariorum.* Part II. *Collectors A–D.* Reg. Veg. 2.

Lawrence, G. H. M. 1951. *Taxonomy of vascular plants.* New York: MacMillan Co.

Lindsey, A. A., R. O. Petty, D. K. Sterling, and W. VanAsdall. 1961. Vegetation and environment along the Wabash and Tippecanoe rivers. Ecol. Monogr. 31: 105–156.

Lipps, E. L., and H. R. DeSelm. 1969. The vascular flora of the Marshall Forest, Rome, Georgia. Castanea 34: 414–432.

Little, E. L., Jr. 1971. *Atlas of United States trees.* Vol. 1. *Conifers and important hardwoods.* U.S.D.A. Forest Serv. Misc. Publ. 1342.

———. 1976a. *Atlas of United States trees.* Vol. 3. *Minor western hardwoods.* U.S.D.A. Forest Serv. Misc. Publ. 1314.

———. 1976b. *Trees and shrubs of the United States. A bibliography for identification.* U.S.D.A. Forest Serv. Misc. Publ. 1336.

———. 1977. *Atlas of United States trees.* Vol. 4. *Minor eastern hardwoods.* U.S.D.A. Forest Serv. Misc. Publ. 1342.

Lobeck, A. K. 1950. Physiographic diagram of North America. New Jersey: The Geographical Press, Hammond and Co.

Löve, A. 1954. Cytotaxonomical evaluation of corresponding taxa. Vegetatio 5-6: 212–224.

Löve, A., and E. Kjellquist. 1974. Cytotaxonomy of Spanish plants III. Dicotyledons: Salicaceae–Rosaceae. Lagascalia 4: 32.

Löve, A., and D. Löve. 1942. Chromosome numbers of Scandinavian plant species. Bot. Not. 94 (1942): 19–59.

———. 1975. IOPB chromosome number reports 49. ed. A. Löve. Taxon 24: 504–507.

Majovsky, J. et al. 1974. Index of chromosome numbers of Slovakian flora. Part 4. Acta Fac. Rerum Nat. Univ. Comenianae, Bot. 23: 1–23.

———. 1976. Index of chromosome numbers of Slovakian flora. Part 5. Acta Fac. Rerum Nat. Univ. Comenianae, Bot. 25: 1–18.

———. 1978. Index of chromosome numbers of Slovakian flora. Part 6. Acta Fac. Rerum Nat. Univ. Comenianae, Bot. 26: 1–42.

Mang, F. 1980. Wird bei Gattungen und Arten die zur Hybridisierung neigen eine Zusatzkartierung erforderlich? Göttingen Floristische Rundbriefe 13: 106–119.

Marshall, H. 1785. *Arbustum americanum.* Philadelphia: Crukshank.

Massey, A. B. 1961. Virginia flora. An annotated catalog of plant taxa recorded as occurring in Virginia. Virginia Exp. Sta. Tech. Bull. 155: 1–258.

Massey, A. B., and C. R. Ball. 1944. The willows of Virginia. Bull. Virginia Polytech. Inst. 37(9): 1–31.

Massey, J. R., E. L. Core, J. W. Hardin, A. E. Radford, and L. S. Radford. 1981. Revised contributor's guide for The Vascular Flora of the Southeastern United States. Mimeographed.

Maxwell, J. A., and M. B. Davis. 1972. Pollen evidence of Pleistocene and Holocene vegetation on the Allegheny Plateau, Maryland. Quaternary Res. 2: 506–530.

McGregor, R. L., and T. M. Barkley, eds. 1977. *Atlas of the flora of the Great Plains.* Ames: Iowa State University Press.

McLeod, K. W., and J. K. McPherson. 1973. Factors limiting the distribution of *Salix nigra.* Bull. Torrey Bot. Club. 100: 102–110.

Mehra, P. N., and T. S. Sareen. 1973. Cytology of West Himalayan Betulaceae and Salicaceae. J. Arnold Arbor. 54: 412–418.

Meikle, R. D. 1975. *Salix* L. In *Hybridization and the flora of the British Isles,* ed. C. A. Stace. New York: Academic Press.

Merrill, E. D. 1949. *Index Rafinesquianus.* Jamaica Plain: Arnold Arboretum.

Mohlenbrock, R. H. 1980. *Illustrated flora of Illinois. Flowering plants. Willows to mustards.* Carbondale and Edwardsville: Southern Illinois University Press.

Muhlenberg, G. H. E. 1803. Über die Nordamerikanischen Weiden. Ges. Naturf. Freunde Berlin II. 4: 233–242.

Nakai, T. 1928. Une nouvelle systématique des Salicacées de Corée. Bull. Soc. Dendrol. France 66: 37–51.

Nakajima, G. 1937. Cytological studies in some dioecious plants. Cytologia Fujii Jubilaei Volumen, 282–292.

Neumann, A., and A. Polatschek. 1972. Cytotaxonomischer Beitrag zur Gattung *Salix.* Ann. Naturhist. Mus. Wien 76: 619–633.

Nuttall, T. 1842. *The North American Sylva.* Vol. 1. Philadelphia: Townsend Ward.

Paiero, P., and M. Schoepf. 1981. Corologia del genere *Salix* L. in Alto Adige. Webbia 35: 1–62.

Palmer, E. J., and J. A. Steyermark. 1938. New varieties and forms from Missouri. Ann. Missouri Bot. Gard. 25: 769–773.

Payne, W. W. 1978. A glossary of plant hair terminology. Brittonia 30: 239–255.

Pennell, F. W. 1942. Botanical collectors of the Philadelphia local area. Bartonia 21: 38–57.

Perret, P., and H. M. Burdet. 1981. Les "Plantae Vapincenses" de Dominique Chaix et les traveaux floristiques de Dominique Villars en Dauphine. Candollea 36: 400–408.

Petrovsky, V. V., and P. G. Zhukova. 1983. The chromosome numbers, morphology, ecology, and taxonomy of the willows of northeast Asia. [Russian]. Bot. Zurn. (Moscow & Leningrad) 68: 29–38.

Porsild, A. E., and W. J. Cody. 1980. *Vascular plants of continental Northwest Territories, Canada.* Ottawa: National Museum of Natural Sciences.

Pursh, F. 1814. *Flora americae septentrionalis.* Vol. 1. London: White, Cochrane, and Co.

Rechinger, K. H. 1964. *Salix.* In *Flora Europea,* eds. T. G. Tutin et al., Vol. 1: 43–54. Cambridge: Cambridge University Press.

Rehder, A. 1949. *Bibliography of cultivated trees and shrubs.* Jamaica Plain: Arnold Arboretum.

Remington, C. L. 1968. Suture-zones of hybrid interaction between recently joined biotas. Evol. Biol. 2: 321–427.

Roer, H. F., and R. Elven. 1975. Key to the genus *Salix* L. in Norway. [Norwegian.] Blyttia 33: 151–194.

Ross, H. H. 1965. Pleistocene events and insects. In *The Quaternary of the United States,* eds. H. E. Wright and D. G. Frey, 583–595. Princeton: Princeton University Press.

Rouleau, E. 1970. *Guide to Index kewensis and its supplements i–xiv.* Montreal: published by the author.

Rousseau, C. 1974. *Géographie floristique du Québec-Labrador.* Trav. Doc. Centr. Etudes Nordiques. 7.

Sargent, C. S. 1913. *Trees and shrubs: illustrations of new or little known ligneous plants.* Vol. 2. Boston: Houghton Mifflin Co.

Schneider, C. 1918. A conspectus of Mexican, West Indian, Central and South American species and varieties of *Salix.* Bot. Gaz. 65: 1–41.

———. 1919a. Notes on American willows. IV. Species and varieties of section *Longifoliae.* Bot. Gaz. 67: 309–346.

———. 1919b. Notes on American willows. V. The species of the Pleoandrae group. J. Arnold Arb. 1: 1–31.

———. 1920. Notes on American willows. IX. a. The species of section *Discolores.* b. The species of section *Griseae.* J. Arnold Arbor. 2: 1–25.

———. 1921a. Notes on American willows. XI. a. Some remarks on the species of section *Cordatae.* b. Some remarks on the geographical distribution of American willows. J. Arnold Arbor. 2: 185–204.

———. 1921b. Notes on American willows. XII. a. Systematic enumeration of the sections, species, varieties and forms of American willows. b. Some remarks on the hybrids hitherto observed among American willows. c. Some remarks on the geographical distribution of the American willows. d. Analytical keys to the species of American willows. J. Arnold Arbor. 3: 61–125.

Schramm, J. R. 1966. Plant colonization studies on black wastes from anthracite mining in Pennsylvania. Trans. Amer. Phil. Soc. N. S. 56: 6–194.

Seymour, F. C. 1969. *The flora of New England.* Rutland, Vermont: C. E. Tuttle Co.

Skvortsov, A. K. 1968. Willows of the USSR. A taxonomic and geographic revision. [Russian]. Moscow: Publishing Office "Nauka."

———. 1969. *Salix.* In *Flora Iranica. Flora des Iranischen Hochlandes und der umrahmenden Gebirge.* by K. H. Rechinger, 12–45. Graz: Akademische Druck- und Verlagsanstalt.

———. 1971. Euro-amerikanische verwandtschaftliche Beziehungen zweier mediterraner *Salix*-Arten. Feddes Repert. 82: 407–420.

———. 1973. Present distribution and probable original range of the crack willow (*Salix fragilis* L.). [Russian]. Problems in biogeocoenology, geobotany and botanical geography, 263–280. Leningrad: "Nauka."

Small, J. K. 1903. *Flora of the southeastern United States.* New York.

———. 1913. *Florida trees.* New York.

———. 1933. *Manual of the southeastern flora.* New York.

Small J. K., and A. M. Vail. 1893. Report on the botanical exploration of southwestern Virginia during the season of 1892. Mem. Torrey Bot. Club 4: 93–201.

Smith, E. B. 1978. An atlas and annotated list of the vascular plants of Arkansas. *Salix,* 307–310. Fayetteville: Department of Botany, University of Arkansas. Multilith.

Smith, F. F., D. K. Smith, and G. W. Argus. 1978. Willows for pleasure and benefit. American Hort. 57 (2): 22–25, 32.

Smith F. F., R. E. Webb, G. W. Argus, J. A. Dickerson, and H. W. Everett. 1983. Willow beaked-gall midge, *Mayetiola rigidae* (Osten Sacken), (Diptera: Ceciomyiidae): Differential susceptibility of willows. Environmental Entomology 12: 1175–1184.

Soper, J. H., and M. L. Heimburger. 1982. *Shrubs of Ontario.* Toronto: Royal Ontario Museum.

Stafleu, F. A., and R. S. Cowan. 1976. *Taxonomic literature.* Vol. 1. Utrecht: Bohn, Scheltema & Holkema.

Stearn, W. T. 1966. *Botanical Latin.* London and Edinburgh: Nelson.

Steyermark, J. A. 1959. Vegetational history of the Ozark forest. Univ. Missouri Stud. 31: 1–138.

———. 1963. *Flora of Missouri.* Ames: Iowa State University Press.

Suda, Y. 1960. The number of chromosomes in some Japanese salicaceous plants. III. Sci. Rep. Tohoku Univ., Ser. 4, Biology. 26: 25–36.

———. 1963. The chromosome numbers of salicaceous plants in relation to their taxonomy. Sci. Rep. Tohoku Univ., Ser. 4, Biology. 29: 413–429.

Suda, Y., and G. W. Argus. 1968. Chromosome numbers of some North American *Salix*. Brittonia 20: 191–197.

Sudworth, G. B. 1934. Poplars, principal tree willows and walnuts of the Rocky Mountain region. *Salix,* 48–97. U.S.D.A. Tech. Bull. 420.

Thorne, R. F. 1949. Inland plants on the Gulf Coastal Plain of Georgia. Castanea 14: 88–97.

———. 1951. Vascular plants previously unreported from Georgia. Castanea 16: 29–48.

———. 1954. The vascular plants of southwestern Georgia. Amer. Midl. Naturalist 52: 257–327.

Tomlinson, P. B. 1980. *The biology of trees native to tropical Florida.* Allston, Massachusetts: Harvard University Printing Office.

Torrey, J. 1847. *Flora of the state of New York.* Vol. 2. *Salix,* 204–213. Albany.

Tucker, A. O., N. H. Dill, C. R. Broome, C. E. Phillips, and M. J. Maciarello. 1979. *Rare and endangered vascular plant species in Delaware.* Newton Corner, Massachusetts: United States Fish and Wildlife Service.

Vachova, M., and J.. Chmelar. 1976. IOPB chromosome number reports 53. ed. A. Löve. Taxon 25: 490.

Vachova, M., and V. Ferakova. 1978. IOPB chromosome number reports 61. ed. A. Löve. Taxon 27: 382–383.

Vachova, M., and J. Majovsky. 1976. IOPB chromosome number reports 53. ed. A. Löve. Taxon 25: 489–490.

Villars, D. 1786. *Histoire des plantes de Dauphiné.* Vol. 1. Grenoble.

Voss, E. G. (Chairman editorial committee). 1983. *International code of botanical nomenclature.* Reg. Veg. 111.

Ward, D. B., ed. 1979. Plants. In *Rare and endangered biota of Florida.* ed. P. C. Pritchard, 1–175. Gainsville: University Presses of Florida.

Watts, W. A. 1969. A pollen diagram from Mud Lake, Marion County, north-central Florida. Geol. Soc. Amer. Bull. 80: 631–642.

———. 1971. Postglacial and interglacial vegetation history of southern Georgia and central Florida. Ecology 52: 676–690.

Watts, W. A., and M. Stuvier. 1980. Late Wisconsin climate of northern Florida and the origin of species-rich deciduous forest. Science 210: 325–327.

Wherry, E. T., J. M. Fogg, Jr., and H. A. Wahl. 1979. *Atlas of the flora of Pennsylvania.* Philadelphia: Morris Arboretum of the University of Pennsylvania.

Whitehead, D. R. 1967. Studies of full-glacial vegetation and climate in southeastern United States. In *Quaternary Paleoecology,* eds. E. J. Cushing and H. E. Wright, 237–248. New Haven: Yale University Press.

Wilcox, W. 1973. A survey of the vascular flora of Crittenden County, Arkansas. Castanea 38: 286–297.

Wilkinson, J. 1944. The cytology of *Salix* in relation to its taxonomy. Ann. Botany (London) II. 8: 269–284.

Willdenow, C. L. 1806. *Caroli a Linné species plantarum.* 4th ed., Vol. 4, Part 2. Berlin.

Wood, C. E., Jr. 1944. Notes on the flora of Roanoke County, Virginia. Rhodora 46: 69–85.

Zhukova, P. G., A. A. Korobkov, and A. D. Tikhonova. 1977. Chromosome numbers of some plant species in the eastern arctic Jakutia. [Russian]. Bot. Zurn. (Moscow & Leningrad) 62: 229–234.

ACKNOWLEDGMENTS

In making this study I have enjoyed meeting and learning from a large number of people, many of whom contributed significantly to my understanding of the willows of the Southeastern States. I wish to thank them all for their help and for making this work such a pleasure. In addition I wish to single out a number of people who made special efforts to assist me.

I am indebted to James R. Massey, Chapel Hill, North Carolina, who started me on this study by requesting that I prepare a treatment of *Salix* for *The Vascular Flora of the Southeastern United States.* I am also very indebted to Carroll E. Wood, Jr., Cambridge, Massachusetts, who first interested me in *Salix* of the Southeastern States and who, in providing me with a careful and critical review of this manuscript, helped me to avoid many pitfalls. All errors that still linger in this work, however, are solely my responsibility. I also thank Christiane Anderson, Ann Arbor, Michigan, and Sidney McDaniel, Mississippi State, Mississippi, for their helpful comments.

I thank the curators of more than 50 herbaria who kindly provided specimens for study, made study space available, and extended many other courtesies: A, APCR, APSC, ASU, AUA, BALT, CAN, CFR, CITA, CLEMS, CM, DAO, DHL, DOV, DUKE, DWC, EKY, F, FLAS, FSU, FUGR, GA, GEO, GH, K, KNK, KY, LFCC, LSU, LSUS, LTU, LYN, MARY, MEM, MISS, MO, MOAR, MTJB, MUHW, MUR, NA, NCSC, NCU, NLU, NO, NY, O, P, PAC, PH, S, TENN, UARK, UNA, UNCC, US, USF, VA, VDB, VPI, VSC, WCUH, WILLI, WIS, WKU, WVA, WVW. These herbarium acronyms follow Holmgren et al. (1981) with the exception of CFR (personal herbarium of Clyde F. Reed, Baltimore, Maryland) and WKU (herbarium of Western Kentucky University, Bowling Green, Kentucky). Although the names of many of the curators do not appear here, I very much appreciate their help. Other curators made special searches for type specimens or provided bibliographic materials: K. Bremer, Stockholm, Sweden; F. Butzin, Berlin, Germany; P. Lassen, Lund, Sweden; A. Lawalrée, Meise, Belgium; A. Maule, Copenhagen, Denmark; N. Morin, St. Louis, Missouri; R. Vickery, London, England; G. Wagenitz, Göttingen, Germany; and J. Wells, Bloomfield Hills, Michigan. I am indebted to R. D. Meikle, Royal Botanic Gardens, Kew, for information on the synonymy and taxonomy of *Salix cinerea* subsp. *oleifolia.*

This research and its publication were supported financially by the National Museum of Natural Sciences, Ottawa, Canada. Field work in 1984 was partially funded by the Ontario Ministry of Natural Resources, Tree Improvement and Forest Biomass Institute.

Field assistance was provided by Eric Argus, Mary Argus, and John Argus in 1981, 1984, and 1985, respectively; John Argus helped with proofreading. Special collections and/or help with field studies were provided by R. K. Godfrey, Tallahassee, Florida; K. Perkins, Gainesville, Florida; A. Gholson, Chattahoochee, Florida; W. Judd, Gainesville, Florida; and A. Tucker, Dover, Delaware. Critical comments on nomenclature were offered by E. G. Voss, Ann Arbor, Michigan, and R. D. Meikle, Kew, England. Assistance in recording data and mapping was provided by V. Tanaguchi and D. Farhar, Ottawa. Bibliographic material was provided by R. Long and J. Brennan, New York Botanical Garden. In addition the following persons provided special kinds of information, helped in the location of critical specimens or collecting localities, or in other ways made my work easier or more complete: T. Ager, Reston, Virginia; J. Baskin, Lexington, Kentucky; E. Browne, Jr., Memphis, Tennessee; M. Canoso, Cambridge, Massachusetts; J. Campbell, Lexington, Kentucky; N. Coile, Athens, Georgia; S. Cooper, Tallahassee, Florida; C. Douglass, Clemson, South Carolina; W. Duncan, Athens, Georgia; J. Dobberpuhl, Elkins, West Virginia; J. Eckenwalder, Toronto, Ontario; G. Frank, Blacksburg, Virginia; K. Gordon, Jackson, Mississippi; W. Grimé, Chicago, Illinois; P. Holmgren, New York, New York; A. Hotchkiss, Louisville, Kentucky; A. Howden, Ottawa, Ontario; R. McGregor, Lawrence, Kansas; P. Mazzeo, Washington, D. C.; J. Mears, Philadelphia, Pennsylvania; J. Moore, Raleigh, North Carolina; L. Morse, Arlington, Virginia; A. Mosseler, Toronto, Ontario; W. Overlease, West Chester, Pennsylvania; J. Pittillo, Cullowhee, North Carolina; C. Reed, Baltimore, Maryland; J. Rettig, Little Rock, Arkansas; L. Rogers, Greenville, South Carolina; P. Sommers, Nashville, Tennessee; J. Thieret, Highland Heights, Kentucky; L. Uttal, Blacksburg, Virginia; D. Ward, Gainesville, Florida; T. Wieboldt, Blacksburg, Virginia; D. Windler, Towson, Maryland; and Zhao Shidong, Shenyang, China.

The line drawings were made by Susan Laurie-Borque, Ottawa, Ontario. Finally I would like to thank I. M. Brodo and K. Pryer, National Museum of Natural Sciences, Ottawa, Canada, for reading and commenting on the manuscript.

APPENDIX I

Chromosome Numbers of Native and Naturalized Species of Salix of the Southeastern United States

S. alba, 2n = 76 (Blackburn & Harrison 1924; Wilkinson 1944; Federova-Sarkissova 1946; Suda & Argus 1968; Mehra & Sareen 1973; Majovsky et al. 1976, 1978; Chmelar 1979).

S. amygdaloides, 2n = 38. Based on specimens from Saskatchewan (Suda & Argus 1968), Wyoming (Dorn 1975), and Manitoba (Löve 1954).

S. babylonica, 2n = 76 (Nakajima 1937; Bowden 1945; Almeida 1946; Suda 1960, 1963; Majovsky et al. 1974; Mehra & Sareen 1973; Vachova & Chmelar 1976; Chmelar 1979).

S. bebbiana, 2n = 38. Based on specimens from Ontario (Dorn 1976), Manitoba (Löve 1954), Alberta and Wyoming (Dorn 1975), and USSR (Zhukova et al. 1977; Petrovsky & Zhukova 1983).

S. caprea, 2n = 38 (Blackburn & Harrison 1924; Wilkinson 1944; Nakajima 1937; Löve & Löve 1975; Vachova & Ferakova 1978). The report by Wilkinson (1944) of 2n = 76 (specimen of unknown provenance) may be based on *S. muscina* Dode (Neumann & Polatschek 1972) or on *S. cinerea* s.l. The report of a triploid (2n = 57) by Löve & Löve (1942) may also be based on a misidentification.

S. cinerea subsp. *cinerea,* 2n = 76 (Blackburn & Harrison 1924; Wilkinson 1944; Federova-Sarkissova 1946; Neumann & Polatschek 1972; Majovsky et al. 1976; Chmelar 1979).

S. cinerea subsp. *oleifolia,* 2n = 76 (Wilkinson 1944; Neumann & Polatschek 1972).

S. discolor, 2n = 76, 95, 114. All three chromosome numbers have been reported from Saskatchewan (Suda & Argus 1968; Suda, unpubl.). The hexaploid level (2n = 114) occurs in Ontario (Dorn 1976).

S. eriocephala, 2n = 38. Based on specimens from Ontario (Dorn 1976), as *S. rigida,* and plants cultivated in Europe of unstated provenance (Majovsky et al. 1974; Vachova & Chmelar 1976; Chmelar 1979), as *S. cordata* and *S. rigida.* The report of 2n = 44 (Wilkinson 1944), as *S. cordata,* is uncorroborated and probably in error.

S. exigua, 2n = 38. Based on specimens from British Columbia (Suda & Argus 1968), Wyoming (Dorn 1975), and cultivated plants of unknown provenance (Neumann & Polatschek 1972; Chmelar 1979).

S. fragilis, 2n = 76 (Blackburn & Harrison 1924; Neumann & Polatschek 1972; Majovsky et al. 1974; Chmelar 1979). The reports of diploids (2n = 38) by Federova-Sarkissova (1946) and Majovsky et al. (1974), and of hexaploids (2n = 114) by Marklund (in Löve & Löve 1942) and Suda (1963) are unusual departures not only in this species but in the group of tetraploid species that make up section *Salix.*

S. humilis var. *humilis,* 2n = 76. Based on specimens from Ontario (Suda & Argus 1968; Dorn 1976).

S. lucida, 2n = 76. Based on specimens from Ontario (Dorn 1976) and cultivated plants of unknown provenance (Blackburn & Harrison 1924).

S. nigra, 2n = 38. Based on specimens from Ontario (Dorn 1976) and cultivated specimens of unknown origin (Chmelar 1979).

S. pentandra, 2n = 76 (Blackburn & Harrison 1924; Wilkinson 1944; Federova-Sarkissova 1946; Vachova & Majovsky 1976).

S. purpurea, 2n = 38 (Blackburn & Harrison 1924; Holmberg 1931; Wilkinson 1944; Federova-Sarkissova 1946; Suda 1963; Löve & Kjellquist 1974; Majovsky et al. 1974, 1978; Vachova & Chmelar 1976; Chmelar 1979).

APPENDIX II

Infrageneric Classification of Salix in the Southeastern United States and Numerical List of Taxa

Salix Linnaeus

Subgenus *Salix*

Section *Humboldtiana* Pax in Engler & Prantl

1. *S. caroliniana* Michaux
2. *S. nigra* Marshall
3. *S. amygdaloides* Andersson
4. *S. floridana* Chapman

Section *Salicaster* Dumortier

5. *S. lucida* Muhlenberg
6. *S. pentandra* Linnaeus (naturalized)

Section *Salix*

7. *S. alba* Linnaeus (naturalized)
8. *S. fragilis* Linnaeus (naturalized)

Section *Subalbae* Koidzumi

9. *S. babylonica* Linnaeus (naturalized)

Section *Longifoliae* Pax in Engler & Prantl

10. *S. exigua* Nuttall

Subgenus *Vetrix* (Dumortier) Dumortier

Section *Vetrix*

Subsection *Substriatae* Görz

11. *S. bebbiana* Sargent

Subsection *Laeves* Camus & Camus

12. *S. discolor* Muhlenberg
13a. *S. humilis* Marshall var. *humilis*
13b. *S. humilis* var. *microphylla* (Andersson) Fernald
14. *S. caprea* Linnaeus (naturalized)
15a. *S. cinerea* Linnaeus subsp. *cinerea* (naturalized)
15b. *S. cinerea* subsp. *oleifolia* (Smith) Macreight (naturalized)

Section *Vimen* Dumortier

Subsection *Arbusculae* Hayek

16. *S. sericea* Marshall

Section *Cordatae* J. Barratt ex Hooker

Subsection *Lutea* Dorn

17. *S. eriocephala* Michaux

Subgenus *Helix* Dumortier

Section *Helix*

Subsection *Purpurea* Hayek

18. *S. purpurea* Linnaeus (naturalized)

APPENDIX III

Cultivated Taxa of Salix in the Southeastern United States

Willows have long been cultivated in the Southeastern States for ornamental purposes (Smith et al. 1978). Some are grown for their conspicuous or early staminate flowers (*S. cinerea, S. caprea,* and *S. gracilistyla*), and others for their weeping branches and graceful appearance (*S. babylonica* and *S. alba* cv. pendula). The branchlets in some provide a colorful (*S. alba* var. *vitellina* and cv. britzensis) or an unusual contorted display (*S. matsudana* cv. tortuosa and *S. udensis* cv. sekka). Some of these willows are highly susceptible to infestation by the willow beaked-gall midge (Smith et al. 1983). This is especially true of *S. cinerea* and *S. muscina* and the hybrids *S. aurita* × *purpurea, S. caesia* × *myrsinifolia, S. caprea, S. caprea* × *?elaeagnos,* and *S. purpurea* × *silesiaca.*

Introduced species that have become naturalized in the Southeastern States, including *S. pentan-*

dra, S. alba, S. fragilis, S. babylonica, S. caprea, S. cinerea, S. purpurea, and some of their hybrids, are discussed above. There are also a large number of species, cultivars, and hybrids that are known only in cultivation. The following list does not include all taxa that are or may be in cultivation but only those that are documented by herbarium specimens, usually deposited in the United States National Arboretum Herbarium or the National Herbarium of Canada.

Introduced Taxa of Salix Known Only in Cultivation in the Southeastern United States

S. aegyptiaca Linnaeus
S. alba cv. britzensis
S. alba f. *argentea* Wimmer
S. apennina A. Skvortsov
S. appendiculata Villars × *elaeagnos* Scopoli
S. appendiculata × *purpurea* Linnaeus
S. arbuscula Linnaeus × *herbacea* Linnaeus (*S.* ×*simulatrix* F. B. White)
S. aurita Linnaeus × *cinerea* Linnaeus (*S.* ×*multinervis* Doell)
S. aurita × *purpurea* Linnaeus (*S.* ×*dichroa* Doell)
S. aurita × *silesiaca* Willdenow (*S.* ×*subaurita* Andersson)
S. babylonica Linnaeus × *fragilis* Linnaeus (*S.* ×*blanda* Andersson)
S. babylonica Linnaeus cv. crispa
S. caesia Villars
S. caesia × *purpurea* Linnaeus
S. caesia × *myrsinifolia* Salisbury
S. caprea Linnaeus × *cinerea* Linnaeus (*S.* ×*reichardtii* A. Kerner)
S. caprea × ?*elaeagnos* Scopoli (*S.* ×*seringiana* Gaudin)
S. caprea × *viminalis* Linnaeus (*S.* ×*smithiana* Willdenow)
S. caprea × *repens* Linnaeus
S. cinerea Linnaeus × *lapponum* Linnaeus (*S.* ×*laestadiana* Hartman)
S. cinerea × *purpurea* Linnaeus (*S.* ×*pontederana* Willdenow)
S. daphnoides Villars
S. dasyclados Wimmer
S. dasyclados × *viminalis* Linnaeus
S. elaeagnos Scopoli
S. elaeagnos × *purpurea* Linnaeus (*S.* ×*wichurae* A. Kerner)
S. elaeagnos × *silesiaca* Willdenow
S. elegantissima Koch
S. fragilis Linnaeus cv. bullata
S. gilgiana Seemen
S. glabra Scopoli
S. gracilistyla Miquel var. *gracilistyla*
S. gracilistyla var. *melanostachys* (Makino) C. Schneider
S. integra Thunberg
S. lapponum Linnaeus
S. matsudana Koidzumi cv. tortuosa
S. medemei Dode
S. mielichhoferi Sauter
S. miyabeana Seemen
S. muscina Dode
S. myrsinifolia Salisbury
S. myrsinifolia × *phylicifolia* Linnaeus (*S.* ×*tetrapla* Walker ex K. Koch)
S. myrsinifolia × *retusa* Linnaeus (*S.* ×*cottetii* Lagger)
S. purpurea Linnaeus cv. eugenei
S. purpurea cv. gracilis
S. purpurea × *repens* Linnaeus (*S.* ×*doniana* Smith)
S. purpurea × *silesiaca* Willdenow (*S.* ×*siegerti* Andersson)
S. purpurea × *udensis* Trautvetter & G. Meyer
S. purpurea × *viminalis* Linnaeus (*S.* ×*rubra* Hudson)

S. repens Linnaeus
S. rosmarinifolia Linnaeus
S. triandra Linnaeus
S. udensis Trautvetter & G. Meyer
S. udensis Trautvetter & G. Meyer cv. sekka
S. variegata Franchet
S. viminalis Linnaeus

In addition to these introductions from Europe and Asia, some extraterritorial North American species are cultivated in the Southeastern States. The only documented cases of cultivation are for *S. cordata* Michaux and *S. glaucophylloides* Fernald, but other species, such as *S. lucida* and *S. discolor,* are or were in cultivation and others are to be expected.

INDEX TO NUMBERED COLLECTIONS EXAMINED

All collections are listed by principal collector. See Argus (1985) for a detailed list of collections examined; for ordering information see p. 5.

Abel, S. *S. babylonica:* 98.
Achuff, P. *S. caroliniana:* 77.
Adams, J. *S. eriocephala:* 2389.—*S. nigra:* 2258, 4757, 4758.
Adams, T. *S. nigra:* 60.
Ahles, H. *S. babylonica:* 39759.—*S. caroliniana:* 616, 10256, 10257, 10459, 11017A, 11747, 17966, 22025, 22196, 22196A, 22367, 23368, 23665, 23749, 24007, 24009, 24246, 39749, 39989, 41381, 51146, 76572.—*S. eriocephala:* 58299.—*S. humilis* var. *humilis:* 4008A, 12764, 13777, 16462, 21851, 26119, 30758, 38393, 43654, 46177, 55007, 58463, 59944.—*S. humilis* var. *microphylla:* 10759, 11537, 54392.—*S. nigra:* 10661, 11017, 11390, 11484, 11484A, 12503, 13251, 14857, 14976, 15298, 21577, 21597, 21793, 22731, 23087, 23205, 23205A, 23992, 24313, 24554, 24750, 25152, 38968, 40385, 40866, 40909, 50204, 50612, 53223, 53224, 53228, 54012, 54636, 55442, 56115, 56479, 57881, 59160.—*S. purpurea:* 42589.—*S. sericea:* 10055, 15947, 39605, 51981, 52002, 53182, 53183, 53184, 53186, 53227, 54007, 54723, 58300, 83185.
Akens, M. *S. caroliniana:* 73.
Alcorn, J. *S. nigra:* 193, 459.
Alexander, E. *S. exigua:* 189.
Allard, H. *S. alba:* 9893.—*S. caroliniana:* 9359A, 9390.—*S.* ×*conifera:* 10087.—*S. discolor:* 9084, 9087, 9499, 9505A, 9505B, 9536, 21677.—*S. eriocephala:* 9359B, 10061, 11078, 11577, 20829.—*S. eriocephala* × *sericea:* 19692.—*S. humilis* var. *humilis:* 212, 4518, 6480, 7608, 7645, 8125, 9099, 9325, 9327, 9329, 9451, 9764, 9809, 9811, 9847, 9847A, 9847B, 10087, 10200, 10201, 10869, 11174, 11175, 12050.—*S. humilis* var. *microphylla:* 7608, 9901.—*S. nigra:* 434, 1034, 1124, 3649, 5030, 7238, 7354, 7396, 7401, 8518, 9295, 9313, 9318, 9361, 11677, 20328.—*S.* ×*rubens:* 20432.—*S. sericea:* 2398, 5322, 7984, 9083, 9094, 9294, 9296, 9316, 9342, 9391, 9746, 9756, 9760, 9763, 9816, 10201, 11206, 11462, 12053, 21671.
Allen, C. *S. humilis* var. *microphylla:* 940.—*S. nigra:* 6526, 8014.
Allen, J. *S. alba:* 119.
Allison, A. *S. nigra:* 886, 887, 888, 889.
Alman. *S. caroliniana:* 2, 54.
Amberg, L. *S. nigra:* 60.
Anderson, J. *S. caroliniana:* 3.
Anderson, M. *S. purpurea:* 46.
Anderson, S. *S. nigra:* 1396.
Antozzi, W. *S. caroliniana:* 54.
Applegarth, P. *S. exigua:* 102.
Appler, P. *S. nigra:* 443, 534.
Arce, G. *S. caroliniana:* 5.
Arcuri, A. *S. caroliniana:* 74, 122.
Argus, G. W. *S. alba:* 10820, 11540.—*S. babylonica:* 10827, 10935, 11375, 11546.—*S. caroliniana:* 11663, 10822, 10823, 10825, 10889, 10890, 10891, 10892, 10893, 10894, 10895, 10896, 10899, 10900, 10901, 10903, 10904, 10906, 10908, 10909, 10911, 10914, 10915, 10916, 10917A, 10918, 10919, 10920, 10921, 10922, 10923, 10924, 10925, 10926, 10927, 10928, 10930A, 10964B, 10965, 11392, 11393, 11394, 11395, 11396, 11397, 11398, 11399, 11400, 11401, 11402, 11403, 11404, 11405, 11406, 11407, 11408, 11409, 11440, 11443, 11444, 11445, 11446, 11450, 11451, 11452, 11456, 11457, 11460, 11461,

11462, 11463, 11463A, 11464, 11472, 11473, 11474, 11475, 11475A, 11483, 11494, 11496, 11650, 11651, 11652, 11653, 11654, 11658, 11663.—*S. caroliniana* × *nigra:* 11389, 11415, 11492A, 11493A, 11495, 11496A, 11500A. —*S. eriocephala:* 10821, 10828, 10829, 10832, 10833, 10952, 10974, 10975, 10976, 11511, 11512, 11513, 11514.—*S. eriocephala* × *sericea:* 10830.—*S. exigua:* 10838, 10840, 10843, 10859, 11526, 11527, 11534, 11535, 11537, 11539, 11544, 11547.—*S. floridana:* 11453, 11455, 11458, 11459, 11477, 11478, 11479, 11480, 11481, 11482.—*S. humilis* var. *humilis:* 10849, 10863, 10864, 10865, 10866, 10936, 10937, 10941, 10942, 10946, 10953, 10954, 10955, 10956, 10957, 10958, 10959, 10963, 11465, 11466, 11467, 11468, 11469, 11470, 11471, 11476, 11484, 11485, 11486, 11487, 11488, 11519.—*S. humilis* var. *microphylla:* 10852, 10853, 10854, 10855, 10856, 10867, 10868, 10869, 10870, 10871, 10872, 10873, 10938, 10939, 10940, 10943, 10944, 10945, 10971, 10972, 10973. —*S. lucida:* 10948, 10950.—*S. nigra:* 10819, 10824, 10826, 10834, 10835, 10837, 10839, 10841, 10842, 10844, 10847, 10848, 10850, 10861, 10883, 10884, 10885, 10887, 10888, 10890, 10902, 10905, 10907, 10910, 10912, 10913, 10913A, 10913B, 10917B, 10929, 10930B, 10931, 10951, 10960, 10962, 10964A, 10965, 10966, 10967, 10968, 10969, 10970, 11377, 11379, 11380, 11381, 11384, 11385, 11386, 11387, 11388, 11390, 11391, 11410, 11411, 11412, 11413, 11414, 11416, 11417, 11418, 11419, 11420, 11421, 11422, 11423, 11424, 11425, 11426, 11427, 11428, 11429, 11430, 11431, 11432, 11433, 11434, 11435, 11436, 11437, 11438, 11439, 11441, 11442, 11447, 11448, 11449, 11452A, 11489, 11490, 11491, 11492, 11493, 11497, 11498, 11499, 11500, 11501, 11502, 11503, 11504, 11505, 11506, 11507, 11508, 11509, 11510, 11515, 11516, 11517, 11518, 11520, 11521, 11522, 11523, 11524, 11525, 11528, 11529, 11530, 11531, 11532, 11533, 11536, 11538, 11541, 11545, 11649, 11655, 11656, 11659, 11660, 11664.—*S. purpurea:* 10933.—*S.* ×*rubens:* 10934.—*S. sericea:* 10831, 10836, 10862, 10882, 10932, 10947, 10949, 10960, 10961, 11376, 11378, 11382, 11383.

Arsene, G. *S. nigra:* 12011, 12190.

Artz, L. *S. alba:* 1203.—*S. babylonica:* 1200.—*S. humilis* var. *microphylla:* 624, 923.—*S. nigra:* 1202.— *S. sericea:* 1201.

Ashe, W. *S. caroliniana:* 412.

Aswell, J. *S. nigra:* 6.

Athey, R. *S. humilis* var. *microphylla:* 1708.

Atkinson, A. *S. babylonica:* 50.

Austin, D. *S. nigra:* 3672.

Avery, G. *S. caroliniana:* 1485.

Bailey, C. *S. babylonica:* 80.

Baker, D. *S. babylonica:* 32.

Baldo, B. *S. nigra:* 75.

Baldwin, J. *S. caroliniana:* 129.—*S. humilis* var. *humilis:* 93, 94, 95.

Ball, C. *S. alba:* 3, 6, 22, 34, 35, 2603, 2706, 2708, 2744, 3002.—*S. babylonica:* 4, 5, 30, 36, 2601.—*S. caprea:* 1608, 1609.—*S. caroliniana:* 9, 10, 55, 56, 64, 65, 85, 86, 689, 690, 691, 2598, 2605, 2608, 2711, 2733, 2734, 2735, 2737, 3025.—*S. cinerea* subsp. *cinerea:* 72.—*S. eriocephala:* 2594, 2604, 2710, 2712, 2745, 3019, 3021, 3022, 3026, 3027.—*S. exigua:* 46, 61, 425.—*S. fragilis:* 52.—*S. humilis* var. *humilis:* 38, 39, 41, 53, 54, 102, 103, 1605, 1606, 1607, 2716.—*S. humilis* var. *microphylla:* 1602.—*S. nigra:* 1, 2, 3, 11, 13, 14, 15, 16, 17, 18, 19, 24, 31, 32, 33, 44, 45, 55, 62, 63, 64, 97, 377, 399, 477, 480, 645, 1751, 1752, 1753, 1754, 1755, 2443, 2589, 2590, 2591, 2592, 2593, 2606, 2707, 2709, 2714, 2731, 2732, 2736, 2737, 2743, 2747, 3003, 3006, 3010, 3011, 3014, 3023, 3060.—*S. pentandra:* 43, 2762, 3016.—*S. purpurea:* 67, 3015.—*S.* ×*rubens:* 12, 29, 37, 2735.—*S. sericea:* 20, 25, 28, 42, 56, 57, 58, 104, 105, 106, 107, 562, 563, 564, 1601, 1744, 2243, 2602, 2701, 2702, 2703, 2704, 2705, 2708, 2713, 2715, 2741, 2742, 2746, 2747, 2763, 3012, 3017, 3018, 3020, 3024, 9760.

Balogh, P. *S. nigra:* 345.

Balton, L. *S. nigra:* 459, 527, 688.—*S. sericea:* 528.

Baltzell, L. *S. caroliniana:* 48, 324, 867, 1596, 1671, 1877, 2001, 2718, 3657, 4751, 5780, 5955, 6092, 7312, 8344, 9665.

Banks, W. *S. humilis* var. *humilis:* 16750.

Barans, A. *S. caroliniana:* 210.—*S. nigra:* 38, 196, 590, 713, 725.

Barber, G. *S. caroliniana:* 204.—*S. exigua:* 945.—*S. nigra:* 911.

Barbour, P. *S. nigra:* 1141, 1142.

Barclay, F. *S. nigra:* 93.—*S. sericea:* 92, 92B.

Barkley, L. *S. nigra:* 75.

Barnes, J. *S. caroliniana:* 36.

Barnes, W. *S. nigra:* 4, 349, 360B, 920.—*S. sericea:* 360A.

Barrick, J. *S. nigra:* 79.

Bartholomew, E. *S. alba:* 43.—*S. babylonica:* 827.—*S. caroliniana:* 866, 926, 927.—*S. eriocephala:* 13, 130, 137, 174, 181, 3206.—*S. exigua:* 12, 31, 58, 98, 105, 106, 211, 244, 249, 271, 826, 951, 952, 1234, 1588,

1728, 16874.—*S. humilis* var. *humilis:* 31, 53, 66, 866, 3244, 4143.—*S. humilis* var. *microphylla:* 52, 4142.—*S. nigra:* 7, 8, 9, 27, 31, 60, 72, 794, 929, 950, 4113.—*S. sericea:* 6, 49, 118, 266, 812, 1093, 1119, 1538, 1676.

Bartram, E. *S. caroliniana:* 944, 945, 947, 965.

Baskin, J. *S. humilis* var. *microphylla:* 241.

Batchelder, C. *S. caroliniana:* 4789, 4790, 4791, 4792.

Bates, C. *S. nigra:* 24.

Batson, W. *S. humilis* var. *humilis:* 754.—*S. humilis* var. *microphylla:* 752, 753.—*S. nigra:* 756.

Battin, K. *S. babylonica:* 41.

Baum, V. *S. caroliniana:* 285, 329.

Beach, C. *S. cinerea* subsp. *cinerea:* 107.

Beard, L. *S. caroliniana:* 392.—*S. nigra:* 396, 403.

Beaven, G. *S. nigra:* 237.

Beck, C. *S. caroliniana:* 1, 115.

Beckett, M. *S.* ×*ehrhartiana:* 332, 388.—*S. nigra:* 329, 661, 712.

Beckner, J. *S. nigra:* 1013.

Bell, C. *S. caroliniana:* 1619, 1706, 2422, 4856, 11124, 11483, 11498, 12981.—*S. humilis* var. *humilis:* 14486.—*S. humilis* var. *microphylla:* 6101.—*S. nigra:* 2242, 2645, 2662, 2807, 2858, 3598, 5554, 5920, 6291, 6546, 6825, 11658, 12503.—*S. sericea:* 2053, 3228, 14618.

Benedict, J. *S. babylonica:* 4387.—*S. cinerea* subsp. *cinerea:* 4846.—*S. humilis* var. *humilis:* 3200, 4846.

Benner, W. *S. nigra:* 7104, 8568.—*S.* ×*rubens:* 8629.

Bergeron, G. *S. exigua:* 103.

Berkley, E. *S. nigra:* 587.

Bernstein, L. *S. humilis* var. *humilis:* 73.

Berry, M. *S. caroliniana:* 179.

Bessey, E. *S. caroliniana:* 5, 101.

Bethshares, S. *S. caroliniana:* 56.

Bidwell, J. *S. nigra:* 30.

Biernacki, L. *S. humilis* var. *humilis:* 631, 632.

Biltmore Herbarium. *S. humilis* var. *microphylla:* 1187, 1187B, 1188B.—*S. nigra:* 1230.—*S. sericea:* 1186, 1186B.

Bishop, H. *S. exigua:* 92.

Blackerby. *S. nigra:* 61.

Blair, A. *S. caroliniana:* 263, 468.

Blake, S. *S. caroliniana:* 10578.—*S. eriocephala:* 10578.—*S. humilis* var. *humilis:* 9497, 10559.—*S. humilis* var. *microphylla:* 10827.—*S. sericea:* 10560.

Blanton, F. *S. caroliniana:* 6499, 6500, 6945, 6946.

Blanton, P. *S. caroliniana:* 29.

Blomquist, H. *S. caroliniana:* 10952A, 14350, 15484, 15485.—*S. humilis* var. *humilis:* 2917, 2920, 2921.—*S. nigra:* 2922.—*S. sericea:* 2925.

Blum, K. *S. caroliniana:* 3483.—*S. nigra:* 417, 3248, 3308, 3417, 3420, 3488.

Boddie, R. *S. nigra:* 839.

Boettcher, F. *S. nigra:* 2938.

Bonar, J. *S. cinerea* subsp. *cinerea:* 1.—*S. eriocephala:* 2.—*S. exigua:* 1.—*S. humilis* var. *humilis:* 1.

Boole, J. *S. nigra:* 579.

Boom, B. *S. nigra:* 335.

Boone, W. *S. alba:* 550, 805.—*S. caroliniana:* 440.—*S. eriocephala:* 437, 587.—*S. humilis* var. *microphylla:* 624, 1158. *S. nigra:* 541.—*S. sericea:* 271.

Bostick, L. *S. babylonica:* 290.

Bostick, P. *S. nigra:* 22414.

Botanical Expedition. *S. humilis* var. *humilis:* 2054.

Boufford, D. *S. caroliniana:* 21449.—*S. caroliniana* × *nigra:* 9002.—*S. humilis* var. *humilis:* 12464, 12465, 14217, 21997.—*S. nigra:* 2820, 12611, 18123, 18570, 18595.—*S. sericea:* 18207.

Bougere, L. *S. nigra:* 739, 873.

Boulware, M. *S. babylonica:* 53.

Bowers, F. *S. humilis* var. *humilis:* 45060.

Boyce, S. *S. caroliniana:* 503, 523, 589, 619.

Bozeman, J. *S. caroliniana:* 1119, 2544, 2627, 3907, 4067, 4123, 4676, 4990, 5085, 5565, 5637, 5755, 6045, 6069, 6229, 6447.—*S. nigra:* 4067, 4216, 4710, 4861, 5244, 5303, 5362, 5628, 6025, 6518, 6533, 8829, 8859.—*S. sericea:* 9111.

Bracewell, G. *S. nigra:* 52.

Bradley, T. *S. cinerea* subsp. *cinerea:* 7736.—*S. eriocephala:* 6359, 7398.—*S. humilis* var. *humilis:* 7728.—*S. nigra:* 4069, 7436, 7446, 7454, 7617, 10457.—*S. purpurea:* 10456.—*S.* ×*rubens:* 5641.

Brantley, B. *S. babylonica:* 38.—*S. nigra:* 34.

Brass, L. *S. caroliniana:* 14493, 14493A, 14565, 14566, 14711A, 14814.

Braun, E. *S. alba:* 3977.—*S. caroliniana:* 2476, 4125.—*S. eriocephala:* 2744, 4411.—*S. exigua:* 4307, 4311, 4546, 4694. *S. humilis* var. *humilis:* 214, 218, 1156, 1540, 2475, 3409, 3465, 3483, 3801, 3872, 4216, 4952, 6122, 21817, 23140, 24236, 24569, 24666, 24877, 50161, 80991, 89106.—*S. humilis* var. *microphylla:* 1380, 2182, 3458.—*S. nigra:* 1797, 2988, 3863, 4468.—*S. sericea:* 323, 2850, 3809, 3819, 3820, 4066.

Breeland, R. *S. nigra:* 246.

Bretting, P. *S. exigua:* 82.—*S. nigra:* 81, 83, 178.
Bright, J. *S. caroliniana:* 341, 342.
Briley, T. *S. nigra:* 590.
Britton, J. *S. nigra:* 68.
Brock, H. *S. sericea:* 4255.
Brode, T. *S. nigra:* 3, 18.
Brown, C. *S. babylonica:* 1910.—*S. exigua:* 5187, 5188.—*S. nigra:* 1859, 2048, 5185, 5186, 18263.—*S. sericea:* 156.
Brown, D. *S. babylonica:* 70.—*S. nigra:* 71, 174, 313.—*S. sericea:* 169.
Brown, F. *S. humilis* var. *humilis:* 437, 20799.—*S. nigra:* 436.
Brown, J. *S. caroliniana:* 507.—*S. nigra:* 21.
Brown, L. *S. nigra:* 61.
Brown, R. *S. humilis* var. *humilis:* 3021.—*S. sericea:* 172.
Browne, E. *S. caroliniana:* 4658.—*S. exigua:* 4834.—*S. nigra:* 4779, 5503, 7184, 7290, 7301, 7389.
Bruhn, M. *S. humilis* var. *humilis:* 76.—*S. humilis* var. *microphylla:* 1A, 199.
Brumbach, W. *S. caroliniana:* 5686, 5687, 6143, 7415, 7423, 7424, 7428, 7429, 7430.
Brunton, V. *S. babylonica:* 141.—*S. caroliniana:* 191.
Bryson, C. *S. nigra:* 499.
Buckner, R. *S. nigra:* 70.
Buddell, G. *S. exigua:* 101, 102.
Buell, M. *S. caroliniana:* 3506.—*S. humilis* var. *humilis:* 2233.—*S. sericea:* 206A, 2002, 2069, 2219.
Bunke, C. *S. cinerea* subsp. *cinerea:* 13.
Burk, C. *S. sericea:* 28.
Burkhalter, J. *S. caroliniana:* 5588, 5878.
Burns, P. *S. exigua:* 89.
Burton, D. *S. nigra:* 87.
Bush, B. *S. caroliniana:* 405, 15244, 15762, 15879, 15954.—*S. humilis* var. *humilis:* 2584.—*S. nigra:* 152, 5654, 5929.
Byam, L. *S. exigua:* 150.—*S. nigra:* 112.
Byrd, H. *S. babylonica:* 56.
Cagle, C. *S. nigra:* 99, 121.
Cain, S. *S. humilis* var. *humilis:* 491, 618.—*S. humilis* var. *microphylla:* 4372.—*S. nigra:* 24, 1382.—*S. sericea:* 21, 139, 142, 1005.
Caldwell, D. *S. babylonica:* 65.
Camp, W. *S. humilis* var. *humilis:* 1695.
Canby, W. *S. exigua:* 167.
Caplendor, D. *S. caroliniana:* 574, 603.
Carden, F. *S. cinerea* subsp. *cinerea:* 2609.
Carnes, E. *S. nigra:* 21.
Carpenter, I. *S. babylonica:* 4125.
Carroll, F. *S. nigra:* 305.—*S. sericea:* 371.
Carroll, N. *S. nigra:* 86, 1357.
Carson, R. *S. caroliniana:* 5.
Carter, J. *S. nigra:* 101, 481.
Carter, M. *S. nigra:* 1005.
Cash, J. *S. nigra:* 227.
Cassen, P. *S. caroliniana:* 269, 270.
Catignan, G. *S. nigra:* 24.
Caughey, M. *S. nigra:* 732.
Chamberlain, E. *S. caroliniana:* 21, 1239, 1713.
Chandler, A. *S. nigra:* 4206.—*S. sericea:* 4619.
Chandler, G. *S. nigra:* 703.
Chapman, E. *S. nigra:* 20.
Chase, A. *S. humilis* var. *humilis:* 2164.
Cheevers, M. *S. exigua:* 24, 25.
Chester, E. *S. caroliniana:* 494, 803, 2114, 3292, 4399.—*S. exigua:* 2837.—*S. humilis* var. *humilis:* 55.—*S. humilis* var. *microphylla:* 4263.
Chester, W. *S. caroliniana:* 1687.—*S. nigra:* 2013, 3292.
Christensen, R. *S. eriocephala:* 34.
Chrysler, M. *S. humilis* var. *humilis:* 75.—*S. nigra:* 75.
Churchill, J. *S. caroliniana* × *nigra:* 4279.
Clanton, W. *S. humilis* var. *microphylla:* 24.
Clark, R. *S. babylonica:* 2496.—*S. caroliniana:* 4031, 7419, 9407, 9928, 11809, 13074, 15170.—*S. eriocephala:* 2569, 9547, 9663, 9664, 10282, 12559, 14002, 14597, 14776, 14999, 15298, 15992, 16105, 16346, 16614, 16714, 16957, 17107, 17458, 17619, 17978, 18149.—*S. humilis* var. *humilis:* 572B, 16621A, 17149.—*S. nigra:* 972, 2661, 5521, 10279, 10479, 10605, 10696, 12895, 13229.—*S. sericea:* 561.
Clarke, C. *S. humilis* var. *humilis:* 99.—*S. nigra:* 370.—*S. sericea:* 371.
Clarkson, R. *S. alba:* 949.—*S. cinerea* subsp. *cinerea:* 14.—*S. humilis* var. *humilis:* 185, 866, 2111.—*S. sericea:* 19, 1978, 2392.
Clausen, R. *S. caroliniana:* 3752.—*S. nigra:* 3752.
Clebsch, A. *S. caroliniana:* 691.—*S. humilis* var. *humilis:* 2795.
Clebsch, E. *S. humilis* var. *humilis:* 4766, 14275.—*S. humilis* var. *microphylla:* 9062, 9073, 14277, 14353, 14370.
Clemenson, G. *S. cinerea* subsp. *cinerea:* 40.
Cleveland, H. *S. babylonica:* 59.
Cleveland, R. *S. nigra:* 517.
Clevenger, K. *S. babylonica:* 68.
Clovis, J. *S. floridana:* 1373.—*S. sericea:* 1992.
Cochrane, D. *S. nigra:* 2.
Cocks, R. *S. exigua:* 122, 3627.—*S. nigra:* 91, 95, 101, 107, 111, 113, 115.
Coile, N. *S. nigra:* 299, 1683.
Coker, W. *S. caroliniana:* 2.—*S. humilis* var. *humilis:* 1, 106, 107.—*S. humilis* var. *microphylla:* 2, 4982.—*S. nigra:* 1, 2.

Cole, V. *S. nigra:* 208.
Coleman, J. *S. nigra:* 48386.
Collins, M. *S. babylonica:* 18.
Commons, A. *S. eriocephala:* B.—*S. lucida:* A.
Conrad, J. *S. exigua:* 602, 1230.—*S. nigra:* 136, 751, 843.
Cooley, G. *S. caroliniana:* 770, 1035, 1064, 1126, 2494, 5550, 5640, 5873, 7280, 12608, 12626.—*S. humilis* var. *humilis:* 4602.—*S. nigra:* 3302, 3339, 3820, 3887.
Cooper, A. *S. caroliniana:* 2834, 2945.—*S. sericea:* 22, 247.
Cooper, B. *S. nigra:* 53.
Cooper, M. *S. nigra:* 111.
Cooper, W. *S. nigra:* 90.
Copeland, E. *S. nigra:* 73.
Corcoran, C. *S. nigra:* 1635.
Cormier, K. *S. nigra:* 606.
Corn, L. *S. humilis* var. *humilis:* 178, 179, 180.
Correll, D. *S. caroliniana:* 41562.—*S. humilis* var. *humilis:* 14603, 14623.—*S. humilis* var. *microphylla:* 8712.—*S. nigra:* 6901, 14458, 14633.—*S. sericea:* 14457, 14565, 14579, 14615, 14623, 14631.
Cosentino, P. *S. nigra:* 373.
Counts, R. *S. babylonica:* 15.—*S. nigra:* 47.
Coville, F. *S. nigra:* 175.
Coward, T. *S. nigra:* 31.
Craft, C. *S. nigra:* 145.
Crandall, D. *S. humilis* var. *humilis:* 9346, 9346A.—*S. sericea:* 20497.
Cranfill, R. *S. sericea:* 127.
Creager, F. *S. caroliniana:* 12.
Creear, P. *S. nigra:* 324.
Creel, W. *S. babylonica:* 17.
Crewz, D. *S. caroliniana:* 1230, 1283.
Critchley, J. *S. nigra:* 2.
Cronquist, A. *S. caroliniana:* 4922.
Crooks, A. *S. eriocephala:* 450A.—*S. humilis* var. *humilis:* 410.
Crosby, M. *S. caroliniana:* 4845.—*S. nigra:* 4836.
Crow, C. *S. babylonica:* 387.—*S. exigua:* 399.—*S. nigra:* 75, 203, 462.
Crow, S. *S. nigra:* 183.
Crum, S. *S. babylonica:* 14.
Crum, T. *S. babylonica:* 116.
Crutchfield, P. *S. cinerea* subsp. *cinerea:* 2848.—*S. nigra:* 483, 1314.
Culp, R. *S. caroliniana:* 405.
Curry, M. *S. nigra:* 128, 640, 1475.
Curtiss, A. *S. caroliniana:* 2, 3, 4, 4523, 4523A, 4523B, 4524C, 4524CD, 5826.—*S. caroliniana* × *nigra:* 1, 5826.—*S. humilis* var. *humilis:* 7517.—*S. nigra:* 4524, 4524AB, 4524B, 5826.
Cypert, E. *S. caroliniana:* 407, 430, 431.
D'Arcy, W. *S. caroliniana:* 1483, 2868, 4420.
Daly, E. *S. alba:* 114.—*S. sericea:* 54.
Darwin, S. *S. nigra:* 1340, 1341.
Davenport, L. *S. caroliniana:* 958.—*S. eriocephala:* 1800.—*S. exigua:* 72. *S. nigra:* 313, 1052.
Davidson, J. *S. nigra:* 98.
Davies, P. *S. alba:* 202.—*S. exigua:* 204, 205, 206, 207, 208, 1571.—*S. humilis* var. *humilis:* 201.—*S. nigra:* 209, 210, 211.—*S. sericea:* 203.
Davis. *S. nigra:* 1284.
Davis, H. *S. alba:* 6628.—*S. babylonica:* 3967.—*S. caroliniana:* 6621, 6624, 6629, 16328.—*S. cinerea* subsp. *cinerea:* 2020, 2021.—*S.* ×*conifera:* 15216, 15217.—*S. discolor:* 5058, 6127.—*S. eriocephala:* 2023, 2758, 6622, 6623.—*S. humilis* var. *humilis:* 2759, 2762, 3964, 3965, 5053.—*S. humilis* var. *microphylla:* 2018, 3746.—*S. lucida:* 3652.—*S. nigra:* 5055, 6620, 6625, 6626.—*S. sericea:* 1346, 1611, 6627.
Davis, J. *S. nigra:* 1396.
Davis, R. *S. caroliniana:* 384.—*S. exigua:* 774.—*S. nigra:* 225A.
Dawson, N. *S. nigra:* 115, 1217.
Deam, C. *S. caroliniana:* 58613, 58614, 58615, 58738, 60394, 60707, 60821, 61207, 64065, 64177, 64178, 64185, 64241, 64298, 64305.—*S. humilis* var. *humilis:* 60033.—*S. humilis* var. *microphylla:* 61279.—*S. nigra:* 58871.
Dean, C. *S. caroliniana:* 57620, 64065, 64185.
Dean, J. *S. exigua:* 373.
Dean, R. *S. nigra:* 91, 225, 328.
Dearing, K. *S. nigra:* 45.
Demaree, D. *S. alba:* 11257.—*S. caroliniana:* 2756, 3260, 3312, 3929, 4502, 8509, 8673, 9811, 9878, 9924, 9983, 10197, 10198, 10978, 10997, 14171, 14831, 14835, 14848, 14848A, 15601, 15708, 15747, 15856, 15959, 16007, 16017, 16074, 16200, 16478, 16963, 17226, 17946, 18230, 19836, 20122, 20164A, 20255, 20472, 20575, 20612, 21916, 21966, 22079, 22225, 22259, 22799, 22979, 22991, 23069, 23164, 23393, 23486, 23591, 23977, 25001, 25157, 25379, 25461, 25873, 25975, 25984, 25985, 26022, 26116, 26324, 26567, 27000, 27367, 27728, 27729, 27871, 28836, 28885, 29104, 29233, 30209, 30456, 30496, 30498, 34333, 35327, 45490, 49381, 49520, 55480, 55706, 57703, 57877, 60034, 61694, 63504, 63555, 63569, 64012, 64125, 69649, 70821, 70824.—*S. eriocephala:* 25873.—*S. exigua:* 5087, 8712, 11100, 11360, 11396, 12991, 13823, 14066, 14067, 14234, 14266, 14451, 15225, 15226, 15542, 16620, 16660, 17054, 19690, 19742, 19982, 20377, 20935,

21361, 21555, 23266, 25055, 25462, 25581, 26188, 26326, 27114, 27456, 27615, 27911, 27927, 28572, 28732, 28741, 28892, 28894, 29105, 30208, 30242, 30783, 31719, 33193, 37450, 46415, 53369, 65060, 66955.—*S. humilis* var. *humilis:* 57, 76, 4858, 5964, 6345, 17686, 20768, 22300, 22476, 27089, 37480, 37487, 37617, 41891, 49046, 52629, 56945, 62003.—*S. humilis* var. *microphylla:* 15071, 17640, 18742, 18745.—*S. nigra:* 54, 6782, 6983, 9884, 10034, 11086, 11316, 11320, 13320, 13480, 13486, 13501, 13572, 13582, 13595, 13759, 13770, 13820, 13839, 13956, 14106, 14193, 14235, 14352, 14377, 14390, 14440, 14441, 14455, 14456, 14466, 14567, 14568, 14593, 14606, 14626, 14642, 14727, 14787, 14788, 14832, 14885, 14937, 15211, 15325, 15948, 16075, 16477, 16587, 16660, 16768, 16769, 16826, 16828, 16830, 16868, 16869, 16928, 16972, 17041, 17196, 17300, 17890, 18788, 18881, 18914, 18989, 19063, 19252, 19256, 19624, 19686, 19909, 20123A, 20147, 20343, 20375, 20441, 20587, 20780, 20783A, 20789A, 20824, 20842, 20868, 20876, 20906, 20915, 20918, 20934, 20953, 21096, 21741, 21892, 21916, 21945, 21953, 21961, 22047, 22729, 22965, 22996, 23356, 23394, 23523, 23687, 23700, 23940, 24012, 24502, 24709, 24726, 24767, 25054, 25298, 25574, 25817, 25851, 25854, 25877, 25894, 25895, 26362, 26546, 26569, 26786, 26976, 27022, 27383, 27473, 27585, 27586, 27586A, 27592, 27594, 27595, 27601, 27602, 27607, 27608, 27610, 27611, 27619, 27620, 28166, 28485, 28571, 28717, 28726, 28727, 28738, 28739, 28742, 28755, 28756, 28785, 28786, 28797, 28798, 28799, 28800, 28804, 28874, 28875, 28876, 28877, 28884, 28885, 28886, 28887, 28888, 28893, 28898, 28899, 28905, 28906, 28912, 29329, 29738, 30243, 30351, 30961, 32904, 33762, 34652, 36787, 37083, 37448, 39556, 39669, 48248, 49507, 49521, 49551, 49556, 51952, 52105, 55641, 57628, 57629, 58164, 59811, 63395, 63423, 69383.—*S.* ×*rubens:* 27455.—*S. sericea:* 25872.

DePoe, C. *S. sericea:* 7359.

Deramus, R. *S. caroliniana:* 304.—*S. nigra:* 266, 308, 834.

DeSelm, H. *S. exigua:* 29974.—*S. humilis* var. *humilis:* 1873, 29973.—*S. humilis* var. *microphylla:* 144, 1014, 1066, 1135, 1138, 1873, 1988, 30574.—*S. nigra:* 29976, 29977, 30649.

Dickerson, J. *S. alba:* 13652.

Diggs, G. *S. nigra:* 1496.—*S. sericea:* 1266, 1352.

Dixon, D. *S. nigra:* 1450.—*S. sericea:* 524.

Dodson, C. *S. caroliniana:* 4965.

Dooley, R. *S. exigua:* 53.—*S. nigra:* 38, 56, 79, 101, 540, 633.

Dosher, J. *S. nigra:* 6.

Douglass, M. *S. caroliniana:* 273.—*S. nigra:* 165, 323.

Downs, R. *S. alba:* 7168.—*S. babylonica:* 10673.—*S. caroliniana:* 301, 7005, 8305, 8306, 8429, 8433, 9187, 9441, 9448, 9980, 9987, 11051, 11059, 11921, 12746.—*S. cinerea* subsp. *cinerea:* 5324.—*S. eriocephala:* 5D, 1487, 2821, 7937, 7938, 8035, 8046, 8104, 8308, 10643, 10681.—*S. humilis* var. *humilis:* 181.—*S. nigra:* 801, 971, 1223, 1489, 1618, 1631, 1821, 2561, 2709, 2827, 2948, 3097, 3146, 3287, 3648, 3984, 4107, 4590, 4650, 4986, 5128, 5260, 5851, 5995, 6025, 6127, 6220, 6461, 6490, 6724, 7554, 8030, 8089, 8422, 8430, 8524, 9188, 9443, 9536, 9650, 10123, 10387, 10461, 10586.—*S.* ×*rubens:* 10808.—*S.* ×*sepulcralis:* 7553, 7936A.—*S. sericea:* 6837, 6903, 6949, 7146, 7222, 7298, 7338, 7379, 7544, 7761, 7775, 7961, 7972, 7973, 7974, 8033, 8036, 8304, 8307, 8309, 8310, 9442, 9444, 9445, 9447, 9449, 9450, 10196, 10670, 10715, 10752, 10811, 10847, 10866, 10916, 10917, 11240, 11432, 11866, 12067, 12666, 12704, 77986.

Drummond, T. *S. nigra:* 303.

DuBois, S. *S. caroliniana:* 28, 343.

DuBose, E. *S. humilis* var. *humilis:* 23.—*S. nigra:* 58.

Duckett, S. *S. cinerea* subsp. *cinerea:* 257.

Dudley, T. *S. caroliniana:* 8793.—*S. cinerea* subsp. *cinerea:* 12362.—*S. purpurea:* 14190.—*S.* ×*sepulcralis:* 8929.

Duke, J. *S. alba:* 178.—*S. nigra:* 2.

Dumond, D. *S. nigra:* 1369, 1760.

DuMond, D. *S. humilis* var. *humilis:* 738, 1305.—*S. sericea:* 921, 953, 1326, 1858.

Duncan, M. *S. nigra:* 27.

Duncan, W. *S. babylonica:* 1556, 1838.—*S. caroliniana:* 1959, 2147, 2200, 2244, 2300, 4700, 7435, 8443, 17672, 29068.—*S. humilis* var. *humilis:* 46, 1223, 1456, 1504, 1818, 1819, 1820, 1821, 2618, 2788, 3158, 4361, 5315, 5484, 6278, 6323, 6622, 6906A, 7209, 7267, 7298, 8236, 8327, 8434, 21722, 21772.—*S. humilis* var. *microphylla:* 1479, 5436.—*S. nigra:* 1315, 1441, 1890, 1920, 2120, 2125, 2183, 2184, 2197, 2262, 2284, 2295, 2459, 3005, 4718, 5166, 7992, 9530, 9805, 12802, 15257, 16325, 16455, 16477, 18602, 22493, 22678A.—*S. sericea:* 61, 1879, 1880, 2647, 6397, 7340, 7517.

Dunlop, G. *S. nigra:* 308.

Earle, E. *S. alba:* 4088.—*S. humilis* var. *humilis:* 2245, 5158, 5211.—*S. nigra:* 3225, 3243, 3523, 3541, 5208.—*S.* ×*sepulcralis:* 3232.
Easley, S. *S. cinerea* subsp. *cinerea:* 4.
Eaton, A. *S. caroliniana:* 758.
Ecklund, H. *S. caroliniana:* 133.
Edmunds, M. *S. nigra:* 830.
Edwards, J. *S. nigra:* 30.
Edwin, G. *S. caroliniana* × *nigra:* 352.
Eggleston, W. *S. humilis* var. *humilis:* 4829.—*S. pentandra:* 4555.
Egler, F. *S. caroliniana:* 254.
Elden, F. *S. caroliniana:* 449.
Eller, J. *S. discolor:* 56.
Elliott, C. *S. cinerea* subsp. *cinerea:* 51.
Ellis, W. *S. exigua:* 1341.—*S. nigra:* 28744, 28980.—*S. sericea:* 1640.
Elmore, D. *S. cinerea* subsp. *cinerea:* 25.
Elmore, H. *S. nigra:* 791.
England, B. *S. nigra:* 121.
Enloe, J. *S. nigra:* 660.
Erlanson, C. *S. cinerea* subsp. *cinerea:* 48. —*S. humilis* var. *humilis:* 73, 86.—*S. sericea:* 53.
Ernest, J. *S. nigra:* 14.
Ernsberger, N. *S. babylonica:* 10.
Evans, E. *S. nigra:* 34.
Evans, J. *S. nigra:* 318.
Evans, R. *S. nigra:* 130.
Everett, D. *S. purpurea:* 13560.
Ewan, J. *S. exigua:* 17709, 21283.—*S. nigra:* 17546, 17547, 20066, 20190.
Exploration Party. *S. caroliniana:* 2.
Eyles, D. *S. caroliniana:* 3977, 4243, 4274, 5400.—*S. eriocephala:* 8348, 8523.—*S. exigua:* 7812, 8446, 8527.—*S. nigra:* 3977, 6895, 7158, 8439.—*S. sericea:* 998.
Faircloth, W. *S. caroliniana:* 112, 457, 1560, 2093, 2221, 5907.—*S. eriocephala:* 8222.—*S. humilis* var. *humilis:* 238, 401, 757, 1031, 5864, 6967, 7092, 7151.—*S. nigra:* 748, 2220, 3091, 3878, 4218, 4359, 4615, 6599, 7119, 7489, 7569, 8198.
Fairey, J. *S. nigra:* 524, 674.
Farley, L. *S. nigra:* 4.
Farmer, M. *S. nigra:* 4.
Farrar, L. *S. nigra:* 19.
Farris, R. *S. babylonica:* 56.
Fassett, N. *S. caroliniana:* 20590.—*S. humilis* var. *humilis:* 20925.
Fauntleroy, J. *S. cinerea* subsp. *cinerea:* 665.—*S. eriocephala:* 3, 663.—*S. nigra:* 6.—*S. sericea:* 2, 4, 5.
Felkel, C. *S. nigra:* 57.
Fernald, M. L. *S. caroliniana:* 3886, 4379, 4624, 5285, 5286, 9310, 9567, 14928.—*S. caroliniana* × *nigra:* 4625, 15082.—*S. humilis* var. *humilis:* 3888, 4626, 6980, 6981, 6982, 13311, 14929.—*S. humilis* var. *microphylla:* 5287.—*S. nigra:* 5748, 6185, 8691, 9309, 9909, 14815, 14816, 15083.—*S. sericea:* 3887, 3889, 6580, 6977, 6978, 6979, 10607, 13923.
Findley, D. *S. nigra:* 69.
Fischler, A. *S. nigra:* 58.
Fisher, G. *S. nigra:* 458.
Fleming, G. *S. caroliniana:* 3342.
Fletcher, W. *S. nigra:* 168.
Fogg, J. *S. caroliniana:* 5520, 5521, 22206.—*S. eriocephala:* 13371, 14998.—*S. humilis* var. *humilis:* 12827.—*S. humilis* var. *microphylla:* 12864.—*S. nigra:* 13147, 13457, 14790.—*S.* ×*rubens:* 14970.—*S. sericea:* 5435, 12974, 14954, 16991, 17398.
Ford, E. *S. humilis* var. *microphylla:* 2156.—*S. nigra:* 4505, 5341.
Fosberg, F. *S. caroliniana:* 14666, 14667, 17867, 18062, 36300, 36301.—*S. eriocephala:* 39537.—*S. fragilis:* 30978.—*S. humilis* var. *humilis:* 16895, 16915, 16931, 18617, 18618, 23243, 23244, 39602, 44238.—*S. humilis* var. *microphylla:* 39810.—*S. nigra:* 16716, 16896, 17715, 18855, 30981, 30983.—*S. sericea:* 16916, 16932, 18939.
Fox, W. *S. caroliniana:* 386, 409, 513, 2282, 4629.—*S. humilis* var. *humilis:* 212, 725, 1987, 2152.—*S. humilis* var. *microphylla:* 2771.—*S. nigra:* 1313.—*S. sericea:* 1326, 1340, 2168, 2207.
Frederick, L. *S. nigra:* 42.
Fredholm, A. *S. caroliniana:* 5061.
Freer, R. *S. humilis* var. *humilis:* 181, 1673, 1695, 1724, 1725, 1804.—*S. nigra:* 1628, 1710.—*S. sericea:* 1688, 1693, 1694, 1879.
French, K. *S. nigra:* 109.
French, R. *S. caroliniana:* 504.—*S. nigra:* 347.
Frey, T. *S. exigua:* 108.—*S. nigra:* 13.
Frye, W. *S. humilis* var. *humilis:* 1, 277.—*S. nigra:* 616.
Funk, V. *S. nigra:* 397.
Gaines, X. *S. nigra:* 192, 240.
Garrett, J. *S. nigra:* 717.
Gates, E. *S. nigra:* 35, 177.
Genelle, P. *S. caroliniana:* 555, 615, 1133, 2612, 2632.
Gentry, J. *S. babylonica:* 1094.—*S. caroliniana:* 458.—*S. discolor:* 606.—*S. exigua:* 457.—*S. nigra:* 296.—*S.* ×*sepulcralis:* 628.
Gerson, B. *S. nigra:* 102.
Gibson, D. *S. nigra:* 670.
Gidel. *S. sericea:* 57.
Gilbert, F. *S. humilis* var. *humilis:* 554, 555, 566.—*S. nigra:* 618, 823, 1002.—*S. sericea:* 1001, 1003.

Gilbert, G. *S. nigra:* 823.
Gillespie, J. *S. caroliniana:* 4939.
Gillespie, M. *S. humilis* var. *humilis:* 453.—*S. nigra:* 30B, 34.
Gillett, J. *S. caroliniana:* 9214.—*S. sericea:* 9407.
Gillis, W. *S. caroliniana:* 11216.
Glass, D. *S. sericea:* 36.
Glasscock, I. *S. exigua:* 601.
Glidden, W. *S. humilis* var. *humilis:* 54.
Godfrey, R. *S. caroliniana:* 284, 380, 472, 492, 534, 2374, 6373, 6872, 6874, 7071, 53034, 53131, 53582, 53975, 54449, 54451, 54456, 54462, 54464, 54478, 54479, 54481, 54482, 54511, 54517, 54521, 54526, 54527, 54528, 54660, 54661, 54664, 54665, 54681, 54687, 54688, 54691, 54692, 54716, 54717, 54848, 55212, 55291, 55737, 56490, 57260, 57382, 58274, 60763, 62590, 62655, 63161, 69337, 73695, 74683, 74700, 78495, 80840, 80841, 80843, 80846, 80847, 80850, 80851, 80852, 80853, 80854, 80855, 80856, 80859, 80860, 80861, 80862, 80863, 80864, 80865, 80866, 80867, 80868, 80869, 80873, 80881, 80887, 80888, 80890, 80891, 80892, 80893, 80894, 80895, 80896, 80897, 80899, 80901, 80907, 80908, 80909, 80910, 80911, 80912, 80913, 80914, 80915, 80916, 80917, 80918, 80919, 80920, 80921, 80922, 80923, 80924, 80925, 80926, 81197A, 81260A, 81261A.—*S. caroliniana* × *nigra:* 54449, 54455, 54482, 54517, 54661, 54683, 54686, 81260K, 81260M, 81260N, 81261Q, 81261R, 81261T.—*S. eriocephala:* 79125, 79360, 79419, 80348, 80757.—*S. floridana:* 54973, 55350, 73634.—*S. humilis* var. *humilis:* 7021, 7022, 51855, 53004, 54522, 55997, 56242, 56263, 59862, 60168, 70121, 71212, 72331, 74179, 74979, 75109, 77152, 78451.—*S. humilis* var. *microphylla:* 51874, 69819.—*S. nigra:* 167, 1313, 3584, 3605, 5968, 54447, 54451, 54454, 54455A, 54455B, 54472, 54474, 54476, 54528, 54589, 54621, 54680, 54682, 54683, 54684, 54685, 54689, 54691, 55242, 55261, 55739, 57775, 78917, 80842, 80845, 80848, 80849, 80858, 80872, 80874, 80875, 80876, 80877, 80878, 80879, 80880, 80882, 80883, 80884, 80886, 80889, 80898, 80900, 81197O, 81197P, 81260O, 81261L, 81261M.—*S. sericea:* 59977.
Gonsoulin, G. *S. nigra:* 1735, 2002, 3137.—*S. sericea:* 3579, 3610.
Gough, R. *S. exigua:* 129.—*S. humilis* var. *humilis:* 591.—*S. nigra:* 248, 292.
Gould, A. *S. cinerea* subsp. *cinerea:* 101.—*S. nigra:* 106.
Grace, J. *S. babylonica:* 10.
Grady, B. *S. nigra:* 161.
Graham, W. *S. caroliniana:* 279, 362.
Grant, B. *S. caroliniana:* 12.
Grather, W. *S. exigua:* 130.
Gray, G. *S. babylonica:* 27.
Greaves, J. *S. babylonica:* 1589.—*S. humilis* var. *humilis:* 1731.—*S. nigra:* 1258A.
Greear, P. *S. humilis* var. *humilis:* 6431, 64249.—*S. nigra:* 66324.
Green, J. *S. caroliniana:* 37.
Grimes, E. *S. alba:* 3417.—*S. babylonica:* 4645.—*S. humilis* var. *humilis:* 3276, 3414.—*S. humilis* var. *microphylla:* 3312, 3414.—*S. nigra:* 3352, 3392, 3482, 3487.
Grindstaff, A. *S. sericea:* 3.
Griscom, L. *S. caroliniana:* 21575.
Guhardja, E. *S. nigra:* 102, 129.
Guidroz, T. *S. caroliniana:* 192.—*S. nigra:* 19, 135.
Gunn, C. *S. babylonica:* 1252.—*S. exigua:* 1198.—*S. humilis* var. *humilis:* 2283.—*S. nigra:* 534, 963, 979, 1214, 10986.
Gupton, D. *S. eriocephala:* 3334.—*S. nigra:* 1069.
Guttenberg, G. *S. eriocephala:* 251, 257.
Hackett, H. *S. caroliniana:* 20.
Hale, S. *S. nigra:* 48980.
Hall, C. *S. nigra:* 141.
Hall, J. *S. nigra:* 6.
Hall, W. *S. nigra:* 70.
Hammer, M. *S. caroliniana:* 20.
Hammett, R. *S. nigra:* 128.
Hance, D. *S. nigra:* 49.
Hanna, G. *S. cinerea* subsp. *cinerea:* 58.
Hanna, R. *S. nigra:* 14.
Hansen, B. *S. caroliniana:* 5068, 5196, 5213, 5245, 5541, 5739, 6876, 6927, 6983, 8203.—*S. floridana:* 5226, 5430.
Hansen, D. *S. nigra:* 17.
Harbison, R. *S. cinerea* subsp. *cinerea:* 1041.—*S. cinerea* subsp. *oleifolia:* 1178.
Harbison, T. *S. alba:* 1, 5060, 6317, 7114, 7178.—*S. babylonica:* 14024.—*S. caroliniana:* 1, 1A, 2, 3, 4, 5, 6, 7, 8, 9, 10, 11, 12, 13, 15, 17, 18, 19, 20, 22, 25, 29, 30, 32, 33, 34, 35, 36, 37, 38, 43, 48, 53, 60, 61, 80, 109, 110, 384, 1203, 1204, 1425, 3847, 3853, 3871, 3882, 3884, 3885, 3903, 3905, 3920, 3961, 3962, 3984, 3997, 4001, 4013, 4056, 4081, 4087, 4088, 4172, 4985, 5947, 13848, 13855, 13886, 13893, 13894, 13904, 13985, 13986, 13991, 13992, 14002, 14009, 14057, 14074, 14075, 14091, 14092, 14287, 14413, 15373.—*S. caroliniana* × *nigra:* 2, 14158.—*S. eriocephala:* 13, 39.—*S. humilis* var. *humilis:* 1B, 2, 14, 34, 1417, 4106,

5217, 7132.—*S. humilis* var. *microphylla:* 30, 15242.—*S. nigra:* 1, 2, 3, 4, 5, 6, 8, 9, 10, 11, 13, 14, 15, 16, 18, 19, 21, 22, 23, 26, 27, 38, 44, 1539, 3917, 3918, 3999, 4022, 4023, 4034, 4038, 4044, 4093, 4094, 4095, 4107, 4110, 4126, 4127, 5005, 5061, 5405, 5777, 7012, 7113, 11843, 13892, 13919, 13926, 14076, 14133, 14156, 14160, 14324, 15256.—*S. sericea:* 1, 1B, 2, 3, 4, 5, 8, 29, 35, 46, 915, 3186, 7111, 7193, 14699.

Hardin, J. *S. alba:* 16478.—*S. caroliniana:* 2355, 2356, 2357, 2358, 2435, 2439.—*S. caroliniana* × *nigra:* 2369.—*S. humilis* var. *humilis:* 999, 2415, 2462.—*S. nigra:* 2359, 2369.—*S. sericea:* 2124, 13383, 13616, 13786, 16480.

Harmon, C. *S.* ×*sepulcralis:* 50.

Harper, F. *S. caroliniana:* 1383.—*S. humilis* var. *humilis:* 615.—*S. nigra:* 1548.

Harper, R. *S. caroliniana:* 18, 34, 133, 138, 205, 206, 212, 1499, 3891, 17028.—*S. eriocephala:* 3025, 3907, 3908.—*S. floridana:* 1381, 3634.—*S. humilis* var. *humilis:* 247, 1393, 3811.

Harriman, N. *S. caroliniana:* 13409.

Harris, J. *S. caroliniana:* 1796, 19329, 23222.

Harvill, A. *S. eriocephala:* 14782.

Haskinson, R. *S. exigua:* 149A.—*S. humilis* var. *humilis:* 151.—*S. nigra:* 148, 149.—*S. sericea:* 150.

Hauch, D. *S. caroliniana:* 554.

Haught, O. *S. eriocephala:* 7017.—*S. exigua:* 1145.—*S. humilis* var. *humilis:* 1140.—*S. nigra:* 1132.

Haynes, R. *S. caroliniana:* 2632, 4395, 4464, 4465, 4776.—*S. exigua:* 1550, 4147.—*S. nigra:* 962, 4602, 4645, 5700, 5700A, 5725, 6096, 7133, 7156, 7680.

Hebb, E. *S. nigra:* 409.

Heck, W. *S. babylonica:* 15.—*S. cinerea* subsp. *cinerea:* 21.

Heller, A. *S. alba:* 794, 894.—*S. humilis* var. *humilis:* 1304.

Heller, H. *S. caprea:* 91.

Helms, M. *S. caroliniana:* 1183.

Hendershot, D. *S. nigra:* 15.

Hendricks, R. *S. caroliniana:* 137A, 137B.—*S. nigra:* 92.

Henry, K. *S. nigra:* 55.

Henry, M. *S. humilis* var. *humilis:* 2176, 3382, 5214.—*S. nigra:* 2053, 4034.

Herbruck, N. *S. babylonica:* 4.

Herman, F. *S. nigra:* 11831.

Hermann, F. *S. caroliniana:* 10284.—*S. eriocephala:* 10283.—*S. humilis* var. *humilis:* 10463, 10792, 13752.—*S. nigra:* 10361.

Herrington, R. *S. nigra:* 23.

Hespenheide, H. *S. nigra:* 386.—*S. sericea:* 387.

Hess, W. *S. caroliniana:* 4575.

Hickey, C. *S. nigra:* 272.

Higgins, P. *S. eriocephala* × *sericea:* 1257.—*S. exigua:* 1258.—*S. humilis* var. *humilis:* 1073, 1285.—*S. sericea:* 1073, 1178.

Hightower, K. *S. nigra:* 86.

Hill, S. *S. humilis* var. *humilis:* 8957.—*S. nigra:* 8996.

Hillman, D. *S. nigra:* 132.

Hite, M. *S. caroliniana:* 550.

Hoag, M. *S. nigra:* 16.

Hoener, F. *S. nigra:* 1039, 1234.

Hollerk, G. *S. nigra:* 32.

Hollingsworth, C. *S. caroliniana:* 41.

Hollis, C. *S. nigra:* 55.

Holmes, R. *S. nigra:* 63.

Holmes, W. *S. nigra:* 2559.

Hooten. *S.* ×*sepulcralis:* 223.—*S. sericea:* 196.

Hornback, J. *S. exigua:* 38.

Horsey, R. *S. alba:* 1001.—*S. nigra:* 623, 826, 830, 831, 845, 868, 885, 931, 953, 1011, 1012, 1050, 1130, 1234, 1271, 1703, 1884, 1971, 2183, 2331, 2382.

Horst, J. *S. nigra:* 200.

Horton, J. *S. babylonica:* 842.—*S. humilis* var. *microphylla:* 2111.

Hotchkiss, A. *S. nigra:* 561.

Hotchkiss, N. *S. humilis* var. *microphylla:* 5637.

House, H. *S. humilis* var. *humilis:* 2276.—*S. humilis* var. *microphylla:* 4232.—*S. nigra:* 1822, 1910, 1929.

Howell, A. *S. caroliniana:* 661.

Howton, B. *S. nigra:* 65.

Hubricht, L. *S. caroliniana:* 2418B.

Huck, R. *S. caroliniana:* 641.

Hudnall, S. *S. nigra:* 9.

Hudson, C. *S. nigra:* 117.

Huff, M. *S. caroliniana* × *nigra:* 70.

Hughes, G. *S. caroliniana:* 10A.

Hughes, J. *S. nigra:* 53.

Humphries, B. *S. nigra:* 808.

Hunk, K. *S. caroliniana:* 1219A.

Hunnewell, F. *S. caroliniana:* 5841, 8947, 17950.—*S. eriocephala:* 2623.—*S. humilis* var. *humilis:* 5906, 12454.—*S. humilis* var. *microphylla:* 5092.—*S. nigra:* 2575, 15475.—*S. sericea:* 2634, 3999, 17049.

Hunt, K. *S. caroliniana:* 2, 3, 6, 416E, 516B, 915A, 2924, 2953, 2954, 2991, 2992, 2993, 2994, 2995, 3222, 3232, 3289, 3297.

Hunter, G. *S. nigra:* 1524, 2287, 4017.

Hutchins, R. *S. nigra:* 83.

Hutchinson, R. *S. nigra:* 81, 682.

Hyman, W. *S. sericea:* 11.

Iarussi, P. *S. cinerea* subsp. *cinerea:* 66.

Iltis, H. *S. caroliniana:* 1079, 5111, 5279,

23108.—*S. humilis* var. *humilis:* 2001, 2002, 3388, 19657, 27810.—*S. humilis* var. *microphylla:* 17178.—*S. nigra:* 3410, 3418, 25274.—*S. sericea:* 19656.

Itzig, C. *S. babylonica:* 66.

Jackson, H. *S. nigra:* 72.

Jackson, P. *S. nigra:* 25.

Jack, J. *S. babylonica:* 8245.—*S. nigra:* 2985.

Jacob, H. *S. caroliniana:* 1448.

James, B. *S. nigra:* 10.

James, F. *S. alba:* 6467.—*S. caroliniana:* 2946, 3875, 4173.—*S. eriocephala:* 7893, 8042, 10046.—*S. eriocephala* × *sericea:* 9024.—*S. humilis* var. *humilis:* 2385, 2742, 2817, 3120, 3250, 3604A, 4229, 4297, 4331, 4559, 4755, 5343, 5372, 5429, 5559, 6288, 6640, 6797, 6954, 7145, 7682, 7702, 7874, 8225, 8749, 9867, 13170, 13243, 13506, 13650, 13743, 13904.—*S. nigra:* 174, 2010, 2735, 2921, 2942, 3000, 3072, 4267, 4306, 4345, 4463, 4504, 4538, 4779, 4967, 5295, 5307, 5957, 6122, 6867, 7009A, 7064, 7176, 7329, 7340, 7433, 7478, 8452, 8712, 9036, 9045, 9100, 9336, 9456, 9589, 9845, 9972, 9989, 10056, 10281, 10298, 10682, 10794, 13325, 13331, 13431, 13592, 13698.—*S. sericea:* 1771, 2637, 3068, 7427, 7639, 7689, 7768, 7841, 8140, 8429, 8499, 8904, 8952, 8959, 8974, 9264, 9569, 9576, 9733, 9739, 9759, 9871, 10054, 10211, 10299, 10361, 10506, 10583, 10686, 10775, 10878, 11005, 11103, 11387, 11426, 11475, 11895, 12052, 12233, 13282.

James, R. *S. caroliniana:* 2.—*S. nigra:* 3.—*S. sericea:* 16880.

Jarvis, E. *S.* ×*rubens:* 89.

Jennings, K. *S. nigra:* 250.

Jennings, O. *S. exigua:* 648, 651.

Jennison, H. *S. humilis* var. *humilis:* 3557, 4248. *S. humilis* var. *microphylla:* 180, 1279, 3490, 4196.—*S. nigra:* 147, 1575, 3868.—*S. sericea:* 1014, 2003, 2428, 3098, 3880, 3906, 4061, 4247.

Johnson, M. *S. nigra:* 3353.—*S. sericea:* 3827, 4221.

Jones, E. *S. humilis* var. *humilis:* 2474.

Jones, F. *S. humilis* var. *microphylla:* 2448.—*S. sericea:* 20.

Jones, G. *S. babylonica:* 37783.—*S. humilis* var. *humilis:* 37782.—*S. humilis* var. *microphylla:* 37782.—*S. nigra:* 4003.

Jones, H. *S. nigra:* 431.

Jones, R. *S. babylonica:* 16.—*S. nigra:* 27, 187.—*S. sericea:* 359.

Jones, S. *S. caroliniana:* 23286.—*S. exigua:* 6335.—*S. humilis* var. *humilis:* 1235, 13182, 14453, 18775, 22506.—*S. humilis* var. *microphylla:* 14287, 19389.—*S. nigra:* 62, 436, 2238, 2696, 4049, 5066, 5389, 5841, 5863, 6236, 6591, 6970, 7035, 7266, 7370, 7439, 7536, 7766, 7842, 7999, 8116, 8259, 8490, 8559, 8563, 8762, 9012, 9535, 11198, 11604, 12234, 12477, 12629, 13129, 15569, 15647, 16491, 16990, 18882, 19015, 19723.

Joyce, R. *S. nigra:* 90.

Joye, G. *S. nigra:* 137, 165.

Judd, W. *S. floridana:* 2766, 2778.

Kearney, T. *S. caroliniana:* 1002, 1036, 1089, 1206, 1388, 1457.—*S. humilis* var. *humilis:* 1313.—*S. nigra:* 34.

Keeley, L. *S. nigra:* 184.

Keever, C. *S. babylonica:* 215.—*S. humilis* var. *humilis:* 866.—*S. nigra:* 234, 922.—*S. sericea:* 223, 852.

Kell, D. *S. caroliniana:* 2.

Keller, C. *S. babylonica:* 14.

Kelsey, S. *S.* ×*rubens:* 136.—*S.* ×*sepulcralis:* 41.

Kelso, L. *S. eriocephala:* 5885, 5887, 5906, 5907.—*S. exigua:* 5901, 6238.—*S. nigra:* 6479.—*S.* ×*rubens:* 6697.

Kennedy, J. *S. nigra:* 5, 7, 107.

Kerrick, K. *S. nigra:* 2.

Killar, A. *S. nigra:* 134.

Kiltz, B. *S. eriocephala:* 187.—*S. humilis* var. *humilis:* 13, 14, 224.

Kim, W. *S. caroliniana:* 223, 260.

Kincaid, J. *S. caroliniana:* 33.

King, C. *S. humilis* var. *microphylla:* 353.

Kirby, C. *S. nigra:* 153.

Kirkman, K. *S. babylonica:* 647.—*S. caroliniana* × *nigra:* 133.

Kirkpatrick, B. *S. nigra:* 11.

Kirkpatrick, R. *S. babylonica:* 4.—*S. nigra:* 5, 54, 645, 691.

Knotts, C. *S. nigra:* 46.

Knox, J. *S. nigra:* 208.

Koelz, W. *S. caroliniana:* 13431.

Kologiski, R. *S. sericea:* 160.

Kral, R. *S. alba:* 12251.—*S. babylonica:* 9583.—*S. caroliniana:* 1774, 1974, 2374, 2376, 2566, 2692, 2696, 2760, 5499, 12351, 14668, 38736, 38850A, 42453, 55393, 55468, 57218, 59576, 59582.—*S. eriocephala:* 9823, 12185A, 29967, 38477, 55041.—*S. exigua:* 39752.—*S. humilis* var. *humilis:* 10455, 13670, 23507, 37569, 39740, 40543, 44197, 44625, 47460, 48347, 53105, 55027, 55923, 56323, 62190, 62406.—*S. humilis* var. *microphylla:* 26013, 34354, 35354, 40538, 43758, 47806, 48297, 48346, 53755, 54854.—*S. nigra:* 6046, 8347, 8382, 13622, 16881, 19690, 26176, 27936, 28085, 29922, 29994, 30351, 38476, 41800, 45398, 45774,

54974, 55040, 55083, 55195, 55377, 61538.—*S.* ×*sepulcralis:* 55506.—*S. sericea:* 9842, 12185B, 13048, 23873, 34102, 36375, 38850B, 45590, 46734, 57580, 61556.
Kreager, P. *S. caroliniana:* 2, 3.—*S. nigra:* 1.
Krissinger, W. *S. humilis* var. *humilis:* 175.
Kurz, H. *S. caroliniana:* 5T.—*S. nigra:* A.
Kushlan, J. *S. caroliniana:* 19, 118.
Labisky, R. *S. nigra:* 311.
Lafferty, E. *S. babylonica:* 534.—*S. nigra:* 1144.—*S. sericea:* 624, 681, 773.
Laing, C. *S. sericea:* 18831.
Laing, H. *S. nigra:* 1002.
Laird, P. *S. nigra:* 1498, 2000.
Laird, R. *S. nigra:* 98.
Lakela, O. *S. caroliniana:* 23755, 23817, 23818, 25637, 26718, 26857, 28190, 28397, 28994, 29432, 30665.
Lalanne, G. *S. nigra:* 263.
Lambiotte, J. *S. babylonica:* 46.
Lamb, F. *S. sericea:* 816.
Lane, R. *S. nigra:* 180.
Langlois, A. *S. exigua:* 177.—*S. nigra:* 9.
Lankford, J. *S. babylonica:* 41.
Lapham, V. *S. caroliniana:* 1, 2, 3.
Lasseigne, A. *S. nigra:* 5610.
Lassetter, J. *S. exigua:* 1986.—*S. sericea:* 2404A, 2502.
Lassetter, S. *S. nigra:* 610.
Lassiter, C. *S. nigra:* 239.
Latspeich, C. *S. caroliniana:* 321.
Lawson, D. *S. babylonica:* 218.—*S. caroliniana:* 602, 926, 1592, 2037.—*S. exigua:* 1710, 2102.—*S. nigra:* 841, 1275, 2142.
Leeds, A. *S. humilis* var. *humilis:* 2166.
Lee, S. *S. nigra:* 242, 243, 274.
Leggett, V. *S. exigua:* 441.——*S. nigra:* 141.
Lelong, M. *S. caroliniana:* 5326.—*S. nigra:* 6935.
Lemaire, R. *S. caroliniana:* 117.—*S. nigra:* 985.
Leonard, E. *S. caroliniana:* 93, 238, 315.—*S. cinerea* subsp. *cinerea:* 20159.—*S. nigra:* 20578.
Leonard, S. *S. babylonica:* 1303.—*S. caroliniana:* 6235.—*S. humilis* var. *humilis:* 1329, 3279.—*S.* ×*sepulcralis:* 501.—*S. sericea:* 2378.
Lewis, C. *S. nigra:* 90.
Lewis, J. *S. humilis* var. *humilis:* 350, 661.—*S. nigra:* 402, 412, 1398, 1433, 1510.—*S. sericea:* 366, 373, 411, 1356, 1381, 1514, 3367.
Lewis, L. *S. nigra:* 301, 390, 1111, 2292.
Lelong, M. *S. nigra:* 4208, 4322.
Lieux, M. *S. nigra:* 16.
Lilly, J. *S. sericea:* 52.
Lipps, L. *S. nigra:* 49521.
Lipscomb, B. *S. babylonica:* 293B.—*S. caroliniana:* 293, 293A, 293C.—*S. exigua:* 179, 294.—*S. nigra:* 292, 292A, 294A.
Little, E. *S. caroliniana:* 13840, 14276, 14299, 14300, 14311, 14368.—*S. nigra:* 14629.—*S. sericea:* 14604, 17376, 17377.
Little, F. *S. nigra:* 56.
Lix, H. *S. humilis* var. *humilis:* 559.—*S. nigra:* 560.
Locke, M. *S. exigua:* 584, 2797.—*S. nigra:* 657, 1115, 2391, 2953, 4057.
Loetterle, L. *S. nigra:* 101, 464, 778.
Long, B. *S. bebbiana:* 8035.—*S. eriocephala* × *sericea:* 1533, 1536.—*S. humilis* var. *humilis:* 1052, 1283, 58208, 68355.—*S. humilis* var. *microphylla:* 34055, 54317, 54369.—*S. lucida:* 1329.—*S. nigra:* 3232B.—*S. sericea:* 2552.
Long, J. *S. nigra:* 61.
Long, R. *S. caroliniana:* 2770.
Lord, W. *S. humilis* var. *microphylla:* 163.
Lovell, E. *S. nigra:* 113, 373, 850.
Lowery, D. *S. exigua:* 601.—*S. nigra:* 507.
Lowe, N. *S. nigra:* 22.
Luces, Z. *S. eriocephala:* 10.
Lurry, D. *S. nigra:* 9, 29, 42.
Lycan, I. *S. alba:* 183.
Lycom, J. *S. nigra:* 237.
Lyle, C. *S. nigra:* 19684.—*S. sericea:* 19266.
Lynch, K. *S. nigra:* 36.
Mack. *S. caroliniana:* 60.
MacRoberts, D. *S. exigua:* 1148.—*S. nigra:* 1149.
Magee, K. *S. exigua:* 13.—*S. nigra:* 78.
Mahler, W. *S. nigra:* 4723.—*S. sericea:* 4748.
Malter, J. *S. nigra:* 53935.—*S. sericea:* 52856, 53932.
Manning, R. *S. nigra:* 172.
Mann, I. *S. nigra:* 273.
Mann, M. *S. humilis* var. *humilis:* 339.
Marford, H. *S. nigra:* 29.
Marshall, B. *S. humilis* var. *humilis:* 16.
Marshall, C. *S. humilis* var. *humilis:* 67.
Marsh, D. *S. exigua:* 2024.
Martin, D. *S. nigra:* 64.
Martin, R. *S. caroliniana:* 938, 1486, 1491.—*S. eriocephala:* 2037.—*S. nigra:* 1015, 1632.—*S. sericea:* 1013, 2035.
Marx, P. *S. nigra:* 712, 904, 1818.—*S. sericea:* 2524.
Mason, J. *S. cinerea* subsp. *cinerea:* 41.—*S. nigra:* 2.
Massey, A. *S. caroliniana:* 2391.—*S. cinerea* subsp. *cinerea:* 2406.—*S. eriocephala:* 2324, 2384, 4205, 4658, 13046.—*S. humilis* var. *humilis:* 2407, 3067, 13156.—*S. humilis* var. *microphylla:* 5157.—*S. nigra:* 2385.—*S. sericea:* 2987.

Massey, C. *S. babylonica:* 49.
Masters, W. *S. sericea:* 87.
Matthews, J. *S. babylonica:* 608.—*S. discolor:* 710.—*S. exigua:* 401, 620.—*S. nigra:* 58, 99, 132.
Matthews, W. *S. caroliniana:* 20, 155.
Maxon, W. *S. caroliniana:* 8, 21, 27, 57, 360, 11267.—*S. caroliniana* × *nigra:* 7.—*S. nigra:* 7, 25, 58, 59.—*S. sericea:* 6083.
Mazzeo, P. *S. alba:* 12945.—*S. babylonica:* 1230, 1974.—*S. caroliniana:* 1914A, 1914B, 1945, 2897.—*S. cinerea* subsp. *cinerea:* 10313, 13032.—*S. nigra:* 1301, 1302, 1999.—*S.* ×*rubens:* 13692.—*S.* ×*sepulcralis:* 1200, 18373.
McAllister, B. *S. caroliniana:* 405.
McAnally, B. *S. nigra:* 30.
McAtee, W. *S. caroliniana:* 1719, 1747B.—*S. humilis* var. *humilis:* 3563.—*S. nigra:* 3485, 3569, 3577.—*S. sericea:* 3562, 3570.
McCart, W. *S. caroliniana:* 9513, 9514, 10425, 10494, 10518, 10572, 10577, 10672, 10756, 11304.
McCoy, E. *S. sericea:* 6.
McCoy, J. *S. nigra:* 48, 137.
McCoy, P. *S. humilis* var. *humilis:* 4.
McCoy, T. *S. exigua:* 175.—*S. nigra:* 182.
McCurdy, J. *S. nigra:* 202.
McCurdy, Y. *S. sericea:* 333.
McDade, L. *S. caroliniana:* 461.
McDaniel, J. *S. nigra:* 23.
McDaniel, S. *S. caroliniana:* 7486.—*S. eriocephala:* 22938.—*S. exigua:* 853, 13327.—*S. humilis* var. *microphylla:* 2840.—*S. nigra:* 551, 8649, 13267, 21192.
McDougall, W. *S. nigra:* 1531.
McDowell. *S. nigra:* 322.
McFarland, F. *S. alba:* 3855, 3935, 4078.—*S. caroliniana:* 4582.—*S. eriocephala:* 7, 38, 39, 73.—*S. exigua:* 8, 17, 30, 31, 36, 40, 50, 60, 62, 70, 71, 2764, 3321, 3322, 3325, 3375, 3868, 4085.—*S. fragilis:* 6.—*S. humilis* var. *humilis:* 64, 69, 3576, 3577, 3854, 3856, 4072, 4485, 4536.—*S. humilis* var. *microphylla:* 3185, 3193, 3292, 4060.—*S. nigra:* 1, 4, 5, 9, 16, 26, 29, 35, 37, 51, 59, 61, 63, 65, 2372, 3316, 3324, 3365, 3428, 3591, 4086.—*S. sericea:* 17, 33, 34, 52, 72, 3129, 3323, 3337, 3366, 3590, 3907, 3949.
McFarland, J. *S. nigra:* 1678.
McFarlin, J. *S. caroliniana:* 4363, 4382, 4517, 4533, 4542, 5485.
McGregor, E. *S. nigra:* 505.
McHenry, L. *S. nigra:* 46.
McInteer, B. *S. caroliniana:* 970.—*S. exigua:* 814.—*S. humilis* var. *humilis:* 436, 1294.—*S. humilis* var. *microphylla:* 357.—*S. nigra:* 331, 353, 370, 379, 1539, 1553, 1755, 1903.—*S. sericea:* 10S, 20S, 30S, 812, 912, 1217, 1551, 1920.
McKay, R. *S. nigra:* 192, 264.
McKenzie, F. *S. nigra:* 22.
McKinney, G. *S. nigra:* 17.
McLemore, W. *S. nigra:* 3, 26.
McNeely, J. *S. sericea:* 1299.
McVaugh, R. *S. humilis* var. *humilis:* 4875, 5501, 5510, 5564.—*S. humilis* var. *microphylla:* 4861.—*S. sericea:* 5508, 5655.
McWilliams. *S. nigra:* 218.
Mearns, E. *S. caroliniana:* 2.
Meeks, D. *S. nigra:* 154.
Meijer, W. *S. caroliniana:* 51180.—*S. humilis* var. *microphylla:* 1153.
Mellichamp, T. *S. nigra:* 667.
Mell, C. *S. caroliniana:* 83.
Menzel, R. *S. humilis* var. *humilis:* 400.
Merrell, J. *S. exigua:* 379.
Merrill, G. *S. caroliniana:* 95, 343, 1967, 1979, 2130.—*S. humilis* var. *humilis:* 1689.—*S. nigra:* 1583, 1825.
Metcalfe, F. *S. nigra:* 622, 646.
Meyer, F. *S. alba:* 13681, 14866, 15897, 16579.—*S. babylonica:* 10427, 11744, 14901, 14986, 15113, 15211, 15431, 15432, 15896, 15899, 16415, 17459, 18086, 18456, 18517.—*S. caroliniana:* 10388, 11596, 11633, 11932.—*S. caroliniana* × *nigra:* 10689A, 10689B.—*S. cinerea* subsp. *cinerea:* 9796, 13562, 13688, 14867, 16129.—*S. eriocephala:* 13659F, 13660, 13662, 13697, 18081, 18082.—*S. exigua:* 13671.—*S. fragilis:* 13682, 17699.—*S. nigra:* 9168, 16578.—*S. pentandra:* 14189.—*S. purpurea:* 10315, 12358, 13669, 13677, 13680, 13687, 13690, 13695, 13696, 14881.—*S.* ×*rubens:* 13602.—*S.* ×*sepulcralis:* 15108.
Mikula, B. *S. humilis* var. *humilis:* 4360.—*S. nigra:* 341, 5505, 8362, 8475.—*S.* ×*rubens:* 339, 5506.—*S. sericea:* 6452.
Miller, J. *S. caroliniana:* 203.—*S. humilis* var. *humilis:* 181, 4425.—*S. nigra:* 15, 873.—*S. sericea:* 192.
Miller, M. *S. nigra:* 40.
Millspaugh, C. *S. alba:* 55.—*S. nigra:* 105, 113.—*S. sericea:* 35, 89, 404.
Mitchell, R. *S. caroliniana:* 1018.
Mitchell, T. *S. babylonica:* 15.
Mobbs, B. *S. eriocephala:* 19.
Moertle, R. *S. nigra:* 96.
Mohead, M. *S. babylonica:* 97.—*S. nigra:* 96.
Moldenke, H. *S. caroliniana:* 342.
Montfrans, M. *S. nigra:* 317.
Montgomery, F. H. *S. nigra:* 37, 22493.
Montz, G. *S. exigua:* 669, 2702, 3534, 4053,

4288.—*S. nigra:* 98, 1479, 1605, 1944, 1996, 2683, 2702, 3522, 3920.
Montz, W. *S. nigra:* 34.
Moore, A. *S. sericea:* 2398.
Moore, D. *S. alba:* 5.—*S. caroliniana:* 1, 40, 92, 143, 169, 227, 273, 390, 627, 850, 1079.—*S. eriocephala:* 2, 39, 632, 3002.—*S. exigua:* 35, 168, 345, 515, 524.—*S. humilis* var. *humilis:* 1, 44, 49A, 172, 454, 5351.—*S. nigra:* 34, 37, 434, 593, 4028.
Moore, E. *S. caroliniana:* 143.
Moore, J. *S. caroliniana:* 86, 500.—*S. humilis* var. *humilis:* 989.—*S. nigra:* 651, 1099, 5251.
Moore, K. *S. nigra:* 43.
Moore, M. *S. humilis* var. *humilis:* 68.
Moreland, D. *S. nigra:* 1504.
Morf, H. *S. sericea:* 12.
Morgan, J. *S. nigra:* 966.
Moritz, W. *S. babylonica:* 13.
Morizot, D. *S. nigra:* 57.
Morris, E. *S. nigra:* 26, 50, 15394.
Morton, G. *S. caroliniana:* 4845.—*S. humilis* var. *humilis:* 42659, 42986.—*S. nigra:* 43424.—*S. sericea:* 42673.
Morton, V. *S. alba:* 1814.
Mosier, C. *S. caroliniana:* 109, 338, 347.
Muenscher, W. *S. caroliniana:* 14012, 14198, 14199, 14245.
Mulcahy, D. *S. caroliniana:* 1301.
Muntz, H. *S. nigra:* 8682.
Murphy, G. *S. sericea:* 5135.
Murphy, S. *S. nigra:* 317.
Musselman, L. *S. humilis* var. *humilis:* 3924.
Myers, J. *S. sericea:* 525.
Nash, E. *S. caroliniana:* 134.
Nash, G. *S. caroliniana:* 134.
Nease, F. *S. nigra:* 234, 235, 236, 238, 239, 240.
Nederveld, A. *S. nigra:* 106.
Nee, M. *S. exigua:* 655.
Neff, S. *S. exigua:* 89.
Neighbours, J. *S. nigra:* 26.
Nelson, J. *S. floridana:* 888.—*S. nigra:* 1473.
Netherly, F. *S. nigra:* 86.—*S. sericea:* 32.
Newber, M. *S. sericea:* 440.
Newson, A. *S. caroliniana:* 6, 7.
Newton, D. *S. nigra:* 6.
Nicely, K. *S. eriocephala:* 935.—*S. humilis* var. *humilis:* 1004, 2955.—*S. nigra:* 1799.—*S. sericea:* 627, 831.
Nichols, P. *S. nigra:* 349.
Nixon, C. *S. nigra:* 59.
Noell, I. *S. humilis* var. *humilis:* 49.
Nolen, R. *S. sericea:* 18, 25.
Norris, F. *S. exigua:* 16336.—*S. nigra:* 16348.
Norris, R. *S. nigra:* 786, 812.
Northcutt, C. *S. cinerea* subsp. *cinerea:* 48.—*S.* ×*sepulcralis:* 23.
Norton, J. *S. eriocephala:* 3045.—*S. humilis* var. *humilis:* 3011, 3028, 3032, 3043.—*S. sericea:* 3013, 3018, 3037.
Nor, J. *S. babylonica:* 35.
O'Neill, H. *S. caroliniana:* 396, 414, 591.—*S. humilis* var. *humilis:* 449.
Omar, M. *S. nigra:* 50.
Oosting, H. *S. nigra:* 33103.—*S. sericea:* 2236, 3458, 33102, 34218.
Owens, R. *S. nigra:* 267, 301.
Oxner, M. *S. caroliniana:* 91.
Padgett, M. *S. nigra:* 194.
Padgett, R. *S. babylonica:* 718.
Painter, C. *S. nigra:* 53.
Painter, J. *S. humilis* var. *humilis:* 560, 565, 568, 1502. —*S. humilis* var. *microphylla:* 1471. —*S. nigra:* 1264, 1352, 1372, 1503. —*S.* ×*rubens:* 1246.—*S. sericea:* 562, 563, 564.
Palko, T. *S. nigra:* 23.
Palmer, E. *S. alba:* 16512.—*S. amygdaloides:* 16502.—*S. caroliniana:* 5588, 14308, 17602, 17660, 20499, 23100, 23187, 23211, 23255, 24507, 24736, 25291, 26504, 26567, 27302, 27310, 27518, 29591, 35203, 35414, 38354, 38467, 39306, 39816.—*S. cinerea* subsp. *cinerea:* 39933.—*S. cinerea* subsp. *oleifolia:* 35441.—*S.* ×*ehrhartiana: 15115.*—*S. eriocephala:* 27217, 29651.—*S. exigua:* 7526, 9448, 15058, 17798, 24978, 25013, 25085.—*S. humilis* var. *humilis:* 12627, 17334, 17681, 23007, 23212, 23264, 24529, 25056, 27125, 35279, 35423, 35473, 39783, 42235, 42587, 44479.—*S. nigra:* 1829, 5394, 5627, 6044, 7427, 7429, 7492, 7670, 12669, 16513, 17327, 17556, 17717, 17880, 23070, 23256, 24539, 25149, 26585, 27126, 27226, 35203, 35281, 35414, 35524, 42558.—*S.* ×*rubens:* 17197, 17825.—*S. sericea:* 17703, 35424, 39929, 39944, 42243, 42482, 42559.
Papenoe, J. *S. caroliniana:* 527.
Parker, E. *S. babylonica:* 96.—*S. nigra:* 75, 88.
Parker, H. *S. humilis* var. *humilis:* 89.
Parker, J. *S. nigra:* 415, 648.
Parrish, R. *S. nigra:* 50.
Patrick, T. *S. eriocephala:* 1353.—*S. humilis* var. *humilis:* 382.—*S. nigra:* 210.
Patti, J. *S. caroliniana:* 3.
Patton, E. *S. nigra:* 70.
Pavlu, R. *S. caroliniana:* 1178.—*S. nigra:* 628.
Payne, R. *S. caroliniana:* 54.
Peake, D. *S. humilis* var. *humilis:* 698.—*S. nigra:* 792.—*S. sericea:* 497, 793.
Peattie, D. *S. alba:* 1062.—*S. nigra:* 1626, 2136.—*S. sericea:* 132, 1607, 1802.
Pennell, F. *S. caroliniana:* 12076.—*S. eriocephala:* 1574.—*S.* ×*rubens:* 2052.

Perkins, K. *S. humilis* var. *humilis:* 901, 912, 913.
Perry, L. *S. nigra:* 845.
Philips, D. *S. babylonica:* 11.
Phillips, H. *S. nigra:* 2569.
Pias, P. *S. exigua:* 4090.—*S. nigra:* 24.
Pierrelee, J. *S. exigua:* 149, 648.
Piper, C. *S. caroliniana:* 1, 2.
Pippin, J. *S. nigra:* 52, 140.
Pittillo, D. *S. alba:* 4464.—*S. cinerea* subsp. *cinerea:* 3, 4250, 6531, 7621.—*S. cinerea* subsp. *oleifolia:* 3353, 4389, 4526.—*S. nigra:* 2468, 2967, 2974, 3366.—*S.* ×*sepulcralis:* 3143, 3296.
Plitt, C. *S. alba:* 758.
Plymale, L. *S. alba:* 472.—*S. exigua:* 760.—*S. nigra:* 492.—*S. purpurea:* 470.
Pollard, C. *S. caroliniana:* 874.—*S. nigra:* 23, 37.
Pollock, W. *S. nigra:* A.—*S. sericea:* B.
Porcher, R. *S. caroliniana:* 2A, 1103B, 1320A, 1338, 1352A, 1572, 1656, 1657, 1950.—*S. humilis* var. *humilis:* 1617A.
Porter, C. *S. caroliniana:* 10623, 10650, 10748, 10782.
Porter, D. *S. caroliniana:* 876.—*S. humilis* var. *humilis:* 1967.—*S. humilis* var. *microphylla:* 1968, 2028.
Posey, A. *S. caroliniana:* 30.
Powell, J. *S. caroliniana:* 3.—*S. nigra:* 108.—*S. sericea:* 11.
Prewitt, R. *S. nigra:* 32.
Price, A. *S. nigra:* 672.
Prichard, E. *S. caroliniana:* 558.
Pruski, J. *S. nigra:* 131.
Pullen, T. *S. exigua:* 66323, 66385, 691114.—*S. nigra:* 6489, 8895, 64210, 66512, 66600, 66724, 66793, 66868, 66974, 67525, 67891, 69852, 70381, 70634, 70713, 661046.
Pulley, H. *S. babylonica:* 28.
Pultorak, R. *S. nigra:* 769.—*S. sericea:* 527, 772.
Pyron, J. *S. humilis* var. *humilis:* 25, 1554.—*S. humilis* var. *microphylla:* 524, 1556.—*S. nigra:* 82, 114, 2621.
Quarterman, E. *S. caroliniana:* 4366.—*S. humilis* var. *microphylla:* 1255.—*S. nigra:* 1141.
Quidas, F. *S. nigra:* 42.
Racette, T. *S. caroliniana:* 6.
Radford, A. *S. alba:* 6372.—*S. babylonica:* 32247.—*S. caroliniana:* 20, 40, 6296, 6604, 7671, 7739, 8099, 22040, 22213, 22268, 31777, 31923, 32173, 32406, 33466, 33566, 33728, 33768, 33868, 44616.—*S. caroliniana* × *nigra:* 32404.—*S. cinerea* subsp. *oleifolia:* 7098.—*S. humilis* var. *humilis:* 354, 355, 6141, 10143.—*S. humilis* var. *microphylla:* 549A, 8409.—*S. nigra:* 23, 50, 738A, 7590, 10753, 10860, 11692, 12029, 20139, 20182, 20471, 20791, 20933, 21071, 21079, 21186, 22781, 31983, 32245, 32392, 32587, 33173, 34583, 42829, 43479, 43652.—*S. sericea:* 371A, 375A, 416A, 445A, 471A, 542A, 8867, 9111, 9669, 11822, 32852, 32887, 43442, 43443.
Raines, T. *S. caroliniana:* 76096.
Ramseur, G. *S. humilis* var. *humilis:* 218.—*S. sericea:* 945.
Ramsey, B. *S. nigra:* 39.—*S. sericea:* 39.
Ramsey, G. *S. caroliniana:* 256.—*S. nigra:* 4890.—*S. sericea:* 4832, 4852.
Randolph, F. *S. caroliniana:* 165, 166.—*S. nigra:* 27.
Randolph, L. *S. humilis* var. *humilis:* 252.—*S. nigra:* 92.
Raven, P. *S. caroliniana:* 26322.
Rawlinson, E. *S. humilis* var. *humilis:* 1, 4, 81A.
Ray, J. *S. caroliniana:* 7419, 7796, 10739.—*S. nigra:* 4932, 8313.
Reade, J. *S. humilis* var. *humilis:* 6627.
Redfearn, P. *S. caroliniana:* 2507.—*S. humilis* var. *humilis:* 27929.—*S. nigra:* 2754, 30566, 30652.
Redman, D. *S. discolor:* 531.
Redmond, B. *S. nigra:* 21.
Reed, C. *S. alba:* 61233, 94459, 95892, 101715.—*S. babylonica:* 1687, 36428.—*S. caprea:* 64070, 64071.—*S. caroliniana:* 5145, 20713, 20715, 24972, 28519, 37340, 37396, 37539, 37545, 41608, 42000, 42057, 42754, 49563, 49564, 82304, 94172, 102028, 115114.—*S. cinerea* subsp. *cinerea:* 200.—*S. eriocephala:* 16564, 26505, 79006, 103571, 104354.—*S. exigua:* 15378, 19337A, 44716, 44754, 79975, 81661.—*S. fragilis:* 2504, 10015, 14546, 26209, 44393, 91805.—*S. humilis* var. *humilis:* 28, 1684, 1691, 2124, 6567, 7775, 7924, 10520, 13173, 15907, 16123, 20777, 23723, 31808, 36606, 40626, 40684, 40717, 41682, 41902, 42686, 43856, 46185, 51072, 76322, 77945, 78928, 78929, 80321, 80331, 82623, 83226, 86662, 88494, 89220, 89516, 89682, 89715, 90035, 92287, 92407, 94589, 94591, 95211, 96241, 96482, 96665, 97308, 97683, 98130, 99352, 99832, 99985, 103298, 105007, 105159, 105161, 105904, 105906, 106067, 106358, 115629.—*S. humilis* var. *microphylla:* 2990, 3165, 10606, 13500, 19718, 21104, 23089, 24419, 24422, 75375, 83697, 88176, 89109, 89517.—*S. nigra:* 148, 219, 235, 1669, 1890, 2501, 2540, 2606, 2697, 3089, 3261, 3341, 4919, 7300, 7868, 7931, 8198, 8400, 8837, 8903, 9055, 10612, 12725,

12819, 13144, 14302, 14872, 15372, 15568, 16565, 18873, 18876, 18921, 19334, 19335, 19850, 20508, 23076, 23077, 23354, 24839, 25207, 36291, 43889, 44390, 44763, 62075, 78909, 79041, 87034, 87220, 87221, 88974, 89452, 93950, 98241, 99167, 103696, 109091.—*S. purpurea:* 79009, 79010, 96803, 112976.—*S.* ×*rubens:* 8839, 19857, 79001, 98145.—*S.* ×*sepulcralis:* 110309.—*S. sericea:* 2043, 2486, 3238, 3554, 3572, 4354, 4559, 6179, 8393, 12796, 12804, 12861, 13606, 13930, 13931, 15560, 17063, 17071, 17087, 18394, 21885, 24054, 24418, 27758, 29750, 45491, 78424, 79958, 80103, 80107, 80793, 83605, 83765, 91054, 92588, 92615, 97091, 97472, 98077, 98491, 98588, 99239, 102666, 105925.

Regan, E. *S. nigra:* 158.

Rehder, A. *S. nigra:* 967.

Reid, R. *S. nigra:* 640, 671, 874, 1295, 1654, 2152, 2223.

Reimer, C. *S. nigra:* 58.

Renfroe, R. *S. nigra:* 13.

Reynolds, I. *S. exigua:* 1.

Rhodes, D. *S. exigua:* 5190.—*S. nigra:* 2519, 3827, 5219, 8262, 12008, 12472, 14572, 14766, 15428, 18252.

Richardson, C. *S. babylonica:* 11.

Rich, N. *S. exigua:* 367, 373.—*S. nigra:* 89, 444, 1182.

Ridlehoover, F. *S. babylonica:* 958.

Riedlinger, D. *S. exigua:* 127.

Riggins, R. *S. caroliniana:* 1069.

Roadifer, R. *S. nigra:* 144.

Robert, G. *S. nigra:* 29.

Roberts, J. *S. nigra:* 1005.

Robicheaux, B. *S. nigra:* 334, 888.

Robinette, B. *S. exigua:* 144.—*S. nigra:* 135.

Robinson, A. *S. caroliniana:* 77.

Robinson, B. *S. caroliniana:* 235, 236, 478, 479.—*S. caroliniana* × *nigra:* 477, 479.

Rodenburg, J. *S. caroliniana:* 45.

Roder, L. *S. sericea:* 2002.

Rodgers, C. *S. babylonica:* 69231, 75154, 77002.—*S. caroliniana:* 125.—*S. nigra:* 67056, 73260, 76129, 76211.—*S. sericea:* 61347A, 63001, 69171, 69374, 70084, 73053.

Rodgers, J. *S. caroliniana:* 99.

Rodgers, T. *S. caroliniana:* 76.

Roe, G. *S. nigra:* 989.—*S. sericea:* 1063.

Rogers, H. *S. caroliniana:* 120, 520.—*S. humilis* var. *microphylla:* 1039.—*S. sericea:* 17.

Rogers, K. *S. caroliniana:* 2813A, 45902.—*S. exigua:* 1998.—*S. humilis* var. *humilis:* 2812A, 2812B, 6832.—*S. humilis* var. *microphylla:* 2513, 41201, 41571.—*S. nigra:* 4348.—*S. sericea:* 4068.

Rogillio, H. *S. nigra:* 11.

Rohrer, J. *S. humilis* var. *humilis:* 1712.—*S. sericea:* 1759.

Roper, E. *S. nigra:* 10.

Ross, E. *S. sericea:* 54.

Rossbach, G. B. *S. caroliniana:* 842, 1847, 1897, 2786, 7216.—*S. eriocephala:* 1319, 1619, 4154.—*S. humilis* var. *humilis:* A, 1071, 1072, 1157, 2505, 3929, 3962, 3963, 8353, 8353A, 8353B.—*S. nigra:* 2209, 2575, 4625, 4690.—*S. sericea:* B, 875, 1073, 1342, 2338, 2500, 2605, 4117, 4153, 5855, 6115, 7331, 7353.

Rowley, F. *S. nigra:* 15.

Roycraft, J. *S. nigra:* 178.

Rudd, V. *S. nigra:* 106.

Rugel, F. *S. humilis* var. *humilis:* 2.

Rush, L. *S. nigra:* 44.

Ruth, A. *S. humilis* var. *humilis:* 154.—*S. nigra:* 155, 161, 224, 424.—*S. sericea:* 195.

Rydberg, P. *S. humilis* var. *humilis:* 9147, 9466.

Sabend. *S. nigra:* 62.

Safford, W. *S. caroliniana:* 109, 334.

Sale, J. *S. sericea:* 677.

Salle, E. *S. babylonica:* 23.—*S. nigra:* 148, 220.

Salomon, R. *S. humilis* var. *humilis:* 408.

Sargent, F. *S. caroliniana:* 97, 7259.

Sasnett, P. *S. humilis* var. *humilis:* 38.—*S. nigra:* 44.

Savage, J. *S. nigra:* 311.

Scarborough, K. *S. babylonica:* 892.—*S. nigra:* 787.

Schanbacker, T. *S. babylonica:* 20.

Schipper, A. *S. alba:* 25.

Schlossberg, M. *S. babylonica:* 7.

Schmidt, W. *S. babylonica:* 39.

Schramm, J. *S. babylonica:* 10.—*S. sericea:* 16.

Schrom, L. *S. sericea:* 358.

Schutz, S. *S. nigra:* 162, 438, 452.

Scott, D. *S. nigra:* 38.

Screpetis, D. *S. nigra:* 30.

Scully, F. *S. exigua:* 1320.

Seaman, W. *S. babylonica:* 7035.—*S. humilis* var. *humilis:* 2943, 3122.—*S. nigra:* 3600.—*S. sericea:* 3624.

Searcy, F. *S. nigra:* 639.

Sears, M. *S. caroliniana:* 886.—*S. nigra:* 789.

Seay, J. *S. nigra:* 46.

Seibert, R. *S. caroliniana:* 1192, 1204, 1205, 1393.

Semple, J. C. *S. caroliniana:* 2151.

Sessler, A. *S. nigra:* 21.

Setser, H. *S. nigra:* 492.

Settle, L. *S. caroliniana:* 3.

Seymour, F. *S. nigra:* 16687.

Shacklette, H. *S. exigua:* 934, 936.—*S. humilis* var. *humilis:* 4S.—*S. nigra:* 25, 35, 935.

Shanks, R. *S. caroliniana:* 5035, 5782, 7196.—*S. eriocephala:* 15837, 15898.—*S. humilis* var. *humilis:* 1044, 1067, 3330, 3775, 5549, 5775, 8934, 13756, 14170.—*S. humilis* var. *microphylla:* 2020, 3040, 3589, 5451, 5638, 11339, 14815.—*S. nigra:* 1123, 2100, 4229, 14174, 15459, 15550, 15773, 15791, 15874, 18567. —*S. sericea:* 8837, 15412, 18562, 18566.
Sharp, A. *S. alba:* 4219.—*S. caroliniana:* 346, 395, 397, 845, 1334, 4993, 7579, 9658, 9761.—*S. eriocephala:* 6253, 6303, 6476, 7940, 7958, 8006, 8096, 12141, 16334.—*S. exigua:* 7950, 10408, 10479.—*S. humilis* var. *humilis:* 157, 1242, 1651, 2820, 3718, 3754, 7431, 7795, 7797, 9339, 9636, 11222, 30094.—*S. humilis* var. *microphylla:* 2876, 4750, 9629, 11065, 11331, 22610.—*S. nigra:* 8389, 10283, 12260, 12363, 13041, 25671.—*S. ×sepulcralis:* 8269, 12376, 17042.—*S. sericea:* 25, 11662, 13170, 26984, 28101, 45621.
Sharp, D. *S. nigra:* 1958.
Sharpe, M. *S. nigra:* 309.
Shaver, J. *S. caroliniana:* 2376, 2470.—*S. nigra:* 2433.
Shelby, J. *S. nigra:* 47.
Shell, D. *S. nigra:* 46.
Sherman, B. *S. nigra:* 80.
Short, C. *S. humilis* var. *microphylla:* 7.
Short, D. *S. exigua:* 305.—*S. nigra:* 304.
Shreve, F. *S. caroliniana:* 1014.—*S. humilis* var. *microphylla:* 376.—*S. nigra:* 399, 477, 907.
Shuey, A. *S. caroliniana:* 248, 249.
Shull, G. *S. nigra:* 417.—*S. sericea:* 419.
Sieren, D. *S. caroliniana:* 1537.
Sijam, K. *S. nigra:* 98.
Silberhorn, G. *S. nigra:* 1377.—*S. sericea:* 1365.
Silverstone, P. *S. caroliniana:* 61.
Simmering, R. *S. nigra:* 85.
Simpson, D. *S. nigra:* 5.
Sinclair, C. *S. nigra:* 1755.
Singer, J. *S. nigra:* 50.
Singleton, F. *S. caroliniana:* 5A.
Slack, M. *S. nigra:* 434.
Small, J. *S. caroliniana:* 161, 182, 268, 723, 2898, 2984, 2985, 3076, 3411, 5639, 7313, 8177, 11773A, 12273, 12476, 12715.—*S. humilis* var. *humilis:* 6980.—*S. humilis* var. *microphylla:* 109.—*S. nigra:* 12462.
Smith, B. *S. babylonica:* 1627.—*S. humilis* var. *humilis:* 1623, 1625.—*S. nigra:* 1618.—*S. sericea:* 1617, 1621, 1622.
Smith, C. *S. babylonica:* 945, 1592.—*S. caroliniana:* 89.—*S. cinerea* subsp. *cinerea:* 1411.—*S. exigua:* 551.—*S. nigra:* 319, 370, 428, 498, 550, 566, 622, 1369.
Smith, D. *S. nigra:* 60, 90.—*S. sericea:* 1723.
Smith, E. *S. caroliniana:* 263, 1353.—*S. humilis* var. *humilis:* 26, 508.—*S. nigra:* 65.
Smith, F. *S. caprea:* 19385, 22981.
Smith, G. *S. discolor:* 2628.—*S. humilis* var. *humilis:* 2628, 2632, 2641.—*S. nigra:* 155.
Smith, H. *S. nigra:* 2572.
Smith, J. *S. caroliniana:* 451.
Smith, L. *S. humilis* var. *microphylla:* 4193.—*S. sericea:* 436.
Smith, R. *S. caroliniana:* 916.
Smschek, J. *S. babylonica:* 24.—*S. nigra:* 19.
Smyth, M. *S. eriocephala:* 4242, 4290.—*S. eriocephala* × *sericea:* 5087.—*S. nigra:* 4736.
Sole, J. *S. nigra:* 613, 744.
Solomon. *S. sericea:* 1824.
Solomon, J. *S. humilis* var. *humilis:* 5460.—*S. sericea:* 1625.
Solomon, R. *S. nigra:* 429, 430.
Soltis, D. *S. nigra:* 507.
Spawn, W. *S. nigra:* 1038.
Spencer, R. *S. nigra:* 12.
Stack, J. *S. caroliniana:* 524.—*S. sericea:* 125.
Standley, P. *S. caroliniana:* 12567, 12599, 12952.—*S. humilis* var. *humilis:* 5641.—*S. sericea:* 5483, 5723, 10458.
Stearn, J. *S. caroliniana:* 18.
Stegall, H. *S. nigra:* 36.
Stevens, C. *S. humilis* var. *humilis:* 140.—*S. humilis* var. *microphylla:* 2782.
Stewart, S. *S. caroliniana:* 784.
Stibolt, D. *S. cinerea* subsp. *cinerea:* 70.
Still, G. *S. exigua:* 225.
Still, L. *S. nigra:* 58, 180, 338.
Stimson, W. *S. nigra:* 361.
Stone, D. *S. caroliniana:* 3903.—*S. nigra:* 174, 175, 2588, 3642.
Stone, M. *S. nigra:* 26.
Stone, W. *S. humilis* var. *humilis:* 14013.—*S. nigra:* 46, 105, 14014.—*S. sericea:* 14012.
Strain, R. *S. nigra:* 27.
Strange, T. *S. nigra:* 50.
Stryker, S. *S. babylonica:* 49.
Sturtevant, W. *S. caroliniana:* 93.
Sullivan, D. *S. humilis* var. *humilis:* 389.—*S. nigra:* 401, 422, 425.—*S. sericea:* 394, 402, 413.
Sullivan, G. *S. nigra:* 269.
Sullivan, L. *S. nigra:* 17.
Sundin, T. *S. babylonica:* 690.
Svenson, H. *S. caroliniana:* 9980.—*S. humilis* var. *microphylla:* 7755.—*S. sericea:* 10320.
Takach, P. *S. babylonica:* 3.—*S. humilis* var. *humilis:* 8.
Tanger, L. *S. eriocephala:* 3030, 4252.—*S. nigra:* 3030.
Tannehill, M. *S. nigra:* 264, 428.
Tatnall, E. *S. exigua:* 1.

Tatnall, R. *S. alba:* 2115.—*S. caroliniana:* 1768.—*S. eriocephala:* 1743, 2121, 2141.—*S. humilis* var. *humilis:* 1123, 1344, 1345, 1348, 1666, 2116, 2469, 2879, 4493, 5007.—*S. humilis* var. *microphylla:* 2144.—*S. nigra:* 241, 1757, 2508, 2510, 3293, 4494.—*S. purpurea:* 1440.—*S. sericea:* 1935, 2142.

Tatum, H. *S. caroliniana:* 25.

Taylor, C. *S. babylonica:* 354.

Taylor, J. *S. nigra:* 6364.

Taylor, N. *S. humilis* var. *humilis:* 67, 78.

Taylor, S. *S. nigra:* 7.

Teasley, H. *S. humilis* var. *humilis:* 21B.—*S. sericea:* 2C.

Teer, J. *S. nigra:* 339.

Temple, L. *S. caroliniana:* 6116.—*S. exigua:* 2846, 2850.—*S. nigra:* 2953, 4703, 8219, 8566, 8717.

Terrell, E. *S. caroliniana:* 3496.—*S. nigra:* 4039, 4227.

Terrell, J. *S. fragilis:* 14.

Thieret, J. W. *S. caroliniana:* 34044.—*S. eriocephala:* 50633, 52267.—*S. exigua:* 21032, 21078, 23842, 23970, 24058, 24966, 25112, 25789, 30263, 31450, 53869.—*S. humilis* var. *humilis:* 22975, 31280.—*S. nigra:* 10500, 17042, 17043, 19449, 22827, 22836, 25527, 26263, 26866.—*S. sericea:* 50329.

Thomas, H. *S. nigra:* 501.

Thomas, J. *S. caroliniana:* 627.—*S. nigra:* 811.

Thomas, L. *S. nigra:* 451.

Thomas, R. *S. babylonica:* 6860, 9111, 19708, 22636, 35877, 54137, 59660, 69674, 74810.—*S. caroliniana:* 6173, 6183, 7881, 8013, 8305, 8407, 9494, 14823.—*S. cinerea* subsp. *cinerea:* 12996, 21094, 33411.—*S. exigua:* 6174, 7877, 9324, 14761, 14786, 15468, 16012, 20338, 24222, 28842, 32789, 35700, 36917, 38229, 40615, 41245, 42061, 42118, 42402, 44148, 50516, 52024, 52409, 54534, 56110, 56379, 56388, 57835, 58454, 59071, 59961, 60948, 61811, 64800, 65990, 67504, 67826, 69360, 69717, 72688, 73170, 74743.—*S. humilis* var. *humilis:* 23224, 31770, 49341, 51210, 53237, 56101, 56376, 57723, 65467. —*S. nigra:* 1320, 1873, 2052, 2290, 2308, 2354, 3220, 3306, 3688, 3956, 4629, 4937, 5079, 5212, 5242, 5396, 5601, 5658, 6555, 6677, 7042, 9493, 10218, 10284, 10313, 10380, 12985, 13042, 13061, 13417, 13732, 14047, 14760, 18961, 22589, 22636, 28540, 28706, 28822, 28905, 28974, 29541, 32842, 32937, 33978, 34506, 34991, 35329, 35750, 36857, 38100, 38242, 41203, 41247, 42669, 43197, 45041, 48284, 49869, 52660, 55663, 57351, 57729, 61105, 61143, 63422, 64644, 66003, 67225, 68736, 72073, 72520, 72731.—*S.* ×*sepulcralis:* 10209, 76654.—*S. sericea:* 13297, 13327, 22817, 24341, 32967, 39856, 57899, 73584.

Thomas, S. *S. babylonica:* 14.—*S. caroliniana:* 34.

Thompson, G. *S. nigra:* 704.

Thompson, H. *S. nigra:* 193.

Thompson, M. *S. nigra:* 36.

Thompson, R. *S. caroliniana:* 435, 460.—*S. nigra:* 266, 640.

Thorne, R. *S. caroliniana:* 1497, 2766, 3264, 3557, 3750, 7603A, 26321, 44243, 44272.—*S. eriocephala:* 3147, 3799.—*S. floridana:* 2917, 2958, 2972, 3179, 7090, 8581.—*S. humilis* var. *humilis:* 2744, 8778, 11223, 18059, 18060.—*S. nigra:* 2416, 2581, 2646, 2835, 3146, 3621, 3650, 3831, 3984, 4115, 9351, 16576.—*S. sericea:* 11344, 11846, 12117.

Threlkeld, W. *S. caroliniana:* 17.

Tidestrom, I. *S. alba:* 2349, 2350, 6704, 11698.—*S. babylonica:* 3248, 11695.—*S. caroliniana:* 149, 4518, 6950, 7490.—*S. cinerea* subsp. *cinerea:* 2180, 2351.—*S. fragilis:* 7635.—*S. humilis* var. *humilis:* 2008, 3160, 3170, 4598, 5237, 7062, 7472, 11724, 11750.—*S. humilis* var. *microphylla:* 6630, 7061.—*S. nigra:* 4504, 4518, 7102, 7196, 7492, 7515, 7727, 7912, 8089, 11728, 11765, 11766.—*S. pentandra:* 8300.—*S.* ×*rubens:* 157, 173, 2348, 7783, 7901.—*S. sericea:* 2023, 4551, 7942.

Timmer, D. *S. nigra:* 104.

Tinsley, J. *S. nigra:* 76, 77, 93, 94.

Tollett, K. *S. cinerea* subsp. *cinerea:* 13.

Tosh, J. *S. alba:* 918.—*S. caroliniana:* 917, 922.—*S. nigra:* 876, 919.—*S. sericea:* 93, 916, 920, 921.

Totten, H. *S. caroliniana:* 1543.

Traverse, A. *S. caroliniana:* 583.

Travis, M. *S. nigra:* 482.

Trott, L. *S. caroliniana:* 49.

Tseng, J. *S. caroliniana:* 4.

Tucker, G. *S. babylonica:* 4468.—*S. caroliniana:* 3565, 4166A, 4166B, 4362, 7212, 7936, 8531, 10413.—*S. eriocephala:* 6927.—*S. humilis* var. *humilis:* 2403, 2836, 2916, 9448, 9949.—*S. nigra:* 3783, 3910, 5002, 7804, 11253.—*S. sericea:* 2053, 3261.

Tucker, R. *S. caroliniana:* 222.—*S. exigua:* 221.

Underwood, J. *S. humilis* var. *humilis:* 2925, 4087.

Underwood, L. *S. nigra:* 44.

Uphof, J. *S. caroliniana:* 166.

Upton, G. *S. babylonica:* 21.

Utley, J. *S. caroliniana:* 6.
Uttal, L. *S. caroliniana:* 8515.—*S. eriocephala:* 6345.—*S. humilis* var. *humilis:* 6363, 7068, 9659.—*S. sericea:* 6885, 7680, 11808.
Van Drunen, T. *S. caroliniana:* 192.
Varner, J. *S. nigra:* 11571.
Venable, D. *S. nigra:* 233.
Venard, H. *S. humilis* var. *humilis:* 2134.
Wade, C. *S. babylonica:* 22.
Walker, E. M. *S. caroliniana:* 615, 1735.—*S. humilis* var. *humilis:* 174, 1550, 2788, 3299, 3300.—*S. nigra:* 598, 2421, 5552, 5564.—*S. sericea:* 599, 2731, 2732, 5562.
Walker, J. *S. nigra:* 13, 959, 16642.
Walker, W. *S. nigra:* 15.
Wallace, G. *S. caroliniana:* 189.
Walton, L. *S. alba:* 3597.—*S. babylonica:* 35147.—*S. humilis* var. *humilis:* 3561.—*S. nigra:* 3596, 34149.—*S. sericea:* 3546.
Warden, J. *S. nigra:* 168, 205.
Ward, D. *S. caroliniana:* 1246, 1295, 2395, 3283, 3301, 4659, 5496, 5530.—*S. floridana:* 3430, 3431, 3449.—*S. nigra:* 6640, 7162.
Ward, L. *S. caroliniana:* 2108.—*S. nigra:* 126.
Ware, D. *S. caroliniana:* 3960B.—*S. nigra:* 2865, 5347.—*S. sericea:* 3525, 4576.
Warner, J. *S. nigra:* 5.
Warrington, S. *S. eriocephala:* 9, 86, 172.—*S. exigua:* 173.—*S. nigra:* 9, 86.
Wascom, E. *S. nigra:* 506.
Waters, S. *S. babylonica:* 30.
Watson, N. *S. nigra:* 39, 88, 95.
Watters, B. *S. nigra:* 1171.
Watters, R. *S. nigra:* 76.
Weatherby, C. *S. caroliniana:* 16517.—*S. humilis* var. *microphylla:* 452, 1240, 6253, 6341.
Weaver, R. *S. caroliniana:* 2265.
Webb, D. *S. humilis* var. *humilis:* 803.—*S. humilis* var. *microphylla:* 256.—*S. nigra:* 897.
Weiss, O. *S. caroliniana:* 444.
Wenzlaff, M. *S. nigra:* 32.
Westerfeld, W. *S. alba:* 5052.—*S. caroliniana:* 5341, 6025, 6026, 6050, 6051, 6094.—*S. nigra:* 2166A, 5328, 5340.—*S. purpurea:* 1910, 5051.
West, J. *S. nigra:* 53.
Wharton, M. *S. alba:* 5685.—*S. caroliniana:* 8662.—*S. fragilis:* 10276, 10335.—*S. humilis* var. *humilis:* 2664, 3078, 4475, 5803, 5972, 6014, 6178.—*S. humilis* var. *microphylla:* 1815, 1835, 2525B, 3113.—*S. nigra:* 605, 763, 1078, 1430, 1878, 2081, 2866, 4135a, 4207, 4280, 4479, 5686, 6123, 9277, 10145, 10281, 10981, 11000.—*S.* ×*rubens:* 10276.—*S. sericea:* 49, 1249, 2589B, 5909, 6187.
Whatley, J. *S.* ×*sepulcralis:* 50.
Whatley, N. *S. nigra:* 53.
Wheeler, H. *S. eriocephala:* 1357.
Wherry, E. *S. humilis* var. *humilis:* 2656, 2700.—*S. sericea:* 2656, 2657, 2757.
Whetstone, R. *S. caroliniana:* 4265.—*S. nigra:* 4094.
Whetstone, W. *S. eriocephala:* 29.—*S. humilis* var. *humilis:* 53.—*S. nigra:* 37.
White, P. *S. babylonica:* 222.—*S. nigra:* 93.
White, R. *S. babylonica:* 19.—*S. sericea:* 9.
Whitlock, P. *S. sericea:* 94.
Whitman, K. *S. nigra:* 58.
Wickland, D. *S. nigra:* 700.—*S. sericea:* 654, 1531.
Wieboldt, T. *S. humilis* var. *humilis:* 4083.—*S. humilis* var. *microphylla:* 272.—*S. nigra:* 175M.
Wiegand, K. *S. humilis* var. *microphylla:* 929.—*S. nigra:* 927.
Wiersema, J. *S. humilis* var. *humilis:* 71, 72.—*S. nigra:* 1633.
Wiggins, I. *S. caroliniana:* 19263, 19264, 19309, 19399, 19581.
Wihle, M. *S. caroliniana:* 105.
Wilbur, R. *S. caroliniana:* 5458, 5537, 5658, 6450, 6913.—*S. humilis* var. *humilis:* 7116.—*S. nigra:* 5755, 19666, 19676.—*S. sericea:* 3612.
Wilcox, W. *S. eriocephala:* 1505.
Wilkens, H. *S. nigra:* 4553.
Williams, G. *S. exigua:* 1170.
Williams, L. *S. nigra:* 685.—*S. sericea:* 674.
Williams, N. *S. nigra:* 46, 365.
Williams, R. *S. babylonica:* 135.—*S. humilis* var. *humilis:* 62.—*S. nigra:* 28, 229, 522.
Williams, S. *S. nigra:* 19.
Williamson, B. *S. nigra:* 155, 840.—*S. sericea:* 1412.
Williamson, C. *S. caroliniana:* 8.
Willis, F. *S. babylonica:* 5.
Wilson. *S. humilis* var. *microphylla:* 33013.
Wilson, E. *S. caroliniana:* 1866.
Wilson, H. *S. nigra:* 363.
Wilson, L. *S. nigra:* 344, 1462.
Wilson, M. *S. humilis* var. *microphylla:* 1945.
Windler, D. *S. exigua:* 1265.
Wingate, S. *S. babylonica:* 33.
Withers, E. *S. caroliniana:* 2203.
Witman, H. *S. purpurea:* 2515.
Wofford, B. *S. nigra:* 10298.
Wofford, E. *S. humilis* var. *microphylla:* 50925.—*S. nigra:* 80, 1657, 1688.—*S. sericea:* 77.
Wolfe, A. *S. cinerea* subsp. *cinerea:* 17.
Wolfe, P. *S. nigra:* 41.

INDEX TO SCIENTIFIC NAMES

Accepted names are in roman type; the main entry for each is in **boldface**. Synonyms are in *italics*.